Meet the Faculty Candidate Poster Session 2015 – Sponsored by the Education Division

Topical Conference at the 2015 AIChE Annual Meeting

Salt Lake City, Utah, USA
8-13 November 2015

ISBN: 978-1-5108-1841-5

Printed from e-media with permission by:

Curran Associates, Inc.
57 Morehouse Lane
Red Hook, NY 12571

Some format issues inherent in the e-media version may also appear in this print version.

Printed by Curran Associates, Inc. (2016)

For permission requests, please contact AIChE
at the address below.

AIChE
120 Wall Street, FL 23
New York, NY 10005-4020

Phone: (800) 242-4363
Fax: (203) 775-5177

www.aiche.org

Additional copies of this publication are available from:

Curran Associates, Inc.
57 Morehouse Lane
Red Hook, NY 12571 USA
Phone: 845-758-0400
Fax: 845-758-2634
Email: curran@proceedings.com
Web: www.proceedings.com

TABLE OF CONTENTS

Alexandria University
Faculty of Engineering
Chemical Engineering Dept.

Study of the performance characteristics of a stirred tank reactor suitable for diffusion controlled liquid-solid catalytic reactions

Supervisors:
Prof. Dr. A. Konsowa
Dr. Y.O. Fouad
Prepared by:
Eng/ Mahmoud M. Taha
2014

Outlines:

1. Introduction
2. Aim of the present work
3. Apparatus
4. Variables studied
5. Results and discussion
6. Summary and conclusion

Introduction

PART – I –

Introduction

This work is concerned with studying liquid-solid diffusion-controlled reactions such as:

1. Catalytic reactions
2. Electrochemical reactions
3. Biochemical reactions
4. Photocatalytic reactions

Introduction

This reaction kind conducted in many reactors such as:

- Slurry reactors

There are many shortcomings:

1. Catalyst attrition
2. Erosion of impeller blades
3. Low rate of mass transfer
4. Separation of the final product is a troublesome process

Introduction

Several attempts are carried out to study the performance by using:

- Flat plates

  (flat plate fixed to the wall of square agitated vessels)

- Raschig rings

  (packed bed of Raschig rings fixed to the wall or rested at the bottom of agitated vessels)

Introduction

- Expanded metals and screens

<u>Previous work on screens involve:</u>

i. Fixed screens
   (horizontal and vertical)
ii. Screens in motion
   (rotating and vibrating)

Aim of The Work

Developing a high space-time STR suitable for diffusion controlled L-S catalytic reactions.

The work divided into two parts:

1. Evaluating the performance of the reactor ***(k).***
2. Examining the reactor performance in a diffusion controlled process.

Experimental Part

PART – II –

Apparatus

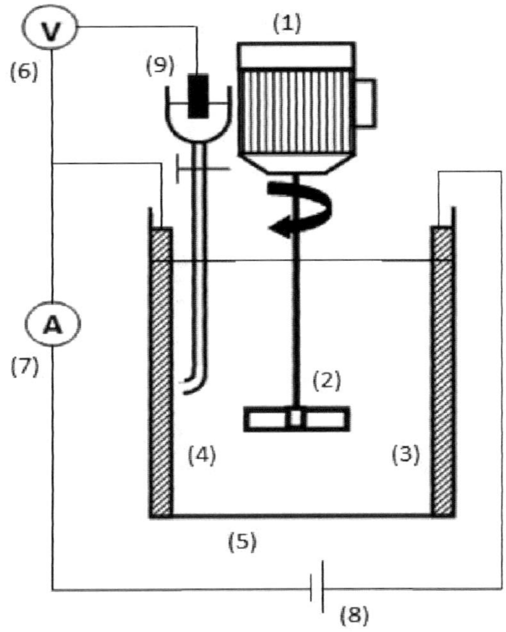

1) Electric motor
2) Isolated impeller shaft & Impeller
3) Stainless steel screens "anode"
4) Nickel plated St. St. screens "cathode"
5) Plexiglass square tank
6) High impedance voltmeter
7) Ammeter
8) 12 V d.c power supply
9) Nickel wire placed in luggin tube

Determination of performance characteristics of new reactor

Technique used:

The electrochemical technique was used because of its *simplicity* and *accuracy*.

Configuration used:

Flow-by electrode configuration.

Principle:

- Mass transfer coefficient (k) is obtained using the limiting current (I_L) values determined from potential-current curves.

$$\frac{I_L}{Z\,F} = k\,A\,C$$

Potential-current curve

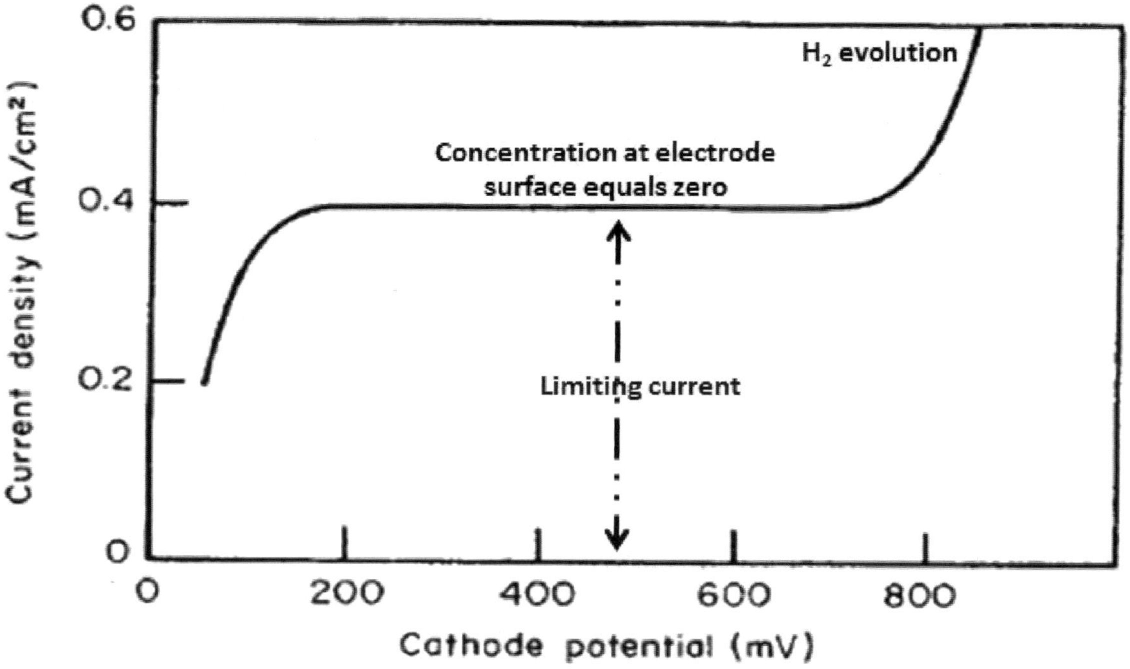

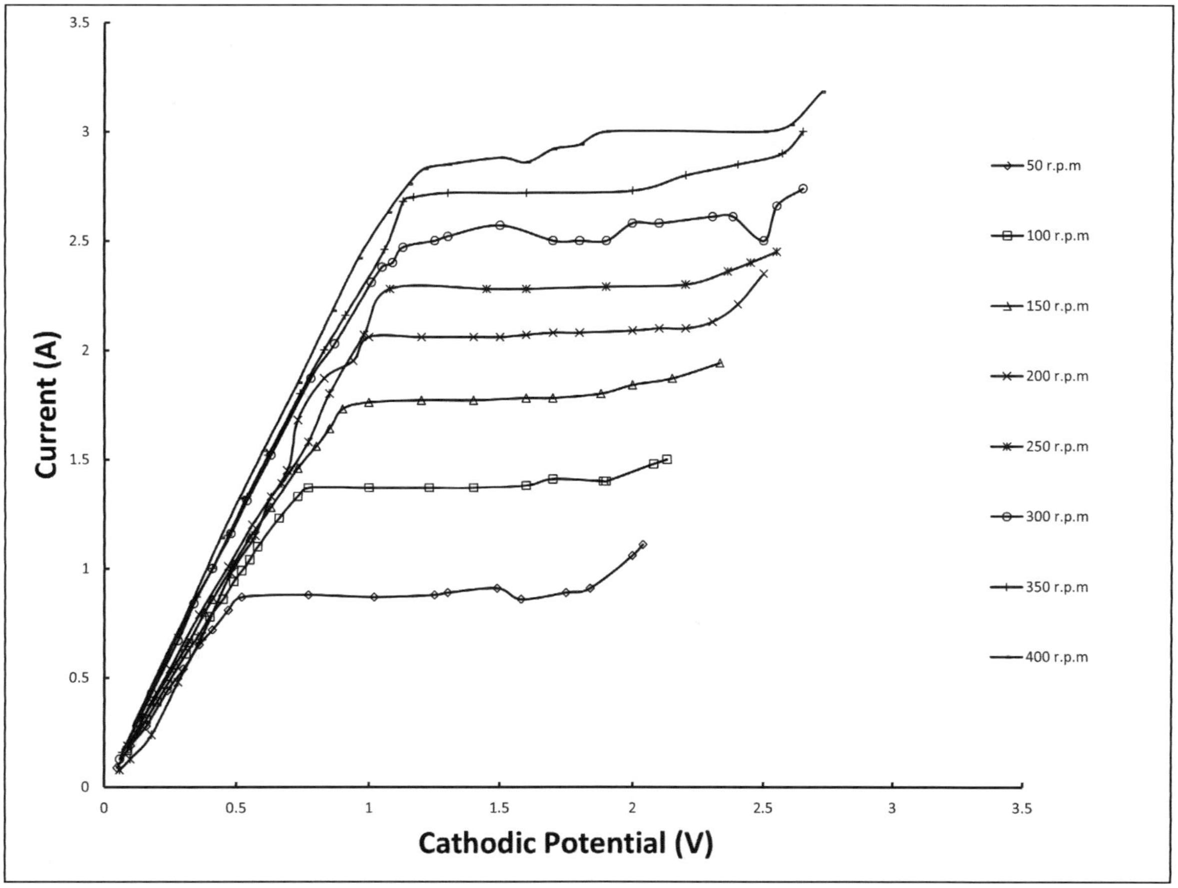

Experiments Conditions

1. Solutions used:

   0.025 M $K_3Fe(CN)_6$ + 0.1 M $K_4Fe(CN)_6$ dissolved in an excess of NaOH solution namely; 1, 2 and 3 N of NaOH solution.

   Reactions involved:

   Anode: $Fe(CN)_6{}^{-4} \rightarrow Fe(CN)_6{}^{-3} + e$

   Cathode: $Fe(CN)_6{}^{-3} + e \rightarrow Fe(CN)_6{}^{-4}$

Experiments Conditions "continued"

2. Room temperature 25±1°C.
3. High ratio of anodic to cathodic area.
4. Electrodes activation in 5% NaOH solution to eliminate chemical polarization and remove any dirts.
5. The solution must kept away from light.

Variables Studied

1. Impeller rotational speed (ω = 50:400 r.p.m)
2. Impeller geometry (45° pitched blade turbine and 90° flat blade turbine)
3. Screen geometrical parameters (mesh number and wire diameter – 5, 10 & 14 mesh no.)
4. Solution physical properties (ρ, μ and Ɗ)
5. Number of closely packed screens forming the reaction surface (from 1 to 5 screens)

Results and Discussion

1. Effect of impeller rotational speed (ω)

$$k = f(\omega)$$

These parameters expressed in terms of dimensionless groups;

$$Sh = f(Re)$$

where;

$$\mathbf{Sh} = \frac{\mathbf{k\ d_w}}{\mathbf{D}} \qquad \mathbf{Re} = \frac{\boldsymbol{\rho}\ \mathbf{n\ d_i^2}}{\boldsymbol{\mu}}$$

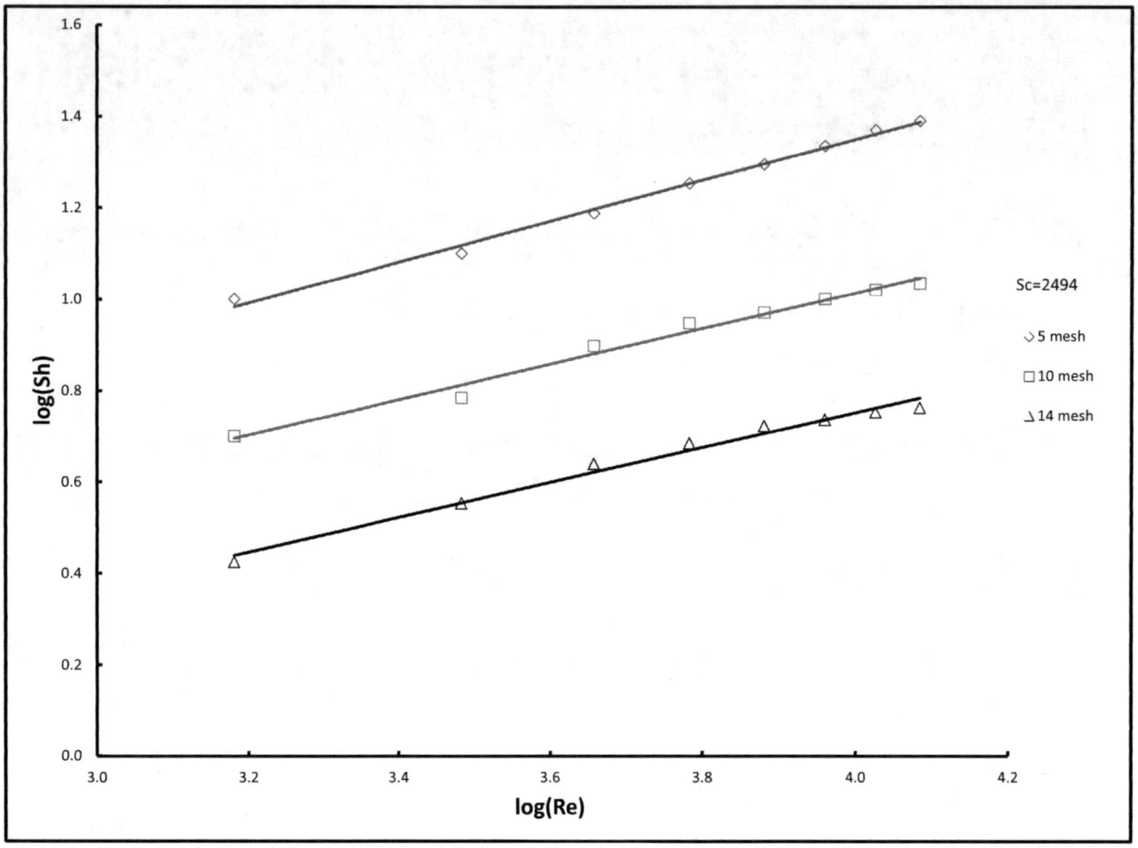

Results and Discussion "continued"

- For 45° pitched blade turbine (axial flow)

$$Sh = a_1 Re^{0.51}$$

- For 90° flat blade turbine (radial flow)

$$Sh = a_2 Re^{0.4}$$

Results and Discussion "continued"

2. Effect of screen geometrical parameters (i.e, mesh number and wire diameter)

$$\text{mass transfer coefficient} = f\left(\frac{1}{\text{mesh no.}}\right)$$

These parameters expressed in terms of dimensionless groups;

$$Sh = f\left({}^{r_h}/_{d_w}\right)$$

Results and Discussion "continued"

Where;

r_h = screen hydraulic radius

$$r_h = \frac{\varepsilon}{a}$$

d_w = screen wire diameter

$\left({}^{r_h}/_{d_w}\right)$ express geometrical parameters of screens

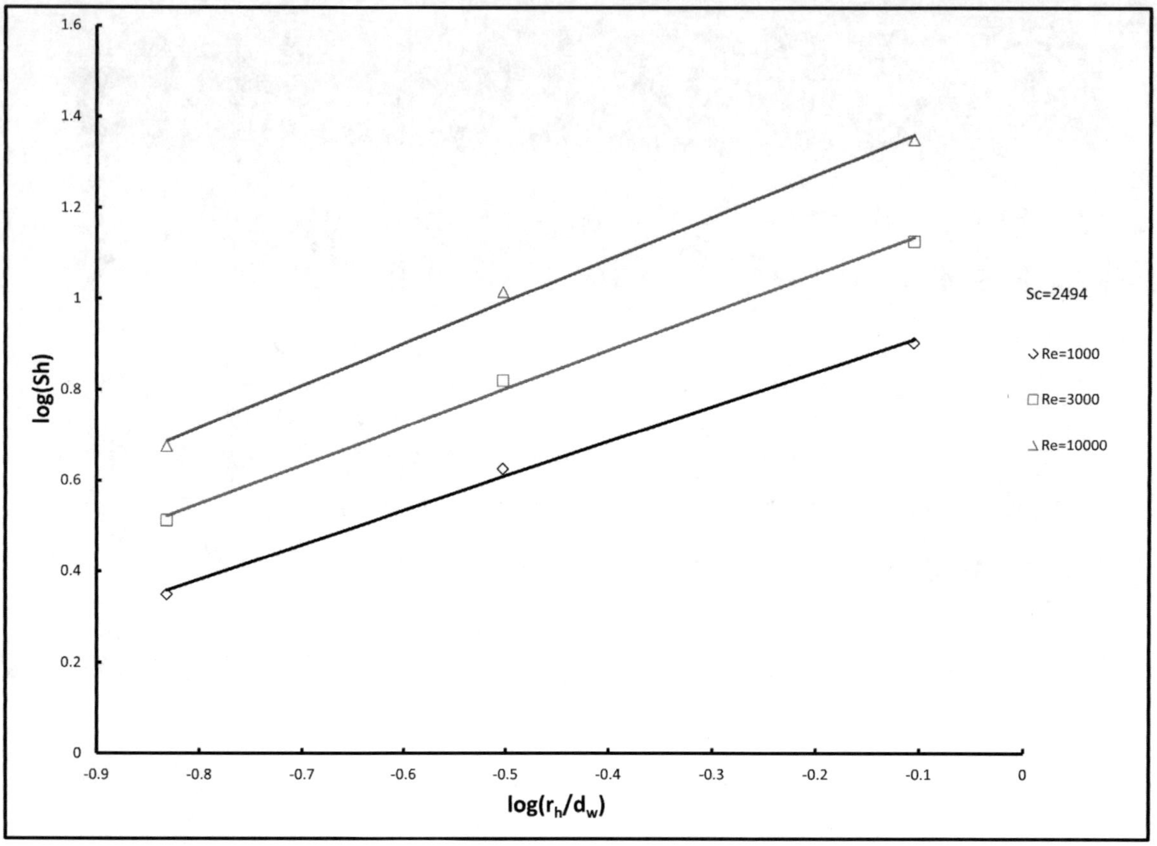

Results and Discussion "continued"

- For 45° bitched blade turbine (axial flow)

$$Sh = a_1 \left({}^{r_h}/_{d_w}\right)^{0.72}$$

- For 90° flat blade turbine (radial flow)

$$Sh = a_2 \left({}^{r_h}/_{d_w}\right)^{0.75}$$

Results and Discussion "continued"

3. Effect of impeller geometry

There are two impellers used:

- 45° pitched blade turbine (axial flow)
- 90° flat blade turbine (radial flow)

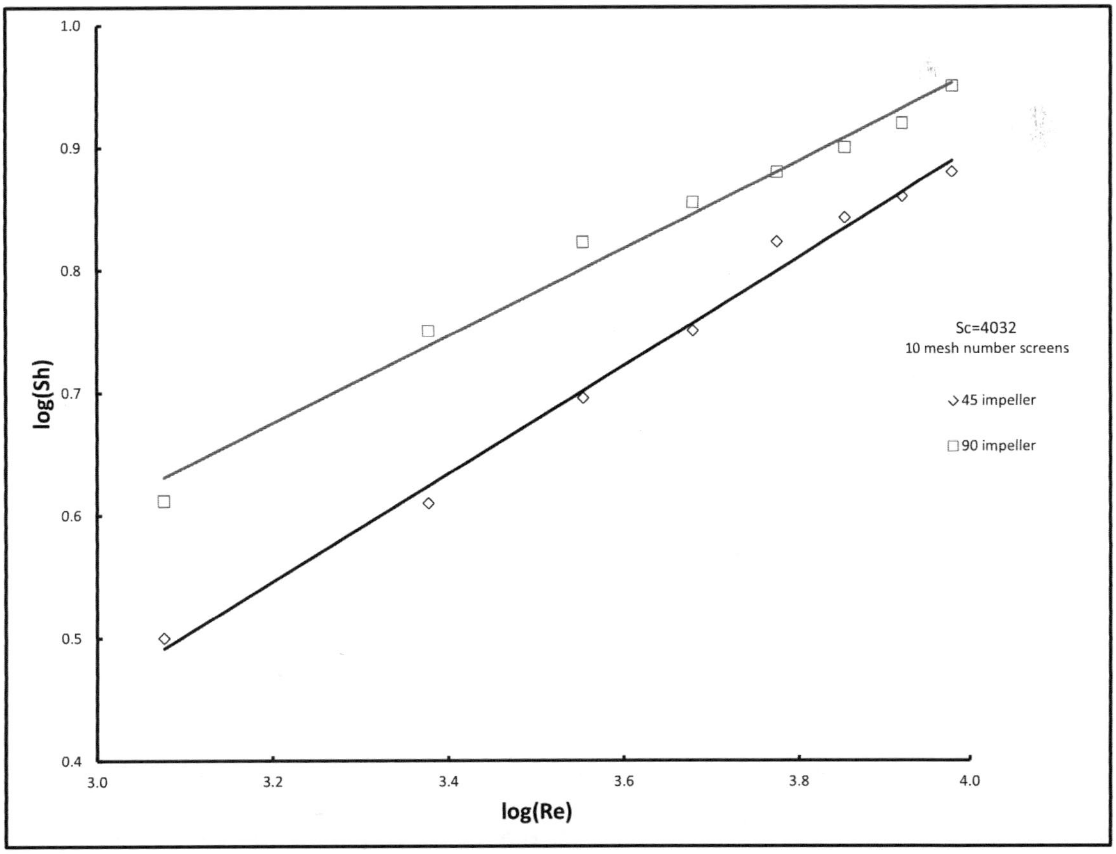

Results and Discussion "continued"

- By dimensional analysis, it is found that

$$Sh = a\, Sc^{\propto}\, Re^{\beta} \left({}^{r_h}/_{d_w}\right)^{\gamma}$$

Where;

Sh	Sherwood number	**Dimensionless groups**
Re	Reynold number	
Sc	Schmidt number	

$$\boldsymbol{Sc} = \frac{\mu}{\boldsymbol{\rho\, D}}$$

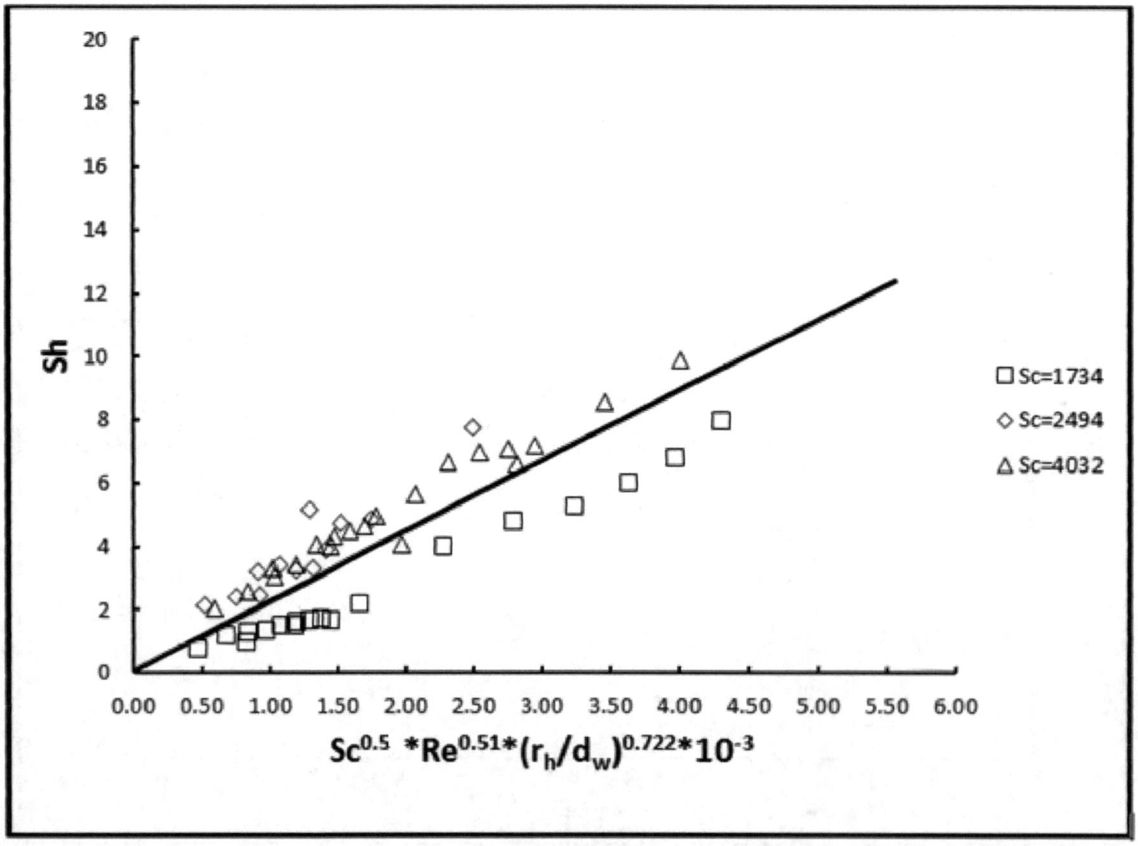

Results and Discussion "continued"

- For 45° pitched blade turbine (axial flow)

$$Sh = 0.0022\ Sc^{0.5}\ Re^{0.51}\ \left({}^{r_h}/_{d_w}\right)^{0.72}$$

- For 90° flat blade turbine (radial flow)

$$Sh = 0.00779\ Sc^{0.5}\ Re^{0.4}\ \left({}^{r_h}/_{d_w}\right)^{0.75}$$

Results and Discussion "continued"

5. Comparison between screen and flat plate electrodes

Results and Discussion "continued"

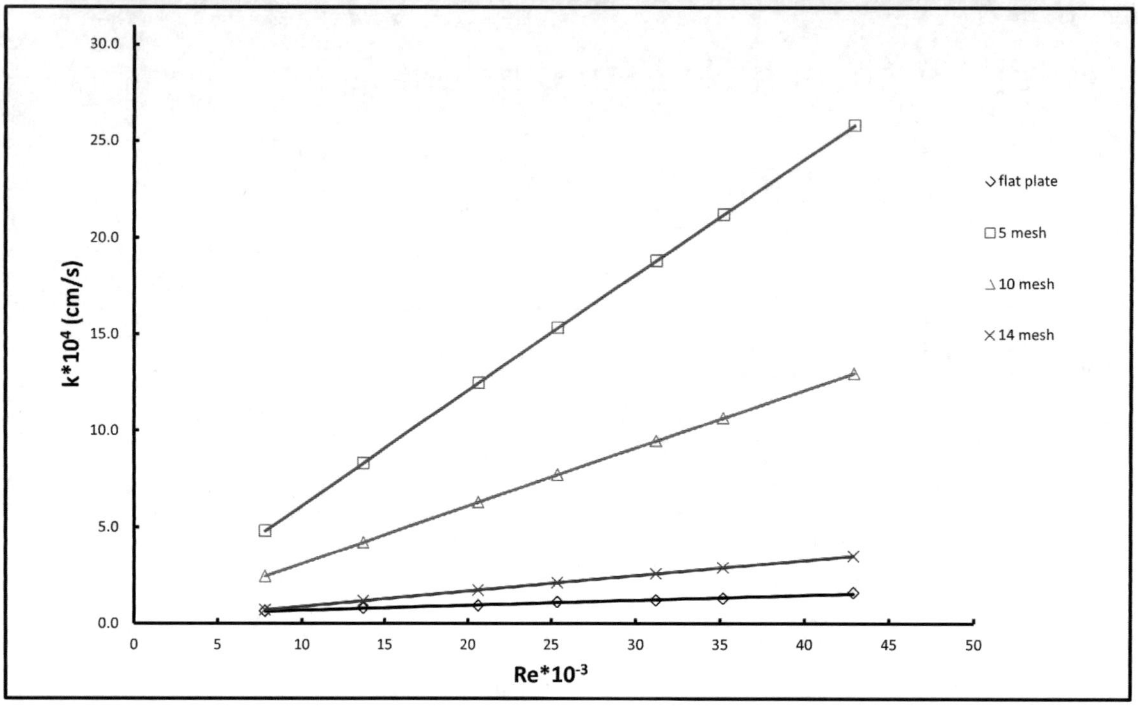

Results and Discussion "continued"

- It is found that;

$$k_{screen} \gg k_{flat\ plate}$$

By a factor ranging from **1.09** to **16.25**

Results and Discussion "continued"

6. Mass transfer at multiscreen electrodes

To increase the proposed reactor productivity, expressed as volumetric mass transfer coefficient (kA)

Results and Discussion "continued"

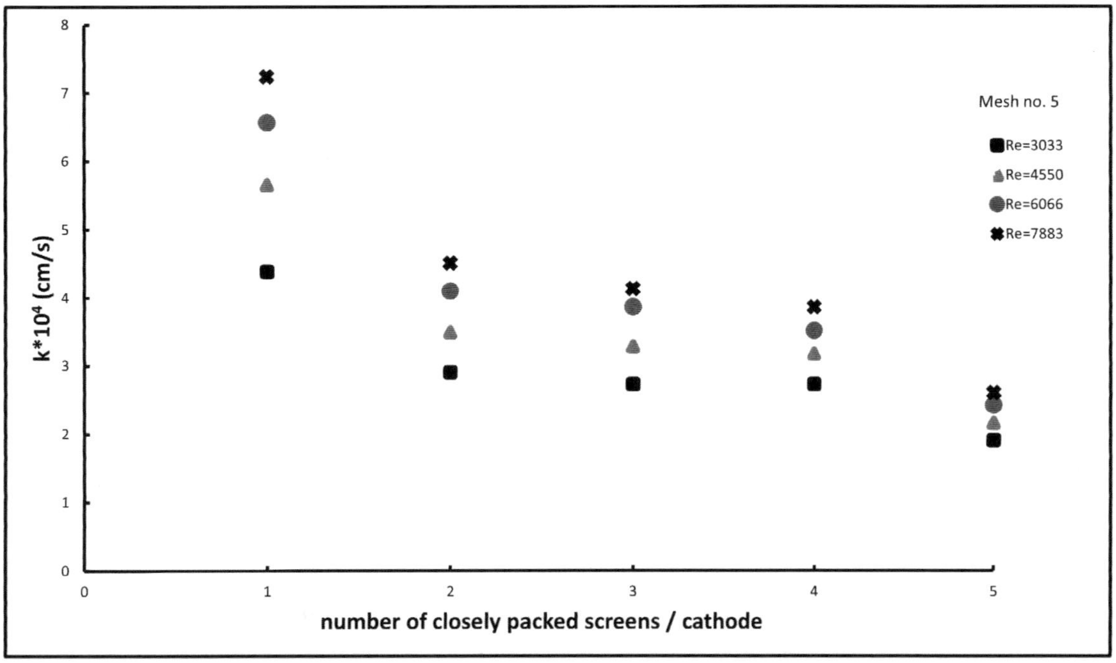

Results and Discussion "continued"

- Form previous data, it shown that:

$$k = f\left(\frac{1}{\text{number of screens per cathode}}\right)$$

- ***Reasons***
  - *i. Turbulence damping inside screen bed.*
  - *ii. Active area decrease at points of contact between screens.*

Results and Discussion "continued"

- Comparison with flat plate previous work

The highest enhancement obtained at the following conditions:

$$\frac{k\,A_{\text{multiscreens}}}{k\,A_{\text{flat plate}}} = \mathbf{46.15}$$

1. Array of four number of screens
2. 5 mesh number
3. ω = 350 r.p.m

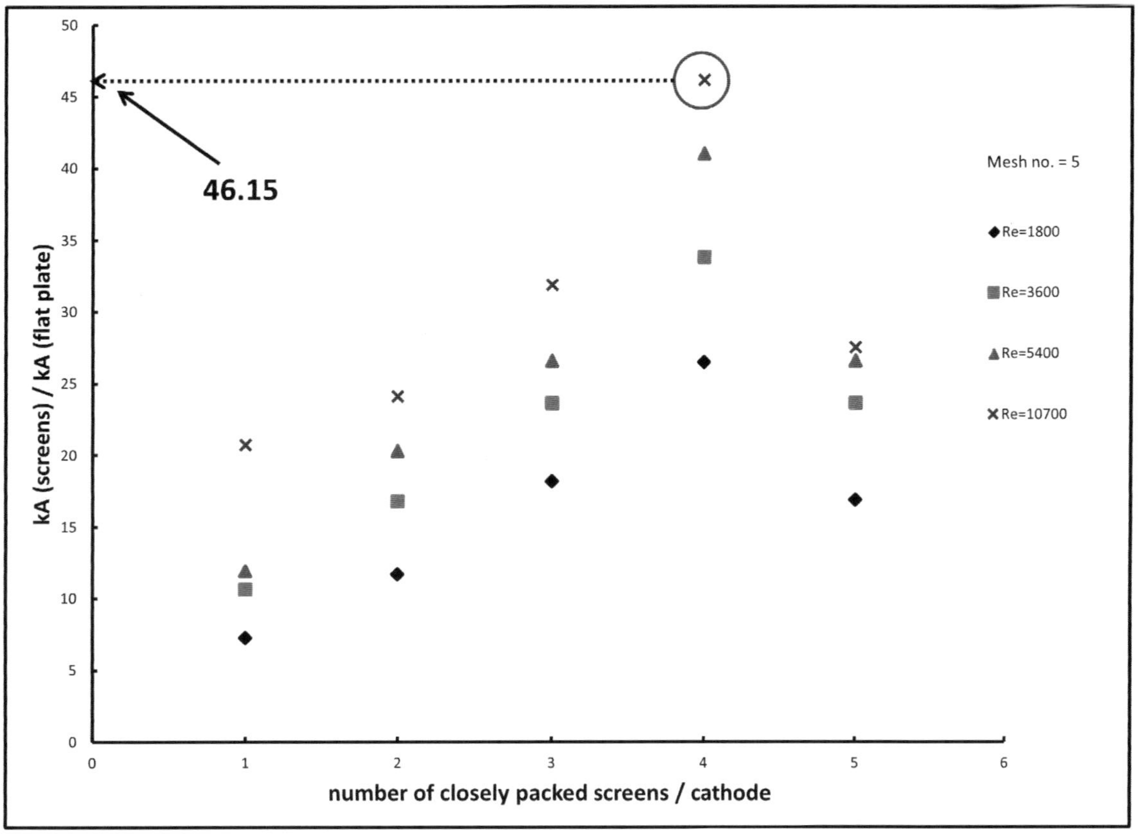

Results Summary

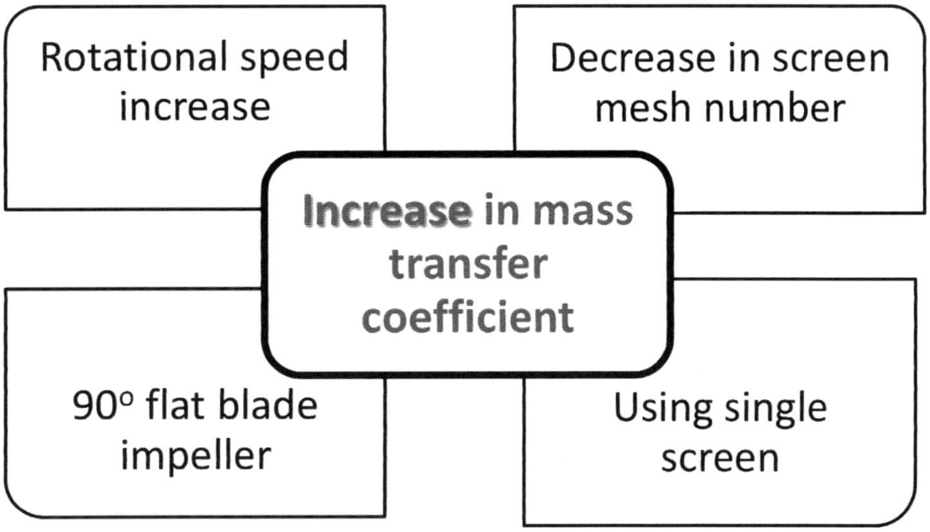

Application of present reactor

The present reactor used for conducting diffusion-controlled reactions involving dilute solutions.

➢ Such as

Removal of heavy metals , <u>Copper ions</u>, from waste solutions.

Apparatus

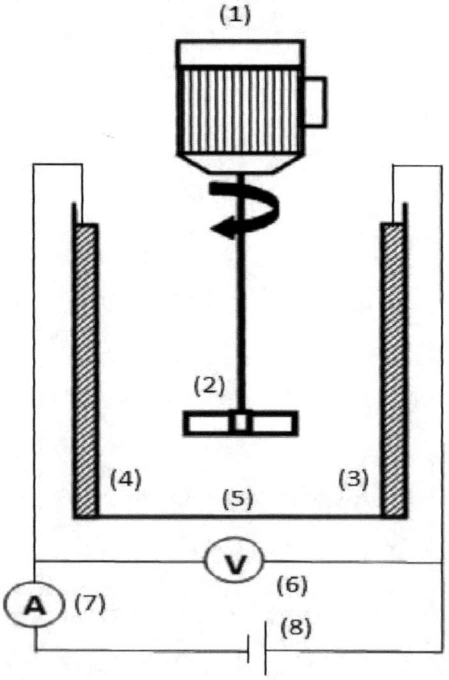

1) Electric motor
2) Impeller shaft
3) **Insoluble lead sheet "anode"**
4) **St. St. screens "cathode"**
5) Plexiglass square tank
6) High impedance voltmeter
7) Ammeter
8) 12 V d.c power supply

Examination of Reactor Performance

Technique used:

Metal deposition in acidified solution.

Principle:

Mass transfer coefficient (k) is determined from reactor design equation by measuring the solution concentration (C) at different time intervals.

The rate of copper deposition in a batch reactor is given by the following equation:

$$\frac{-Q}{A}\frac{dc}{dt} = k\,c$$

By integration;

$$Q \ln\frac{C_o}{C} = k\,A\,t$$

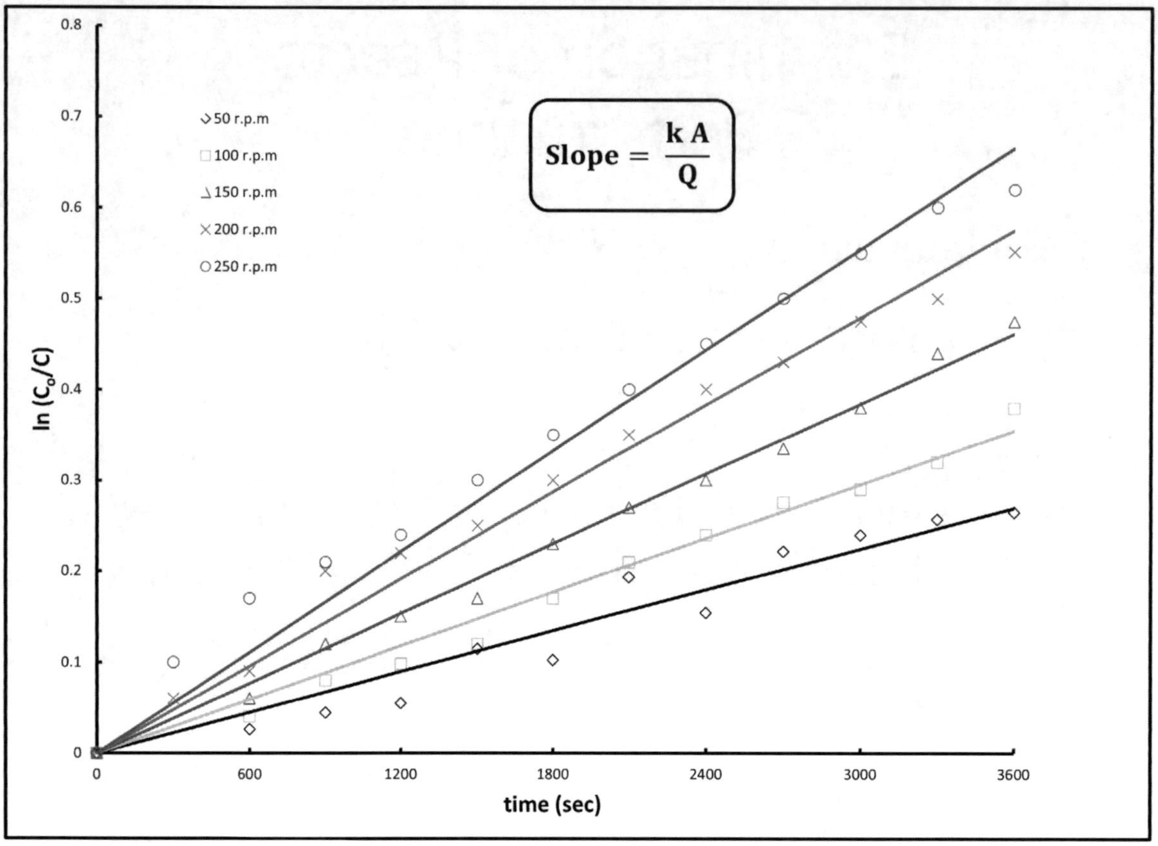

Experimental Section

1. Solutions used:

$$0.05\ M\ CuSO_4 + 1.5\ M\ H_2SO_4$$

Reactions involved:

Anode: $H_2O \rightarrow 2H^+ + \frac{1}{2}O_2 + 2e$

Cathode: $Cu^{++} + 2e \rightarrow Cu$

2. 60 minutes of electrolysis

Results and Discussion

- % removal of copper ions ranged from
  - For 45° pitched blade turbine (axial flow)

    23.26 to **46.18 %** **(Per hour)**
  - For 90° flat blade turbine (radial flow)

    33.90 to **54.71 %** **(Per hour)**

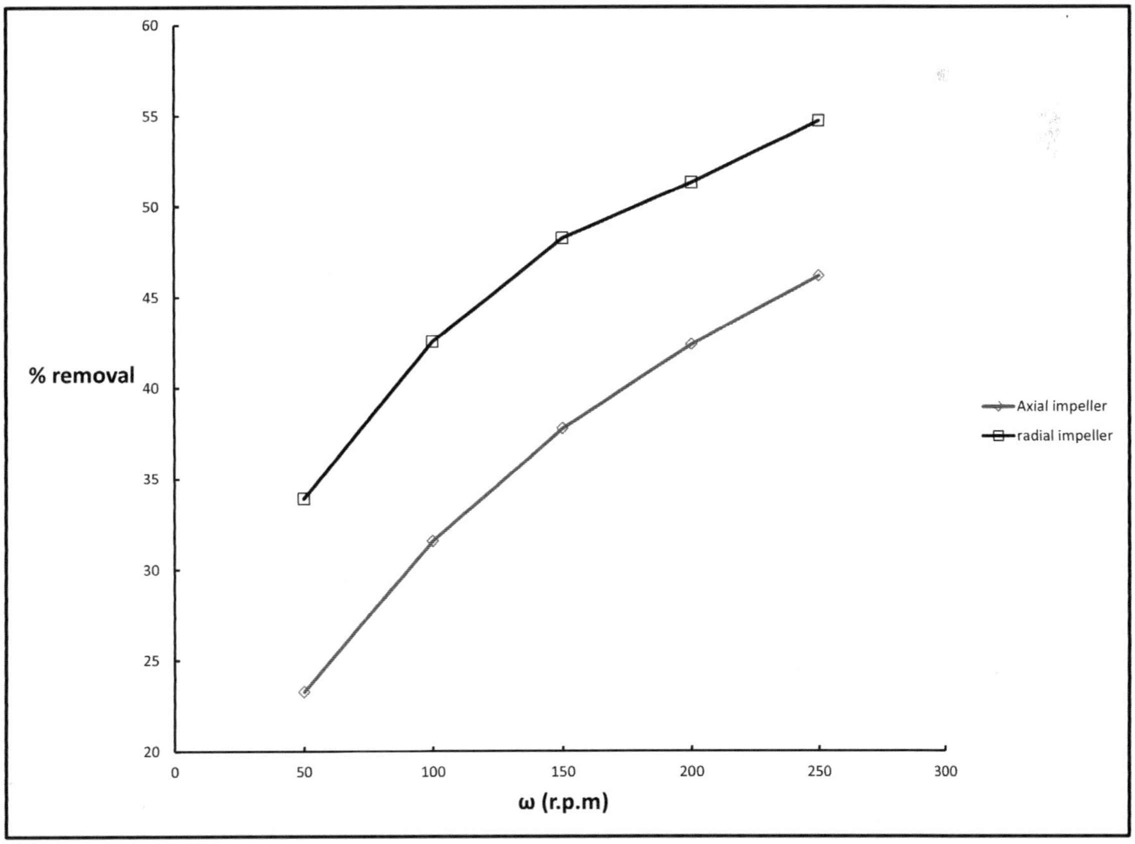

Results and Discussion "continued"

$k_{exp.} > k_{theo.}$ "Maximum deviation = 3.5%"

1) Surface roughness caused by deposition of copper powder
2) Potential is shifted to negative values so H_2 bubbles evolved

Conclusions

1) The reactor save floor space needed so capital cost is reduced.
2) The present dimensionless mass transfer correlation not only can be used to design batch reactors but also continuous reactors.
3) By analogy, the internal heat transfer coefficient can be calculated using the overall correlation obtained.

Conclusions

4. This reactor offers many advantages:

   - Enhancing rate of mass transfer
   - Enhancing rate of heat transfer
   - High surface area per unit volume
   - Dispersion of sparingly soluble components
     (solids – immiscible liquids – gases)

Thank You …

Mahmoud M.Taha

2014

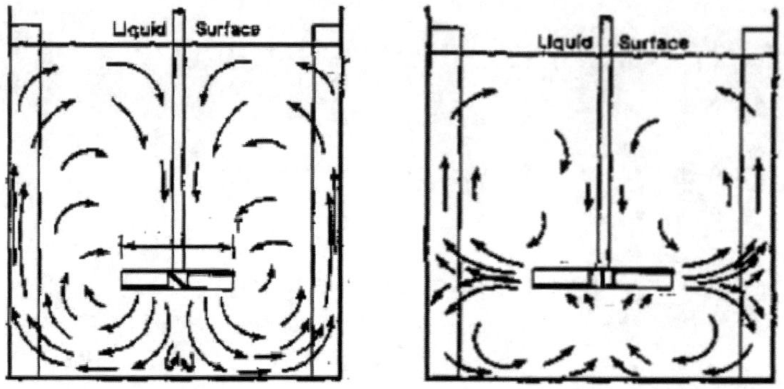

Axial flow impeller **Radial flow impeller**

Flow pattern for different impellers used

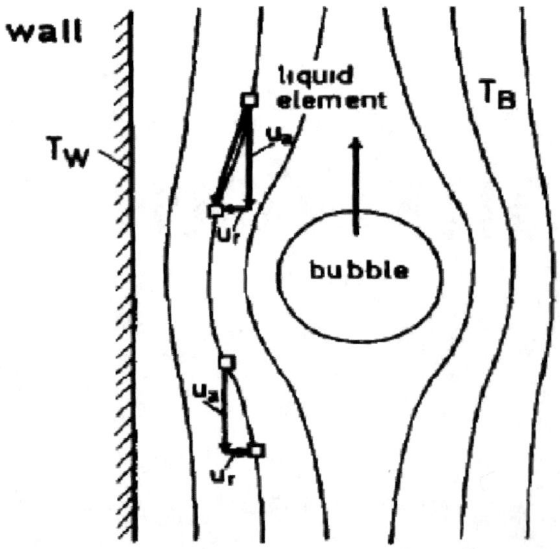

Flow across a rising bubble <u>"Deckwer 1980"</u>

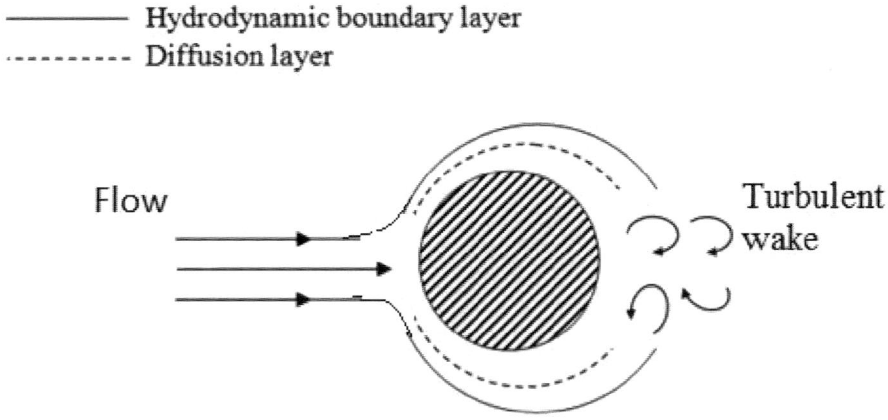

Flow across cylinders or screen wires

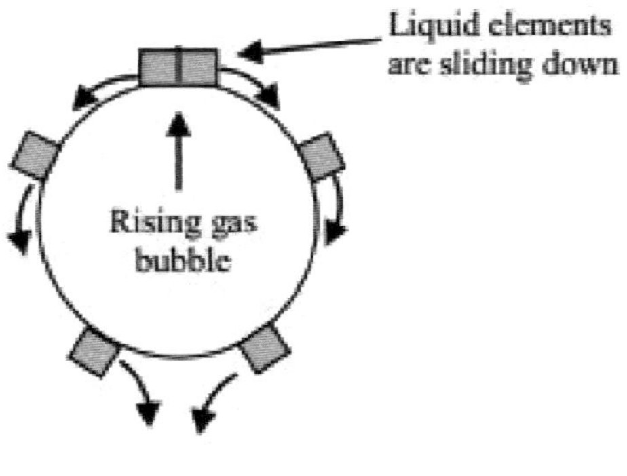

Schematic of Penetration theory

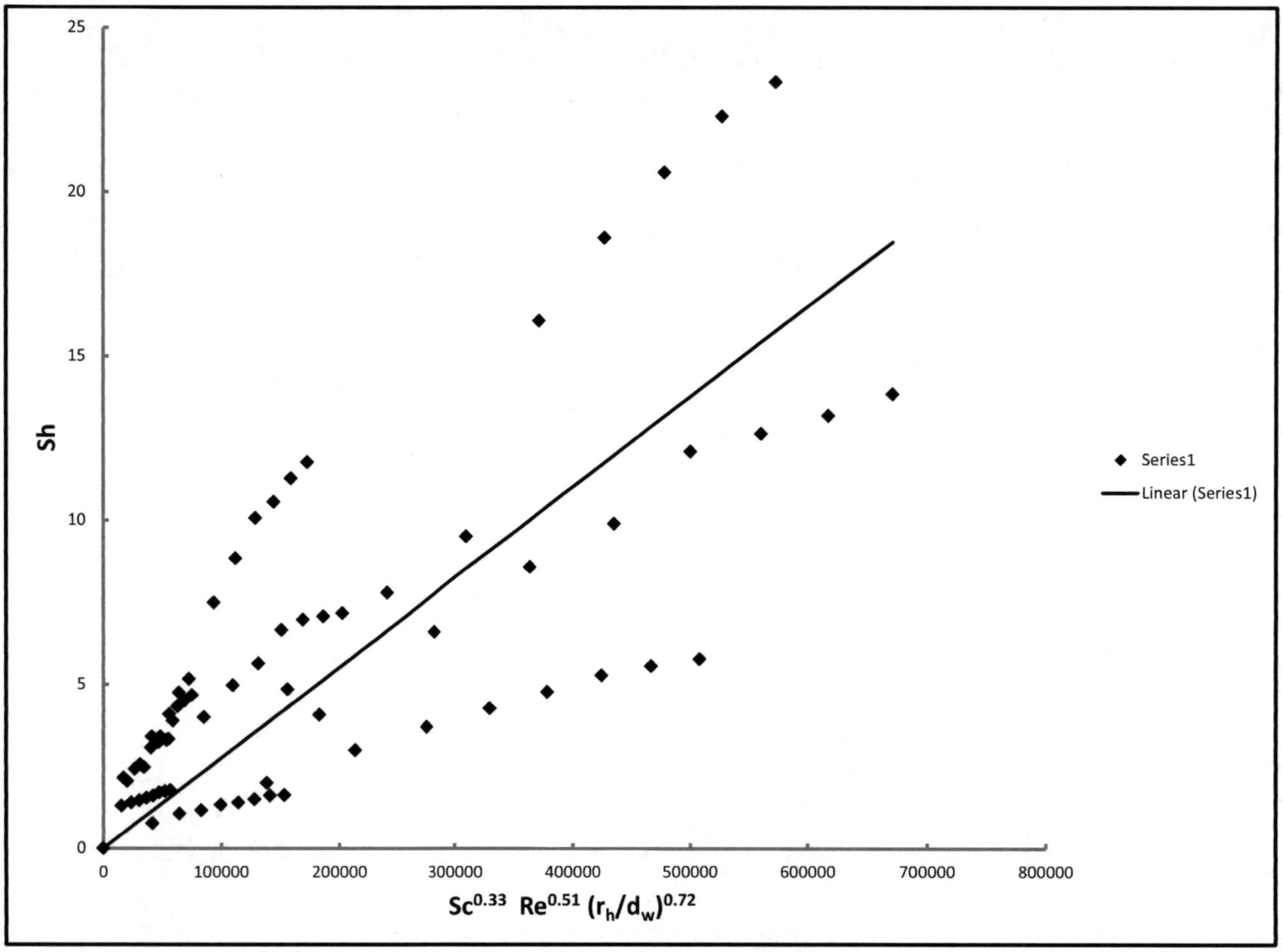
25
20
15
10
5
0
Sh
0
100000
200000
300000
400000
500000
600000
700000
800000
Sc^0.33 Re^0.51 (r_h/d_w)^0.72
Series1
Linear (Series1)

(6it) Robust Adaptive Model Predictive Control of Chemical and Biological Systems
Babaei Pourkargar, D., The Pennsylvania State University
My doctoral research work has contributed to the synthesis of advanced control structures for processes where complex transport phenomena and chemical reactions take place in the chemical and advanced material industries. These processes, called distributed parameter systems (DPSs), can be mathematically modeled by a set of nonlinear partial differential equations (PDEs) and are exemplified by packed and fluidized bed reactors in chemical plants; reactive distillation in petrochemical industries; lithographic processes; chemical vapor deposition, etching processes and plasma discharge reactors in microelectronics manufacturing and complex materials production; crystallization and polymerization processes; and tin float bath processes in glass production industry. In our research, we circumvent the restrictions of the current control approaches tailored for PDE systems via model order reduction by designing low-dimensional output feedback controllers on the basis of reduced order models (ROMs) of the governing PDEs. The successful implementation of the model reduction based approaches depends heavily on the basis functions that are required to construct the ROMs using weighted residual methods. A limitation is that analytical methods to derive a consistent basis function set are not applicable in the presence of unknown parameters, nonlinear spatial derivatives or when the process operates over complex spatial domains. On the other hand statistical methods, such as variants of widely popular proper orthogonal decomposition (POD) method, guarantee basis identification only under strict assumptions that cannot be verified in practice.

Motivated by the above limitations my research focused on

1. Deriving a control-tailored and computationally efficient robust algorithm to recursively compute the optimal set of empirical basis functions required by weighted residual methods to discretize the governing PDEs and construct locally accurate ROMs in the form of ODEs. The algorithm, known as adaptive proper orthogonal decomposition (APOD), circumvents the limitations of online statistical techniques and leads to a three-fold increase in computational speed.
2. Synthesizing advanced nonlinear output feedback control structures that can guarantee closed-loop stability. A wide range of nonlinear Lyapunov-based robust and adaptive controllers and nonlinear dynamic observers were synthesized to address the regulation and tracking problem of nonlinear DPSs as most effective as possible in the presence of time-varying unknown parameters and system uncertainty.
3. Designing a supervisory structure which monitors the controller performance in stabilization and tracking the desired spatiotemporal dynamics, (a) to retune the controller parameters and (b) to revise the ROMs as needed via APOD. Taking advantage of such supervisory control structure, the ROM revisions are minimized which leads to reduction in the required information from spatially distributed sensors to recursively update the empirical basis functions by APOD.

Regarding my future research efforts, I intend to extend my computational work to modeling, estimation, control and optimization of advanced chemical and biological systems with specific applications to catalytic chemical processes, advanced biofuels production and continuous pharmaceutical processes. Specifically, the intellectual objective of my future research is to develop a systematic control and optimization framework tailored for chemical and biological systems with unknown transport-reaction mechanisms when the systems uncertainty plays a vital role. For example, the source of parametric uncertainties and unmeasurable unknown parameters in chemical and biological systems, which necessitates applying such novel control and optimization approaches, is reaction rates, activation energies, fouling factors and microbial growth rates which are only approximately known via complex experiments.

The presence of uncertainty in the form of additive disturbances, state estimation and modeling errors, and the associated topic of robustness are still major challenges in control theory and specifically in model predictive control (MPC) designs. The reason for such challenges is that to solve the optimal control problem in the presence of uncertainty we require feedback from the system; solving an open-loop optimal control problem to determine a control law, as is done in nominal MPC, is not optimal. The natural extension of MPC to address the control problem of uncertain systems would be to optimize over a sequence of control laws not over a sequence of control actions. However this dynamic programming approach may be unacceptable in practical situations. One of the practical methods to cope with such uncertainty in the system model is via robust MPC (RMPC) where the required estimation upper bound of the time-varying uncertainty can be improved as time evolves. Generally, the performance of robust controllers is restricted by the quality of the model and its uncertainty description due to the disability of the control approach in learning the system changes during process evolution. To circumvent such limitation we can employ adaptive MPC (AMPC) to improve closed-loop system performance by updating the model through unknown parameters' estimation based on measurement outputs.

While linear AMPC designs have reached a certain maturity, very few AMPC schemes have been developed for nonlinear systems and only for unconstrained processes. The nonlinear AMPC problem is highly challenging because the robustness to the model uncertainty must be addressed in the presence of constraints while the adaptive estimator is evolving without assuming the separation principle which is only valid for linear systems. I plan to investigate the nonlinear AMPC problem by introducing extra degrees of conservativeness considering robust performance and closed-loop stability. The proposed research in this theoretical subject will introduce systematic computationally efficient process identification methods to estimate system parameters required by MPC to compute the optimal control action subject to a continuous decrease in the identification error. It will also resolve fundamental theoretical and computational issues associated with system identification techniques in control and optimization policies for a wide range of chemical and biological systems.

408819 Preparation and Carbon Dioxide Separation Performance of a Hollow Fiber Supported Ionic Liquid Membrane

Abstract:

In this study, we prepared polyethersulfone hollow fibers using double colaminar flow in a coaxial microfluidic device. The structures of the hollow fibers were adjusted by changing the composition of the dope solution. The produced hollow fibers were used as a support for room temperature ionic liquids to study their potential for carbon dioxide/nitrogen separation. The stability of supported ionic liquid membranes with different structures was evaluated using two different ionic liquids. The effect of the permeate direction and the support structure on the pure gas permeability and ideal selectivity was also studied. The ideal selectivity, represented by the ratio between the permeabilities of carbon dioxide and nitrogen, was compared with data in the literature described by the Robeson upper bound correlation. High selectivity values obtained for carbon dioxide compared with those for nitrogen indicated that the produced supported ionic liquid membranes have good potential for carbon dioxide separation processes.

411237 Shaping Catalysis through Tailored Nanostructures: For Energy Conversion and Storage

Abstract:

Catalyst is a substance that participate chemical reactions by increasing the reaction rate while the catalyst itself is not being consumed. Most catalysts used in industrial processes are based on nanomaterials, because of their high surface-to-volume ratio therefore high material utilization at nanoscale. This presentation covers several recent development of nano-catalysts in energy conversion and storage. The performance of catalysts has been improved through tailoring the sizes, compositions, and morphologies of nanomaterials. This presentation outlines 1) synthesis and preparation of catalytic nanomaterials, including Pt and Pt-alloy catalysts, 2) the connection between shape (morphology) of nanomaterials and the catalytic properties, 3) correlation between electrocatalyst composition profile and electrocatalysis, 4) effective catalyst design strategy, 5) balancing activity and durability by tuning nanostructure, 6) chemical modification of surface to tailor catalytic properties, and 7) superstructures as novel research platform to study catalysis. This presentation emphasizes the novel approaches that improve catalysts performances such as activity and durability through careful tailoring of nanomaterials.

414862 Smart Gating Membranes with K+-Responsive Pore Size and Surface Property

Abstract:

A novel smart gating membrane with K^+-responsive pore size and surface property is prepared by grafting poly(*N*-isopropylacrylamide-*co*-acryloylamidobenzo-15-crown-5) (poly(NIPAM-*co*-$AAB_{15}C_5$)) copolymer chains in the pores of porous nylon-6 membranes with a two-step method combining plasma-induced pore-filling grafting polymerization and chemical modification. Due to the cooperative interaction of host-guest complexation and phase transition of the poly(NIPAM-*co*-$AAB_{15}C_5$) copolymer, not only the membrane pores with grafted gates could spontaneously switch from "closed" state to "open" state, but also the membrane surface is induced the hydrophilic/hydrophobic change of the membrane surface by recognizing K^+ ions in the environment and vice versa; while other ions (e.g., Na^+, Ca^{2+} or Mg^{2+}) can not trigger such an ion-responsive switching function. The positively K^+-responsive gating action of the membrane is rapid, reversible, and reproducible. The proposed K^+-responsive gating membrane provide a new mode of behavior for ion-recognizable "smart" or "intelligent" membrane actuators, which is highly attractive for controlled release, chemical/biomedical separations, sensors, etc.

417644 Design of Injectable Hydrogels for Regenerative Medicine

Abstract:

Stem cell-based therapy is emerging as a viable method to treat various injuries and degenerative diseases. However, stem cell delivery via direct injection results in poor cell survival (typically <5%) due to lack of mechanical protection during injection and lack of biochemical cues post-injection. I have designed highly tunable three-dimensional (3D) hydrogels using building blocks of multivalent protein-polymers to enhance cell survival throughout the injection process and to offer long-term mechanical and biochemical support post-injection. These biomaterials are used both as *in vitro* platforms to study fundamental questions about three-dimensional matrix-cell interactions and as clinical therapies to enhance cell transplantation efficiency and achieve neurovascular regeneration. As a principal investigator, I intend to initiate a multidisciplinary research program combining my expertise in materials design, protein engineering, polymer chemistry, and neural and vascular cell biology for treating neurovascular diseases. My scientific training in these multidisciplinary fields uniquely positions me to synergistically combine peptide and protein motifs with synthetic macromolecules to create new hydrogel systems for understanding cell-biomaterial interactions *in vitro* and addressing therapeutic needs *in vivo*. This program will focus on 1) the design of injectable stem cell niches for drug and cell delivery therapies and 2) the investigation of how niche properties influence cellular function and promote neurovascular tissue regeneration.

414129 Rational Design of Catalytic and Hydrocarbon Trapping Materials to Meet Automotive Emissions Regulations

Abstract:

Rational Design of Catalytic and Hydrocarbon Trapping Materials to meet Automotive Emissions Regulations

As the emission legislation for vehicle pollutants is becoming more stringent worldwide due to increasing concerns of the impact of air pollution on both the environment and public health, much opportunity is created in the fields of catalysis and catalytic materials. Several challenges for which I am particularly well-positioned to make important advancements include the following: (i) Developing catalytic materials that are active for the oxidation of diesel at temperatures as low as 150 ^{o}C, (ii) Studying and understanding the function of hydrocarbon (HC) trapping materials, such as ion-exchanged zeolites with different metals, for both fundamental and applied purposes, (iii) Investigating the direct conversion of methane (natural gas) to ethanol, where ethanol blends (e.g. E85) are provided by most gas station in the US. To-date, availability of direct conversion of methane to ethanol and higher alcohols is still limited. Data such as these will lead to the development of catalysts and processes for more energy efficient vehicles and lower automotive exhaust emissions.

My Ph.D. thesis at the University of South Carolina focused on developing a fundamental understanding of the synthesis of dendrimer-stabilized nanoparticles in solution and on oxide supports. Most researchers had focused on the final catalytic material, and left untouched the interactions of dendrimers with the metal ions in solution. Our goal was to investigate the complexation between dendrimer templates and metal ions and increase the metal uptake by the dendrimers by varying the experimental conditions. By doing so, I developed a new synthetic procedure towards the formation of dendrimer-stabilized Au/Rh nanoparticles and the subsequent synthesis of cost-effective supported metal catalysts. The major novelty of this synthetic route is that pH control helps to regulate the size of Au/Rh nanoparticles formed, when a dendrimer route is used. This project was funded by Toyota Motor Engineer and Manufacturing of North America, Inc. and I have participated in the design and writing of 3 annual proposals that were successfully funded. Diverging from this project – in conjunction with Dr. Michael Amiridis and Dr. John Regalbuto, I also became interested in different synthetic routes of nearly uniform and highly dispersed nanoparticles on high surface area oxide supports. To this end, I synthesized a series of Ag heterogeneous catalysts using the Strong Electrostatic Adsorption (SEA) technique, previously unexplored in the literature, in which the metal nanoscale size is manipulated by altering the experimental conditions.

In my postdoctoral research in Oak Ridge National Laboratory, I developed a method to treat automobile exhausts for the regulated hydrocarbon (HC) emissions resulting from cold-starting engines. The most effective solution is to employ suitable materials which can trap HCs temporarily, such as ZSM-5 and β-zeolites. The silver exchanged ZSM-5 and β-zeolites, exhibited an increased propylene adsorption/desorption in the absence of both H_2O and CO_2. Combination of synthesis and evaluation of such materials gives me tools to better elucidate fundamental understanding in the adsorption of HCs in the automotive exhaust emissions before they pass to the catalytic converter. Recently I have initiated a project for the design of Pd based catalytic materials supported on a ZrO_2-SiO_2 mixed oxide support, aiming to enhanced low-temperature oxidation performance. Under certain conditions, Pd/ZrO_2-SiO_2 outperformed a Pd/Pt commercial diesel oxidation catalyst. The rigorous synthesis of catalytic materials was new for my research group in ORNL. As a result, I now have experience in starting up an academic research lab including designing and constructing laboratory process equipment necessary for my research.

Apparently, catalytic materials with enhanced low-temperature oxidation performance are necessary, with significant attention being paid to their stability under harsh reaction environments, typical to automotive emissions control due to hydrothermal aging and poisoning. Furthermore, HC porous trapping materials for the treatment of automobile exhausts resulting from cold-starting engines need to be developed, ensuring temporary retention of HCs until automotive emission control catalysts are lit off. After resolving those issues, future work will build upon my expertise in catalyst synthesis and HC trapping materials to develop fundamental understanding of structure-activity relationships for the formation of useful materials that will provide a basis for lowering the automotive emissions. Since this is an emerging field, I anticipate this type of studies to extent into the future and I seek to undertake them as a tenure track faculty.

(6ds) Solvent-Based Control over Nanostructure of Midblock Sulfonated Block Copolymers

Mineart, K., North Carolina State University

Block copolymers containing ionic pendant groups, also referred to as block ionomers, have become increasingly popular due to their potential application as fuel cell and water desalination membranes, as well as components in photovoltaic devices and polymeric actuators. Block ionomers hold promise in these contemporary technologies because of their inherent ability to form separate ionic and nonpolar microdomains at nanoscale dimensions. Segregation of ionic and nonpolar segments enables simultaneous water, or ion, transport in conjunction with mechanical robustness. The ionic and nonpolar blocks are, however, highly incompatible, which can lead to long-term metastable morphologies. While thermal annealing is typically used to refine block copolymer self-assembly, most block ionomers possess inaccessible glass transition temperatures, thereby making thermal annealing ineffective. The development of morphological and property design paradigms in block ionomers represents a main challenge preventing widespread use. Here, we present a facile means by which to alter the morphological behavior of midblock-sulfonated pentablock ionomers through the use of solvent-based processing. A combination of transmission electron microscopy and microtomography (TEM/T) and synchrotron small-angle X-ray scattering (SAXS) are used to probe nanostructural features present in bulk films (~100 μm thick) produced from a variety of different preparation strategies. Results indicate that solution-cast morphologies are solvent-templated and that subsequent vapor annealing can either modify geometric parameters within one morphology or convert any morphology to the anticipated equilibrium nanostructure, depending on the selectivity of the solvent. To the best of our knowledge, these results provide the first evidence of morphological control/refinement in block ionomers of commercially relevant molecular weight.

418364 Molecular Interactions and Behavior in Complex Systems

Sunday, November 8, 2015
Exhibit Hall 1 (Salt Palace Convention Center)
Blair Kathryn Brettmann, Institute for Molecular Engineering, University of Chicago, Chicago, IL

In the coming years I aim to establish an independent research program focused on molecular interactions and behavior in complex systems. I have set this direction due to my PhD work on molecular level mixing in pharmaceutical drug products and from my time in the building materials industry, where I noted a large unmet need for deeper understanding of complex products, especially polymer composites. As a future principle investigator, my research program will combine a scientific understanding of behavior and structure on the molecular to micron scale with a drive towards complex systems encountered in realistic applications.

During my PhD at the Massachusetts Institute of Technology with Bernhardt Trout and Allan Myerson, I was part of the MIT-Novartis Center for Continuous Manufacturing. We focused on transitioning from batch to continuous processing of pharmaceuticals, and, as part of the downstream team, I developed an electrospinning-based method of mixing polymer excipients with active ingredients in solution and drying into a film that could be processed into a tablet. Key to adoption of our manufacturing process was a thorough understanding of the mixing of the active ingredient with the polymer and how that affects the properties of the final drug product. Using solid-state NMR, I was able to confirm homogeneous mixing with no phase separation, which provided enhanced stability and drug release behavior. For rapid development and high quality products, studies such as this on molecular to micron scale behavior are essential at early stages of process and product development. This was also evident during my two years at Saint-Gobain, where I worked on projects relating to solution deposition of coatings from polymer and particle mixtures.

Driven by a desire to understand molecular interactions in the types of complex systems I worked with at Saint-Gobain, my post-doctoral research at the University of Chicago with Matthew Tirrell in conjunction with Phillip Pincus at the University of California Santa Barbara has been focused on modeling polyelectrolyte brush collapse in the presence of multivalent ions. This has allowed me to explore the behavior of industrially relevant materials in much greater depth, with the end goal of improving control of their properties and utility for a wider variety of applications.

Going forward, an understanding of the detailed interactions and behaviors of mixtures will become increasingly important as products with higher levels of functionality are developed and require more complex combinations of components, all while maintaining low production costs. This is the case for applications ranging from pharmaceutical products to energy efficient building materials, and my future research will focus on improving the understanding of these systems in order to greatly reduce the time for development of new products and improve the overall performance and quality.

418619 Theoretical Approaches to the Design of Clean-Energy Processes and Materials

Abstract:

Theoretical Approaches to the Design of Clean-Energy Processes and Materials

Peter C. Psarras, Stanford University

Introduction

This past half-century has brought much change to the landscape surrounding the chemical processes that sustain our global socio-economic framework. In the midst of this evolving climate, it has become increasingly apparent that new motivations, such as those for greater energy-efficiency and less water-intensive processes, are not necessarily aligned to those of previous concern. Fortunately, this call to attention has re-energized the field of materials research, whereby promising new technologies can serve to bridge the gap between stale, out-of-date processes and those who would fit better into this new energy and environmentally-conscience landscape. Concurrent to this scientific revitalization has been a related emergence in its own right – the dawn of high performance computing environments. As such, theoretical methods represent a powerful approach to the design of materials aimed at resolving the shortcomings of essential industrial processes through revision and innovation. These methods are often favorable in situations where there exists a need to explore and screen a number of materials at low cost and risk, where conditions prove unsuitable or even unfeasible via conventional methods, or where an understanding is sought at molecular levels or timescales that are incompatible with experimental measurement. This presentation will introduce how different theoretical methods can be used to solve four issues related to climate change: nitrogen-selective membranes for indirect CO₂ capture, nitrogen-functionalized micro/mesoporous carbon sorbents for direct CO₂ capture, the assessment of carbon neutrality for irreplaceable industrial processes, and the effect of late-transition metal doping on methane selectivity in Fischer-Tršpsch (FT) catalysis.

Vanadium-Ruthenium Alloys for Indirect CO₂ Capture

Membranes represent one promising technology for carbon capture, whereby the flue stream is subjected to some material that can selectively remove CO₂ and prevent its release. Many intrinsic properties of the capture material can be investigated theoretically. Here, we propose to investigate alloyed VRu membranes for selective nitrogen capture. In this technology, rather than separating out CO₂ directly, nitrogen is separated from the flue stream. Nitrogen is far more efficient to separate, owing to its greater flue concentration and corresponding pressure differential across the membrane. Consequently, the remaining flue becomes more concentrated after nitrogen removal, CO₂ included, making all downstream capture points more efficient by the same virtue. Naturally, a highly selective membrane with good mechanical stability and high nitrogen permeability is sought. This study will prepare 14 V_aRu_b alloys in concentrations spanning those considered in real application. Vanadium is known to bind nitrogen strongly, making it a good candidate for activation of the N-N triple bond. However, this tight binding inherently slows membrane diffusion. Nitrogen permeability can be represented as a product of its solubility and diffusivity. The former is sought by considering the effect of Ru concentration on relative binding energies and vibrational frequencies of interstitially-bound nitrogen, while diffusivity can be examined by calculating all potential TS-pathways, tabulating their rates, and employing kinetic Monte Carlo methods. Together, the results of these two methods will combine to predict nitrogen permeability as a function of Ru content. These findings can inform on membrane design, whereby theoretically optimal Ru composition can be validated experimentally.

Nitrogen-Functionalized Porous Carbons for Enhanced CO₂ Capture

One of the most mature methods for CO₂ capture involves CO₂ absorption via amine-based solvents; however, this technology is notoriously energy intensive, as amine regeneration requires high regeneration temperatures. Additionally, this process is water-intensive, a factor that is becoming increasingly important in drought-ridden states like California. Solid sorbents represent an alternative method for CO₂ capture, with liquid solvents replaced by (typically) large surface area, carbon-based porous frameworks. These sorbents are generally low-cost, easily fabricated, and are far less energy intensive in terms of regeneration. Further, their design can be customized through the inclusion of surface-functionalized groups. Unfortunately, current sorbents display CO₂ uptakes that are deemed too low to be cost-competitive with traditional solvent-based scrubbing. Nitrogen-functionalization of these porous carbons could result in more efficient CO₂ capture, owing to the same chemistry exploited by basic solvents in capturing acidic CO₂. Similar studies have examined the effects of oxygenated functional groups on CO₂ uptake and selectivity over N₂.

This study will employ grand canonical monte carlo methods to explore the effect of quaternary, pyridinic, pyrollic, and oxidized-N groups on CO₂ uptake in porous carbon sorbents. Doping amounts will be varied to ascertain the optimum coverage for CO₂ capture. Additionally, gaseous mixtures of CO₂/N₂ and CO₂/N₂/H₂O will be examined to assess CO₂ selectivity. Theoretical performance will be validated through comparison with experimentally obtained CO₂/N₂ isotherms over fabricated hierarchial micro/mesoporous N-doped carbon sorbents. The combination of these methods will help to inform on sorbent design by illustrating which groups are most important for enhanced CO₂ uptake. Design to include more of a particular functional group can be achieved through, for example, a change in carbonization temperature.

Assessment of the CO₂ Capture Potential from Irreplaceable Industrial Sources

In 2013, CO₂ emissions from all industrial processes totaled 163.0 MMT, equivalent to the annual CO₂ emissions of roughly 35 million automobiles. This figure excludes indirect emissions associated with electricity usage. The heaviest emitters (coal power plants, natural gas combined cycle plants, etc.) continue to receive the majority of attention and funding in terms of carbon capture projects; however, these major sources may also benefit from less CO₂-intensive alternatives. Unfortunately, many industrial processes fall into a category by which there are no alternative routes to product available. For example, steel and cement production both involve processes that directly emit CO₂ as a by-product (via the oxidation of metallurgical coke and conversion of calcium carbonate to lime, respectively). As there materials constitute the irreplaceable fabric of industrialization, CO₂ emissions from these, as from other irreplaceable industrial processes, are projected to increase unabated. With capture technology in place, these emissions can be diverted instead to viable CO₂ reuse and sequestration opportunities, such as oil refining, enhanced oil recovery, food processing, metal treatment, and fertilizer production.

To assess the capture potential of irreplaceable industry, it will be necessary to geo-reference these sources alongside all current and potential future CO₂ users (sinks), with the goal of making economically sound linkages between source and similar-sized usage markets. This will entail a cost analysis of on-site capture plus additional transport costs (freight versus pipeline, hazmat fees, etc.). Geographic information systems (GIS) mapping will assist in defining the most cost-effective mechanisms for CO₂ delivery. As these costs are inventoried, the financial incentive gap necessary to compel the targeted source-sink pairings to move forward is calculated. This effort will develop a current economic assessment of moving irreplaceable industry toward carbon-neutrality. Though these industries represent a small portion (*ca.* 3%) of total CO₂ emissions, their permanence requires immediate attention. As these low-hanging fruits are tackled, this study may serve as a model for assessing carbon-neutrality in other sectors.

The Effect of Late Transition Metal Doping on Methane Selectivity in Fischer-Tropsch Catalysis (Dissertation Work)

Fischer-Tropsch synthesis has endured a long and industrious career as the premier chemical method for the synthesis of hydrocarbons from non-petroleum-based precursors. Historically, a collection of factors has dictated interest in the FT process, including foreign policy, availability of precursor(s), availability of alternate fuel sources and reserves, and environmental considerations. The effect of late transition metal substitution into Fe(100), Ni(111), and Co(0001) surface analogs is investigated using density functional theory (DFT) methods. The surface is modeled using a 7-atom cluster, with perimeter atoms saturated with hydrogen atoms to approximate surface coordination and mitigate dangling bond artifacts. *Ab initio* calculations are performed to predict adsorption energies and transition state geometries and energies for eight surface adsorbate species: C, CH, CH₂, and CH₃ to represent the hydrogenating steps on surface carbide, while C+ CH, CH + CH, C + CH₃, and CH₂ + CH₂ represent species involved in four competitive coupling pathways. A review of the effect of Cu, Ag, Au, and Pd on the reaction energies and barriers associated with these critical steps is discussed. Methane selectivity can be described by a single descriptor, the carbon binding strength; thus, a comparison of these energies over different surface/promoter analogs is an effective method to screen for promising candidates. In this representation, a lowering of carbon binding strength is equivalent to the lowering of methane selectivity.

419051 Continuous and Oscillatory Multi-Phase Microscale Technologies for Pharmaceuticals, Materials and Energy

Sunday, November 8, 2015
Exhibit Hall 1 (Salt Palace Convention Center)

Milad Abolhasani and Klavs F. Jensen, Department of Chemical Engineering, Massachusetts Institute of Technology, Cambridge, MA

My proposed interdisciplinary research program will include development of multi-phase micro systems for fundamental studies of (a) synthesis of cadmium-free semiconductor nanocrystals (i.e., quantum dots (QDs)) for applications in biological imaging and displays; (b) utilization of light-absorbing nanoparticles for the next generation of energy-efficient carbon capture and sequestration technologies; and (c) different mechanisms involved in single/multi-phase C-C and C-N cross-coupling catalytic cycles. Capitalizing on the enhanced mixing and mass and heat transfer rates enabled by flow segmentation compared to batch and single-phase systems, I will focus on exploring the nucleation and growth mechanisms of cadmium-free QDs. The lack of understanding of these mechanisms has prevented further progress and scalability of cadmium-free QDs in comparison with II-VI QDs. The second direction of my research program will be focused on fundamental heat and mass transfer studies of photo-thermal release of carbon dioxide (CO_2) in the presence of light-absorbing nanoparticles and investigation of CO_2 conversion into minerals as well as renewable energy sources. Finally, building on my recently developed oscillatory multi-phase technique, in the third theme of my research program, I will utilize multi-phase oscillatory flow strategy towards exploring the rate limiting steps and fundamental mechanisms associated with single/multi-phase C-C and C-N cross-coupling reactions. The findings of this research will be applied towards development of more efficient catalysts and optimization of flow chemistry approaches for continuous manufacturing of pharmaceutical targets.

420307 Experimental and Computational Studies of Fluid-Particle Flow Systems

Abstract:

Fluid-particle flow systems, such as fluidized beds, exhibit a rich array of complex physical phenomena and are vital to the pharmaceuticals, oil, chemicals and food industries. While these systems have been used industrially for nearly a century in processes ranging from fluidized catalytic cracking to granulation, they are also important in new applications to solve 21st Century problems, such as chemical looping combustion for efficient carbon capture and sequestration. Currently, the energy and economic efficiency of these systems, as well as the use of these systems in new processes, is limited by a lack of in-depth understanding of the complex physics occurring in these systems over a wide range of length scales, and the relationship between this physics and reaction kinetics. Recently, the application of new experimental techniques and computational models has started to shed new light on these fascinating and industrially important physical phenomena.

Over the course of my research career, I have combined computational and experimental techniques to provide understanding of the fundamental science of fluidized beds beyond that which could be achieved using any individual technique. During my PhD at the University of Cambridge, I created a 3-D cylindrical Euler-Lagrange model of fluid-particle flow systems to simulate fluidized beds in their most commonly used state. I compared simulation predictions with previously obtained experimental results in order to validate the model, shed light on the nature of certain experimental techniques and reveal the origin of pressure oscillations in fluidized beds. Additionally, I developed new experimental techniques in order to conduct the first ever measurements of gas velocity and velocity distribution in a fluidized bed of particles using magnetic resonance imaging (MRI). These measurements provide the first data and details showing the distributions in gas velocities in different states of fluidization: minimum, homogeneous, and bubbling fluidization. I compared these results with classical analytical theory for gas flow in fluidized beds, in order to provide insights on the validity of theories used every day by industrial practitioners. As a post-doctoral researcher at Princeton University, I am bridging my doctoral research by investigating how computational predictions of gas phase dynamics match my experimental results. I am also branching out into new areas of fluid-particle flow by developing computational models to investigate wet fluidized beds, in which a small amount of liquid forms bridges between particles, leading to agglomeration, as well as falling granular jets which exhibit phenomena indicative of an effective surface tension.

As a research faculty member, I plan to expand upon the foundation I have built in order to explore the critical interplay between physics and chemistry in fluid-particle flow systems over multiple length scales. My research program will involve developing and combining multi-scale modeling and multi-scale measurement efforts to provide insights on physical phenomena that occur on length scales varying from one particle diameter to a reactor tens of meters in diameter. This research path will also entail coupling the chemical reactions with the modeling and measurement techniques in a way such that the relationship between the physical and chemical phenomena can be better understood. Finally, my program will involve direct collaboration with researchers and industrial practitioners who build fluidized beds for applications such as combustion and catalysis, in order to aid in the design and measurement of their experiments. My ultimate aim is to enable the next generation of two-phase granular flow systems, which can offer real world solutions to global challenges ranging from clean energy production to food supply to safe and economic production of pharmaceuticals.

I also intend to use my unique educational background, coming from a double major in physics and chemical engineering as an undergraduate at MIT to a PhD student and Gates Scholar at the University of Cambridge to a post-doctoral researcher at Princeton University, to enhance chemical engineering education at a research university. From my experiences, I have seen a number of ways in which the on-campus undergraduate and graduate education experience in the U.S. can be further enhanced. Firstly, as a supervisor in the education system unique to Oxford and Cambridge, I have seen the benefits of students discussing material and problems in 2-to-4-person student groups with a graduate-level supervisor in an organized yet open-ended discussion setting. Secondly, as a Gates Scholar, I have had the privilege of interacting with leading graduate students in diverse disciplines, ranging from international relations to public health to engineering; this experience has shown the exceptional and expansive nature of the graduate education that comes from discussing important research problems that cut across academic disciplines. As a faculty member, I plan to take an active role in education to ensure more chemical engineering students at my university are provided with these valuable learning opportunities.

421058 Functional Polymers for Industrial and Bio-Applications: Synthesis, Properties & Engineering

Abstract:

Abstract: From the ring-opening polymerization of N-carboxy anhydrides (NCAs) yielding polypeptides, to carbon/graphene materials and military-related applications such as the preparation of responsive hydrogels against toxic gases, as well as designed functional materials for internal bleeding and imaging to tissue adhesives/sealants, the challenges are numerous. Organic synthesis is a great tool for designing new monomers, clickable reagents and functionalized organic compounds for coupling reactions, leading to well-defined polymers and materials that can be used for industrial and bio-applications. Particular focus is given on water-soluble polymers that can form responsive hydrogels, leading to swelling/collapsing mechanisms against an agent or higher affinity compounds or temperature, and can be examined through rheology and neutron scattering. Furthermore, injectable materials are designed that can lead to significant reduction in bleeding time in live animals or can be spontaneously polymerized under physiological conditions, adhering tissues or sealing air/lymphatic fluids, supporting the current clinical materials. Finally, collagen-mimetic hydrogels for cosmetics are designed in large scale via polypeptide synthesis, with highly stretching and reforming capabilities, with the aim of commercializing the product.

(6ea) Electrocatalysis and Photocatalysis for Energy Sustainability

McKone, J. R., Cornell University
Abruña, H. D., Cornell University
DiSalvo, F. J., Cornell University

This poster features previous and ongoing work on materials and methods for renewable energy conversion and storage using artificial photosynthesis and battery approaches. Specifically, we have demonstrated electrochemical solar-driven hydrogen evolution using Si and WSe_2 semiconductors that have been decorated with Pt catalysts. We have further developed non-noble alternatives to Pt, including Ni-Mo alloys and metal phosphides, and demonstrated their utility in functional solar hydrogen evolving devices. More recently, we have proposed and developed a unique device for solar photochemical energy storage called the "solar flow battery," which directly interfaces semiconductor absorbers with kinetically facile redox couples. This approach allows for efficient and fully integrated solar energy capture, storage, and release on demand as electricity. Finally, we have explored and validated a promising approach for flow battery electrolytes involving the use a single redox-active organic molecule to facilitate both oxidative and reductive processes. Alongside this work I will discuss plans for future research in energy conversion and storage, with an emphasis on combining the tools and insights of gas phase heterogeneous catalysis and electrocatalysis.

423026 Biomass Pretreatment Using Ionic Liquid and Glycerol Mixtures

Sunday, November 8, 2015
Exhibit Hall 1 (Salt Palace Convention Center)
Joan G. Lynam, Chemical and Materials Engineering, University of Nevada, Reno, Reno, NV

Lignocellulosic biomass is a sustainable resource that can supplement fossil fuel use to produce liquid fuels and chemicals. However, its recalcitrant structure including cellulose, hemicelluloses, and lignin biomacromolecules is challenging to deconstruct. Pretreating biomass to convert it to useful liquids dominates process economics. Many pretreatment methods exist, but most require hazardous chemicals or processing conditions. Many ionic liquids (ILs), salts molten below 100 °C, can be used to deconstruct lignocellulosic biomass and are less hazardous than the volatile organic compounds often used.

While effective and recyclable, ILs are expensive. To reduce costs, dilution with other safe compounds would be desirable, if there is no impact on deconstruction efficiency. Glycerol, a food additive, is inexpensive and becoming more so since it is a by-product of the expanding biodiesel industry. Use of glycerol as an additive or diluent for ILs is presented in this poster.

Rice hulls are an available biomass, with over 100 million tons produced per year, but with little practical use. The IL 1-ethyl-3-methylimidazolium formate (EMIM Form) when mixed with an equal amount of glycerol has been shown to be effective in pretreating rice hulls. Ambient pressure, a pretreatment temperature of 110 °C, and a reaction time of three hours gave rice hulls that could be enzymatically hydrolyzed to give reasonably good glucose and xylose yields considering the recalcitrance of this silica-armored biomass.

The IL EMIM Form was also effective when diluted with an equal amount of glycerol to pretreat loblolly pine, a fast-growing softwood. Loblolly pine was pretreated at 140 °C for three hours to produce a solid enriched in cellulose and hemicelluloses, while a lignin-rich product could be precipitated from the IL-glycerol mixture. Similar products were obtained from pretreatment with a mixture of 75% 1-ethyl-3-methylimidazolium acetate (EMIM Ac) and 25% glycerol. Enzymatic hydrolysis of the pretreated solids gave glucose, mannose, and xylose yields up to 18 times that of the raw pine. Klason lignin in the pretreated pine was reduced.

Viscosity measurements of pure glycerol, EMIM Form, EMIM Ac, and IL-glycerol mixtures were very different at ambient temperature, but were quite similar at typical biomass pretreatment temperatures. Biomass pretreated by mixtures with higher viscosity tended to give better carbohydrate yields after enzymatic hydrolysis. This phenomenon may relate to more energy put into shearing flow of the IL-glycerol-biomass system resulting in biomass particle shearing or stretching that enhanced solvent access into the biomass. Higher excess molar volumes, V_m^E, tended to align with better carbohydrate yields after enzymatic hydrolysis. This suggests that separation of ions by a molecular solvent may faciliatate entrance of the pretreating ions into biomass.

422168 Design and Fabrication of Nanostructred Materails for Energy Applications and Functional Nanocoating

Abstract:

Nanostructured materials have attracted great interest in recent years due to their outstanding properties ranging from small size, high surface area, and outstanding performance in a variety of applications. For my doctoral research, I have been working on designing and synthesizing nanocomposites for application on catalysts and Li-ion batteries electrode materials. I developed a novel instant gelation method for synthesizing of porous metal oxides/sulfides/carbinde/nitride with large surface area. If carbon source is introduced, the metal components can be well embedded in a 3D porous conductive carbon matrix. The nanocomposites are ideal candidates for batteries, catalysts, sensors etc.

As a postdoctoral fellow, I have developed a general approach to prepare various free-standing metal sulfides and carbon mat for next-generation flexible batteries. I also developed a facile and cost-effective electrospray approach for ultrauniform, ultrafast, instantly-drying, and binder-free electrode preparation via utilizing micrometer-size graphene oxide nanosheets. This approach can significantly simplify the tedious traditional preparation techniques – drop cast and doctor blade, and more importantly get better adhesion and high uniformity. With this method, the desired mass loading of electrodes can be achieved with only a few minutes of spraying. The electrodes made can be directly used for battery assembly, no extra drying step required. As an extension of the electrospray technique, I also conducted experiments on functional surfaces nanocoating with desired properties (e.g. anti-corrosive, anti-fingerprint, anti-reflective, self-cleaning properties), in which "lotus effect" property was achieved.

In my independent faculty career, I would like to continue working on nanomaterials with application ranging from lithium batteries (Li-ion, Li-air, Li-sulfur battery), sodium-ion batteries, manganese-ion batteries, supercapacitors to catalysts, fuel cells, and solar cells. I will also be working on functional film coating via electrospray, for example, epitaxial manganese-based perovskite oxides thin film with colossal magnetoresistance effect, anti-corrosive film, and self-clean film coating.

422191 Development of Next Generation of Energy-Efficient Separation Technologies through Advanced Tunable Materials

Abstract:

My research work focuses on design and development of novel high performance hollow fiber membranes and sorbents with complex morphologies out of advanced polymers and nanostructured materials, and their applications in membrane-based gas separation, and renewable energy production. After completing my PhD on biofuels purification by pervaporation-based membrane process at Chinese Academy of Sciences, I worked as an Assistant Professor for more than one year at the same institute and continued to focus on the research in bioethanol and biobutanol production from biomass resources by fermentation/pervaporation integrated process. In 2012 I joined Koros Research Group as a Laboratory Manager and Research Engineer at Georgia Institute of Technology. During my work with Dr. Bill Koros, my research interests in the areas of **Energy Efficiency and Environmental Sustainability** have broadened greatly, but still focus on the advanced membrane materials for low energy intensive large scale separation processes.

Currently, I am a Research Engineer at Georgia Tech, where I am expanding my expertise in membrane-based gas separation by engineering the tunable membrane materials. As a Laboratory Manager, I also provide mentorship and instruction to the undergraduates, PhD students and post-doctoral fellows in a group of over 20 researchers since 2012. This position has allowed me to expend my expertise from dense films and asymmetric flat-sheet membranes to high performance hollow fiber membranes, and also from membrane-based separations to combined hollow fiber sorbents adsorption and membrane-based coupled separations. By using advanced crosslinkable polymers, functionalized polymers of intrinsic microporosity (PIMs), and several other commercially available polymers, I have created high performance dense films and hollow fiber membranes for aggressive sour gas separations, CO_2 capture, and renewable energy production.

My future research will address both fundamental issues and applied research related to membrane-based gas separations, CO_2 capture, and organophilic pervaporation membrane technology for renewable energy production. Using the knowledge of the industrially available material properties developed in the fundamental studies, I will strongly focus on conducting cutting-edge research in engineering the advanced polymer structures and economical processes to produce high performance membranes with the emphasis on global CO_2 reduction, energy conversion, and clean technology which will be met industrial applications.

423148 From DNA to Polymer Membranes: Soft Materials for the 21st Century

Abstract:

My research focuses on theoretical and computational problems in soft matter and complex fluids. I did my PhD work at the University of Minnesota with Kevin Dorfman, where we studied equilibrium and hydrodynamic properties of DNA confined in nanometer scale environments. Such environments are particularly important to nascent genomic mapping technologies, which seek to linearize DNA as a step prior to optical imaging. In our research, we strove to advance both a fundamental understanding of polymer physics and practical considerations of device design. For example, in one study we demonstrated that the configurational fluctuations of a DNA molecule decreased rapidly below a threshold confinement diameter. This insight provided a rationale for a previously unexplained observation that optical mapping of DNA only became feasible below this threshold.

I am currently working as a postdoctoral scholar with Glenn Fredrickson at the University of California, Santa Barbara on modeling complex polymer phase behavior in convective flows. Pertinent industrial examples include (but are not limited to) processes for polymer membrane formation and reactive compatibilization. Experience has shown that numerical tractability is a challenge for such simulations. Recently, we have developed an efficient numerical method for a class of models derived from a general framework based on non-equilibrium thermodynamics and statistical field theory. Work is ongoing to use these methods for a multi-component model of an immiscible polymer solution to examine membrane formation.

Going forward, I am excited to explore new research topics in soft matter theory. A recent perspective on the grand challenges in soft matter theory and computation (Liu et al., Soft Matter, 2015) outlines a significant need for work on problems that are (i) far from equilibrium and (ii) involve a hierarchy of length and time scales. Making progress in these areas will involve the need to use cutting edge numerical methods and to develop new ones as necessary. In doing so, I will focus on bridging the fundamental and the practical in an effort to help develop the next generation of soft materials.

423477 Highly Active and Durable Extended Surface Electrocatalysts

Abstract:

Extended surface nanostructures have been developed as catalysts for a variety of electrochemical processes. The majority of efforts in this field have focused on oxygen reducing electrocatalysts for proton exchange membrane fuel cells and oxygen evolving electrocatalysts for acidic electrolyzers. Preliminary advanced geometry materials, however, have also been developed for alcohol oxidation and anion exchange membrane fuel cells (oxygen reduction, hydrogen oxidation, and alcohol oxidation). The formation of these nanomaterials has included various gas phase (sputtering, atomic layer deposition, and chemical vapor deposition) and liquid phase (direct deposition and spontaneous galvanic displacement) processes. In recent years, spontaneous galvanic displacement has become a focus due to a unique combination of high site-specific activity and surface area commonly found in materials produced by the method.

The principal application of interest with this catalyst type has historically been proton exchange membrane fuel cells, where the amount of platinum in the catalyst layer accounts for a significant portion of fuel cell cost and limits the commercial deployment of the device.[1] Extended surface nanomaterials have previously been developed as electrocatalysts for fuel cells and offer key advantages to nanoparticles: an order of magnitude higher specific activity; long range conductivity; and long term durability.[2] Extended nanostructures, however, are traditionally limited by low surface area. Spontaneous galvanic displacement occurs when a less noble metal template contacts a more noble metal cation and combines aspects of corrosion and electrodeposition. Although catalyst development has included a variety of approaches, catalysts formed by galvanic displacement are ideally situated, being able to take advantage of the specific activities generally associated with the catalyst type while significantly improving upon the surface area.[3]

Efforts in this field initially used silver and copper templates as a learning tool, but have since moved to nickel and cobalt since these metal have lower redox potentials and will not plate the anode during fuel cell operation. Recent developments in platinum-nickel nanowires have produced materials with surface areas in excess of 90 $m^2\ g_{Pt}^{\ 1}$.[4] Post-synthesis processing has created nanostructures with oxygen reduction mass activities 7 times greater than platinum nanoparticles and 5 times greater than the U.S. Department of Energy membrane electrode assembly target in rotating disk electrode half-cells. Following accelerated durability testing (30,000 cycles, 0.6 1.0 V, rotating disk electrode), these materials lose 3% oxygen reduction mass activity and less than 0.5% catalyst mass to dissolution. Remaining barriers, including synthesis scalability and electrode fabrication, are being addressed. Recent improvements in membrane electrode assembly performance suggest that the promise of extended surface catalysts in half-cells can be realized in the device.

[1] D. Papageorgopoulos, in: U.S. Department of Energy (Ed.), http://www.hydrogen.energy.gov/pdfs/review14/fc000_papageorgopoulos_2014_o.pdf, 2014.

[2] M. Debe, in: U.S. Department of Energy (Ed.), http://www.hydrogen.energy.gov/pdfs/review09/fc_17_debe.pdf, 2009.

[3] S.M. Alia, Y.S. Yan, B.S. Pivovar, Catalysis Science & Technology, 4 (2014) 3589-3600.

[4] S.M. Alia, B.A. Larsen, S. Pylypenko, D.A. Cullen, D.R. Diercks, K.C. Neyerlin, S.S. Kocha, B.S. Pivovar, ACS Catalysis, 4 (2014) 1114-1119.

423515 Dynamic Modeling and Design of Colloidal Assembly

Abstract:

Understanding concentrated colloidal dynamics in the presence of different pairwise interactions and external fields provides a basis to predict the temporal evolution of colloidal microstructures in diverse phenomena including suspension rheology and colloidal assembly. However, a microscopic theory of concentrated colloidal dynamics does not yet exist that rigorously includes both statistical mechanical (configuration dependent free energy changes) and fluid mechanical (configuration dependent multi-body hydrodynamic interactions) contributions. This work shows the implementation of a novel approach to model colloidal assembly by means of analyzing the temporal evolution of microstructures, that can be observed in experiments and simulation, with a small number of order parameters that describe the state of the system. This methodology is used to analyze the structure evolution of isotropic and anisotropic colloidal building blocks. Knowledge of these dynamic models enable the fundamental understanding of assembly systems, as well as the optimization, design and control of assembly systems to produce low-defect colloidal crystals.

424686 Towards the Computational Design of Monolayer (Hydroxy)Oxide-Metal Bifunctional Catalysts

Abstract:

Ultrathin (hydroxy)oxide films grown on transition metal surfaces have widespread applications in processes related to electrocatalysis, photocatalysis, and heterogeneous catalysis, among others. These hybrid systems with novel properties are often prepared and characterized under very specific conditions, such as ultrahigh vacuum. When subsequently used in various applications, however, significant structural evolution, which depends strongly on the reaction conditions in-situ, is expected, and this evolution remains largely unexplored. In order to identify structure-property relationships and, ultimately, design/screen new materials with improved performance, the development of such understanding is essential.

The description of complex metal/hydroxyoxide interfaces has long been a challenge for modeling, because of the complications introduced by strong correlation effects in transition metal oxides and by the very rich interface structures. As a consequence, even many classic questions related to these systems, for example the origin of the strong metal support interaction (SMSI) that plays a central role in the catalysis of some oxide-supported metal nanoparticles, remain unanswered.

Recently, we have addressed the major challenge of developing self-consistent and highly accurate description of strongly-correlated transition metal oxides, hydroxides and oxyhydroxides, through synergistic use of a Hubbard U correction, a state-of-the-art dispersion correction, and a water-based bulk reference state for the calculations. The strong performance of our approach is illustrated on a series of bulk transition metal (Mn, Fe, Co and Ni) hydroxides, oxyhydroxides, binary, and ternary oxides, where the corresponding thermodynamics of redox and (de)hydration are described with standard errors of 0.04 eV per (reaction) formula unit.

We have applied these methodologies to investigate the growth of monolayer Ti- and Fe-(hydroxy)oxide films on VIIB group and IB group noble transitional metal surfaces, and have obtained mechanistic understanding of the origin of the strong metal support interaction. This understanding has been extended to predict the structure evolution of the other monolayer transitional metal (Mn, Co, and Ni) (hydroxy)oxide films on a variety of substrates in electrochemical environments. It has been demonstrated that the structure and oxidation state of the films can be systematically tuned by changing the applied electrode potential, the nature of substrates, and/or the size of the film cluster, and the predictions have been confirmed by synchrotron experiment on the monolayer Ni-(hydroxy)oxide/Pt, a hydrogen evolution (HER) electrocatalyst with improved performance in comparison with traditional Pt catalyst in alkaline conditions.

The above analyses have led us to conclude that the Ni (hydroxyl)oxide/Pt interface under hydrogen evolution HER conditions is composed of NiOH monolayers on Pt, and the predicted structure is subsequently used for a detailed structural and electrocatalytic analysis at three-phase boundaries of NiOH films, Pt substrates, and the surrounding water. The kinetic analysis shows that water can be dissociated to adsorbed OH* and H* groups in a bifunctional mechanism that significantly lowers the activation barriers for alkaline HER and provides a clearly measureable acceleration of the catalyst process as compared to mono-functional Pt catalysts. The mechanistic understanding and kinetic analysis also suggests a new catalyst, $NiOH/Pt_3X$, with improved performance in comparison with NiOH/Pt, which is further supported by electrochemical experiment.

The development of highly accurate description of strongly-correlated transition metal oxides, the mechanistic understanding of interface structure evolution and structure-properties relationships, and the successful prediction of new electrocatalyst has provided the foundation towards rational design of novel catalysts with increased activity, improved selectivity and controlled stability.

Selected Publications (20 total, 12 first (co-)author, h-index=10):

1. Z. H. Zeng, M. Chan, Z. Zhao, J. Kubal, D. Fan and J. Greeley, Synergistic effect of dispersion corrections and reference state for first principles-based prediction of electrochemical Pourbaix diagrams (submitted to **Phys. Rev. Lett.)**
2. Z. H. Zeng, J. Kubal, K.-C. Chang, N. Markovic and J. Greeley, Towards Controlling the Structural Evolution of Monolayer (hydroxy)oxide-Metal on Noble Metal Substrate (submitted to **J. Am. Chem. Soc.**)
3. Hemma Mistry, Rulle Reske, Z. H. Zeng, Zhi-Jian Zhao, J. Greeley, P. Strasser, B. R. Cuenya, **J. Am. Chem. Soc.** 2014, *136* (47), 16473-16476.
4. Z. H. Zeng, M. E. Björketun, S. Ebbesen, M. B. Mogensen and J. Rossmeisl, **Phys. Chem. Chem. Phys. 15 (18), 6769 (2013)**
5. M. E. Grass, D. R. Butcher, Z. H. Zeng, F. Aksoy, H. Bluhm, G. A. Somorjai, W. X. Li, B. S. Mun, Z. Liu, **J. Am. Chem. Soc.** 133 (50), 20319 (2011)

425083 Application of the Technique of Chemical Vapor Deposition of Polymers to a Broad Spectrum of Research Projects – from Nuclear Fusion to Water Purification

Abstract:

Application of the Technique of Chemical Vapor Deposition of Polymers to a Broad Spectrum of Research Projects – from Nuclear Fusion to Water Purification

Aravind Suresh

Assistant Professor-in-Residence, Chemical and Biomolecular Engineering Department,

University of Connecticut, Storrs, CT 06269

E-mail: aravindsuresh.nitt@gmail.com

Research: My research background lies at the intersection of chemical engineering and materials science. My doctoral dissertation was on the development of bi-functional high-temperature oxides with catalytic activity for hydrogen generation and mixed protonic-electronic conductivity for hydrogen purification[1, 2]. After my Ph.D., I have been working on developing inter-disciplinary research projects involving initiated and photo-initiated Chemical Vapor Deposition (iCVD and piCVD) of polymers.

My work led to the discovery of auto-polymerization of hexyl acrylate under ultraviolet radiation of wavelength 254 nm[3]. The remarkable feature of the process was that deposition rate increased linearly with the thickness of the polymer film. The phenomenon suggests that the bulk, and not just surface, of the polymer film acts in continuously accelerating the deposition process. This would mean that either the monomer vapor diffuses into the film or free radicals generated by UV radiation migrate through the thickness towards the gas-solid interface. I am now working on elucidating the mechanism of the process along with obtaining the underlying kinetic parameters.

Prof. Daniel Burkey and I are collaborating with a group at the Lawrence Livermore National Laboratory (LLNL); for their research on nuclear fusion at the National Ignition Facility, they require an optically clear polymer with chemical resistance to oxygen and humidity upon aging. We are exploring iCVD as a potentially better alternative to plasma-enhanced CVD (peCVD) that they currently employ at LLNL. Results indicate that the mild operating conditions and low energetics of iCVD in comparison to peCVD result in a lower concentration of defects and free radicals in the polymer film, which enhances its stability against oxidation.

We are also collaborating with a research group at the University of Connecticut (UConn) in exploring the utility of iCVD for functionalizing polyacrylonitrile (PAN) nanofiber mats prepared by electrospinning. We have been able to put down a conformal polymer film on the nanofibers that completely alters the hydrophilic nature of the mats - the contact angle changes from $\sim 50^{o}$ to $\sim 140^{o}$. The modified mats are currently being tested as the active components in membrane distillation.

Teaching: I have been teaching the senior and junior laboratory courses in the chemical engineering department at UConn for four years now. Chemical engineering laboratory education is my passion and I intend to follow it as I move ahead in my career. Having a tenure-track faculty member who is dedicated to and specializes in developing and maintaining a modern and robust laboratory will be an asset to any chemical engineering department and will add value to its mission of promoting both undergraduate education and undergraduate research.

References:

[1] Suresh, Basu, Carter, Sammes, Wilhite, Journal of Materials Science (2010) **45**: 3215-3227

[2] Zhang, Suresh, Carter, Wilhite, Solid State Ionics (2014) **266**: 58-67

[3] Suresh, Anastasio, Burkey, Chemical Vapor Deposition (2014) **20**: 5-7

425365 Nanobioelectronics in Healthcare: From Nanobots to Wearable Biosensors

Abstract:

The 1966 movie Fantastic Voyage captured the world's imagination, portraying a tiny submarine navigating through the human bloodstream and treating life-threatening medical conditions. Artificial micromotors, operating on locally supplied fuels and performing complex tasks, offer great potential for diverse biomedical applications to realize this vision, including autonomous delivery and release of therapeutic payloads and cell manipulation. During my PhD research in professor Joseph Wang's group at UC San Diego, I developed a variety of artificial micro/nanobots based on different propulsion mechanisms including the fastest synthetic nanomachine to date: the polymer-based catalytic tubular microengine displays very efficient propulsion of over 1000 body length/s. The microengines modified with different functional groups or bio-receptors can selectively isolate specific target (such as protein, DNA, bacteria and cancer cells) from complex media. Newly-developed biocompatible and biodegradable microbots can move autonomously in natural biofluids without additional chemical fuels, holding great promise for in vivo biomedical applications: the polyaniline/zinc microrockets display effective propulsion in acidic environments (such as human stomach); Mg-based Janus micromotors, which utilize macrogalvanic corrosion and pitting corrosion processes, can be powered by salt water. The first in vivo study of artificial microbots is demonstrated in a living organism using a mouse model. The acid-driven propulsion in the stomach effectively enhances the binding and retention of the motors as well as of cargo payloads on the stomach wall. The body of the motors gradually dissolves in gastric acid, autonomously releasing their carried payloads, leaving nothing toxic behind. These works are anticipated to significantly advance the emerging field of nano/microbots and to open the door to in vivo evaluation and clinical applications of these synthetic motors.

While nanobots offer great promise to efficiently deliver the drug payloads and perform diagnosis inside human body, wearable devices are able to realizing personalize medicine externally through non-invasive continuous monitoring on the human skin. However, most currently developed wearable devices are capable only in tracking the physical activities of an individual and fail to provide insight into the individual's state of health. My postdoc research in professor Ali Javey's group at UC Berkeley focuses on the wearable biosensors for non-invasive health monitoring. A fully-integrated sensor array was developed and can serve as an ideal platform for a wide range of real-time healthcare monitoring such as exercise-induced dehydration and medical diagnosis.

With such innovations and developments, along with careful attention to key challenges and requirements, nanobioelectronics (from wearable biosensors to autonomous nanobots in human body) is expected to have tremendous impact on diverse biomedical applications from in vivo drug delivery to non-invasive health monitoring, providing unlimited opportunities limited only by one's imagination.

425216 Design and Synthesis of Functional Polymers for Industrial, Environmental and Energy Applications

Abstract:

Chemical engineering research primarily aims to find solutions to the problems impacting the globe as well as to invent new technologies that improve the quality of life and meet the societal needs. To achieve these targets, the role of an interdisciplinary approach that gathers the advancements in different scientific fields, is significant. My PhD study involved polymer synthesis and application of the materials in an ultrafiltration process to successfully remove boron species from aqueous solutions for the purpose of the reclamation of geothermal waters. In my postdoctoral study, I found the opportunity to design synthetic routes and produce tailor-made polymers in various structural forms to reversibly capture carbon dioxide from ambient air. The main purpose is to develop economically and environmentally feasible technologies to prevent the constant increase of its concentration in the atmosphere.

In addition, we have written a research proposal for the recovery of lithium salts from brine by the implementation of the novel chemistries to produce lithium-chelating resins. Since the conventionally used technologies are expensive and energy-intensive, this work is meant to bring about potential economic impacts of great significance. Moreover, I am also planning to produce block copolymers with well-defined architectures using Polymerization Induced Self-Assembly (PISA) technique. Then, based on a metal-polymer binding chemistry, these nano structures can be used as templates to grow metal and metal oxide nano shapes with inherent stability against aggregation due to their polymeric shields. There is a tremendous number of application areas of great potential especially in energy harvesting and storage materials as well as drug delivery applications.

To conclude, based on my expertise in the synthesis, characterization and application of specially tailored functional polymers, I am planning to implement the recent advancements in the polymer chemistry to provide solutions to environmental, industrial and energy-related problems.

(6el) Tuning the Ionic Conductivity of Polymerized Ionic Liquid Homo-, Random, and Block Copolymers

Evans, C. M., University of California-Santa Barbara
Segalman, R., UCSB

Polymerized ionic liquids (PILs) are an emerging class of ion-conducting polymers based on the familiar chemistries of ionic liquids. In contrast to ionic liquids, either the cation or anion is tethered to the polymer backbone in a PIL which minimizes leakage of the charged species into neighboring materials (for example in a thin film transistor). A major advantage of PILs over conventional ion conducting membranes is that they can operate in the absence of water allowing for high temperature operation. In this project, we seek a better understanding of the conductivity mechanisms in PILs, particularly in materials which conduct protons. One of the main structural features of ion-conducting polymers is the aggregation of ionic groups due to the non-polar, low dielectric constant polymer backbone. These aggregates have a pronounced influence on the conductivity, viscosity, and dielectric constant of the material and we seek to tune these properties via two distinct routes. The first is through the incorporation of polar, non-ionic groups into the backbone of a random copolymer PIL. The second is via nanostructuring of PILs within self-assembled block copolymers containing a non-conductive, mechanically robust block.

A variety of comonomers that are either polar or non-polar and either bulky or flexible have been incorporated into random copolymers with the PIL to tune the dielectric constant, mechanical properties, and conductivity. We have found that up to ~ 20 wt% of a non-conductive species can be added into a random copolymer without substantially altering the Tg-normalized conductivity. This leads to improvements in processability and solubility of such materials and the ability to tune the overall material properties by the judicious choice of comonomers.

The role of nanoconfinement on the conductivity of imidazolium based PILs, where the imidazolium can act as a proton donor/acceptor and conduct protons, has also been investigated. In polystyrene-PIL block copolymers, where the PIL is confined by a hard phase, an order of magnitude increase in the conductivity is observed in the block copolymer relative to the pure PIL. We attribute this to a change in the packing of the imidazolium PIL upon confinement which leads to more efficient pathways for proton transport through the phase separated domain. Substantial differences in the hydrogen bond network in the block copolymer are also observed via FTIR experiments. The effect of confinement is substantially reduced if a softer confining block, methyl acrylate, is employed. Finally, the conductivity in PS-PIL block copolymers is shown to scale with the domain size of the PIL.

425250 Solid Formation in Flow: A Kinetic and Fluid Dynamic Approach

Abstract:

Motivation

Crystal formation (from molecules in solution) and solid deposition (from particulate in suspension) are very intriguing topics at a fundamental level, and at the same time they have a concrete impact on the most diverse aspects of every-day life. For instance, kidney stones are formed via crystallization and solid deposition of soot affects the performance of combustion engines in cars. From a more industrial perspective, crystallization/solid formation in reactors may induce clogging and shutdown of the production process. [1,2]

Approach

The underlying physico-chemical pathways (e.g. nucleation, growth, aggregation, breakage, deposition, re-suspension) regulate the rates and types of solids/crystals formed. These mechanisms are strongly affected by the environmental conditions, e.g. temperature, solvent, flow patterns as well as by particulate-specific properties, such as size of the primary particles/crystallites and surface charge. In this frame, I am employing a combination of computational fluid dynamics (CFD) and population balance equations (PBE), to gain a deeper fundamental understanding on the impact of flow on the mechanisms regulating the solid formation kinetics. The modeling tools being developed allow at the same time to safely guide (micro)reactor and process design, a topic of great interest to both the chemical and pharmaceutical industry.

Personal background

During my PhD at ETH Zurich in the group of Prof. Massimo Morbidelli, I worked on several projects in the fields of polymers and colloids, combining experimental and modeling techniques to gain a phenomenological wide-angle view on the processes under investigation. In particular, I developed a significant experience in the field of PBE (both mono- and multidimensional) solved with different techniques. This allowed me to approach problems such as the formation of multi-active polymer chains, the interplay between aggregation and coalescence of colloidal particles, as well as the aggregation and breakage of protein fibrils. After winning a post-doctoral fellowship from the Swiss National Science Foundation, I joined the group of Prof. Klavs Jensen at the MIT. Currently I am investigating the crystal formation, aggregation and the resulting clogging in microreactors by employing PBE and CFD.

Long term plan

After consolidating my experience with CFD, I plan to approach further problems in crystallization and solid formation. In particular, I would like to contribute in developing modeling tools able to describe crystallization in flow in a general way, deepening microreactor clogging while tackling also other types of problems, such as kidney stones formation. Ideally, I would like to combine my passion for science with the one I have for teaching in a professorship position. The years spent between my PhD and Post-doc allowed me to appreciate the perks of studying (and learning) a new topic, while transmitting knowledge to others. Being teaching assistant for 3 years (ETH master course of "Polymer Reaction and Colloid Engineering"), supervising 4 grad and 1 undergrad students, and collaborating with colleagues from different countries and backgrounds was an amazing experience, which strongly motivated me in pursuing an academic career.

Publication list

https://scholar.google.com/citations?user=x_evVCMAAAAJ&hl=en

References

[1] Rimer J.D.; An, Z.; Zhu Z.; Lee M.H.; Goldfarb D.S.; Wesson J.A.; Ward, M.D.,Science **2010**, 330, 6002

[2] Henry C., Minier JP, Lefèvre G., Adv. Colloid Interface Sci., **2012**, 185-186, 34-76

425478 Micro and Nano-Rheological Methods for Interfacial and Bulk-Fluid Systems

Abstract:

Micron-scale methods for measuring the mechanical and rheological properties of bulk and interfacial fluids are especially attractive techniques when sample sizes are small and when relevant length-scales of the material under scrutiny are below those accessible to larger conventional equipment. Both passive and active techniques have been explored and developed recently, and there is great opportunity to utilize current techniques for investigating bulk and interfacial biological systems. There are also opportunities to develop novel experimental techniques to overcome limitations associated with sensitivity limits in passive methods, and opportunities to employ certain active techniques in the treatment of cancer. Trends in this area are towards nano-scale probing of material properties, and with the development of superparamagnetic iron oxide nanoparticles (SPIONs) complexed with quantum dots, there is great opportunity to develop new nano-diagnostic, and nano-rheological methods with a broad range of biological applications.

425620 High Performance Computing for Engineered Human Health Systems

Sunday, November 8, 2015
Exhibit Hall 1 (Salt Palace Convention Center)

Andrew P. Spann, McKetta Department of Chemical Engineering, The University of Texas at Austin, Austin, TX

My research uses parallel simulations to solve biological fluid and solid mechanics problems involving drug design and delivery. High performance computing enables the representation of important physics that can only be faithfully captured by going beyond the approximations needed to make classical theories tractable. The use of high performance computing to enable the modeling of complex behaviors involving multiple layers of physics and many bodies allows us to make better predictions in pharmaceutical design, reducing the development time and cost of new medicines.

During my Ph.D. at Stanford University as a student of Eric Shaqfeh, I used computational simulations to examine the behavior of soft, deformable vesicles in fluid flow. Understanding the physics of vesicles in flow is essential for the use of artificial vesicles as bio-compatible drug delivery agents because it allows us to achieve better targeting of delivery and to improve the yield in the production and processing of these medicines. My research incorporates subdivision surface techniques originally developed for the field of computer graphics to resolve the bending forces on the vesicle mesh with unprecedented accuracy. The use of Loop subdivision surfaces has enabled the investigation of realistic 3D vesicles, which are far from the spherical limit. Most notably, I am able to simulate dumbbell instabilities in long tubular vesicles and match my simulations to previously published experimental literature. I have applied similar boundary integral techniques to suspensions of platelets and red blood cells and demonstrated that the volume fraction of red blood cells influences the margination of platelets and their ability to adhere to the wall of a microchannel. These simulations demonstrate that artificial blood additives can aid the formation of blood clots through mechanisms other than the explicit mimicry of platelet functionality.

As a postdoc at UT Austin with Roger Bonnecaze, I have investigated the behavior of vesicles and cells in highly confined environments such as tight tubes and post arrays. These simulations are used to determine the mobility of leukocytes and cancer cells that will be incorporated into a large scale simulation of cancer cell migration, tumor growth and metastasis. I have also used my expertise in high performance computing to accelerate simulations of the dynamics of 1000s of drops in template filling in UV roll-to-roll nanoimprint lithography. The multi-scale model for this simulation is based on lubrication and membrane theory and includes the capillary elastohydrodynamics to study the coupled fluid-structure interaction of the flexible substrate as drops merge and fill features on the template. By being able to simulate over a thousand drops instead of dozens, my simulations can make predictions about the maximum throughput and residual layer thickness of these nanomanufacturing processes as well as determine optimal drop patterns and quantify the effects of error tolerances in drop placement.

My future research vision uses high performance computing to build multi-scale models that help us understand the physical processes behind cancer metastasis. Furthermore, I will use the insight gained from these models to design synthetic blood additives and to construct lab-on-a-chip medical analysis devices for detecting and manipulating cancer cells and blood borne pathogens. Through the NSF XSEDE supercomputing program, I will continue to be eligible for time on the computer clusters I am currently using when I become faculty at another institution. By bringing in these supercomputing resources, I will be able to establish my new computational research group swiftly and expediently.

425258 Fundamental and Applied Studies of Dynamic Self-Assembled Biomaterials

Abstract:

Dynamic self-assembly of macromolecules is inherently tied to the function of biological systems across multiple scales, including the molecular (e.g., protein folding) and the cellular (e.g., extracellular matrix organization). At this interface of materials and biology, I plan to run an experimental research program that focuses on three broad goals: 1) to fundamentally understand the interactions behind the self-assembly of biologically-relevant polymers in both time and space, 2) to develop well controlled, dynamic, structured materials that allow us to probe temporal changes in cellular behavior, and 3) to merge our fundamental understanding of self-assembly with our engineered materials systems to generate biologically-inspired approaches to solve problems in human health, drug delivery, and environmental sustainability. This research will combine ideas and methods from multiple fields, including materials science, polymer physics, biomaterials, and tissue engineering. In this presentation, I will discuss insights gained from previous and ongoing work, such as controlling the self-assembly of biomimetic polypeptoids with monomer sequence [1,2] and how that inspired the design of a hydrogel biomaterial with a reversible structural mechanism to regulate modulus in the presence of attached cells [3]. In addition, I will discuss examples of future projects related to my overall research goals, including the design of peptide-functionalized polymeric templates with dynamic structure for the biomineralization of advanced composites.

Selected publications:

[1] A.M. Rosales, B.L. McCulloch, R.N. Zuckermann, R.A. Segalman. "Tunable Phase Behavior of Polystyrene-Polypeptoid Block Copolymers." *Macromolecules* 2012, *45*(15), 6027-6035.

[2] A.M. Rosales, H.K. Murnen, S.R. Kline, R.N. Zuckermann, R.A. Segalman. "Measurement of the Persistence Length of Helical and Non-Helical Polypeptoids in Solution." *Soft Matter* 2012, 8, 3673-3680.

[3] A.M. Rosales, K.M. Mabry, E.M. Nehls, K.S. Anseth. "Photoresponsive Elastic Properties of Azobenzene-Containing Poly(ethylene-glycol)-based Hydrogels." *Biomacromolecules* 2015, 16(3), 798-806.

425672 The Physical Cell: Impact of Mechanics and Rheology on Cellular Function

Abstract:

The internal environment of a living cell comprises a uniquely complex physical milieu for the molecular interactions, conformational changes, and transport processes that underly cell function. My research focuses on examining the basic physical phenomena that underly biological processes at the molecular and cellular level. I develop multiscale models, employing both analytical and computational techniques, that interface directly with experimental data. My work draws on concepts from continuum mechanics, fluid dynamics, and statistical physics to address the interplay of structure, transport, and biological function.

As a doctoral student with Prof. Andrew Spakowitz, I focused on studying how the mechanical properties of DNA impact the packaging and accessibility of the genome. I investigated the role of DNA elasticity in mediating long-range cooperativity between DNA binding proteins, in the packaging of DNA into compact chromatin fibers, and in the fluctuation-dependent kinetics of enzymes that access the DNA. My work on the statistics of large DNA-protein complexes spurred the development of a generalized approach for coarse-grained modeling of polymer systems by mapping onto effective elastic chains. This methodology enables simulations of polymer dynamics and statistics on much larger length scales than were previously accessible.

As a postdoctoral scholar with Prof. Julie Theriot, I study the rheology, fluctuations, and mechanics of the cytoplasm in motile cells. I focus on white blood cells (neutrophils), which exhibit rapid motion accompanied by morphological dynamics. I have developed a novel analysis method for passive microrheology studies in a flowing complex fluid, applying this technique to demonstrate that the neutrophil cytoplasm behaves as a viscous fluid. I also investigate the general physical problem of mixing and dispersion in an actively fluctuating membrane-enclosed fluid domain, elucidating the role of whole-cell deformation on the internal motion of intracellular components. This collaborative work combines analytical and computational studies with direct measurement of cell shape dynamics and organelle motion within living cells at high temporal and spatial resolution.

My future work will focus on the collective physical phenomena that underly transport, mechanics, and kinetics on a cellular scale. In particular, I will explore the impact of cytoplasmic fluctuations and spatial heterogeneity on biomolecular transport and interactions within the cell, as well as the emergence of cell mechanical properties from the dynamic cytoskeletal polymer network. I intend to leverage my background in multiscale biophysical modeling to forge novel links between molecular-scale mechanics and large-scale whole-cell phenomena.

(6eu) Modeling Porous Materials and Confined Fluids from Atomistic to Continuum Scales

Gor, G., Rutgers, The State University of New Jersey

Expertise and Vision: As a faculty member, I will establish a world-leading group specializing in developing methods for modeling porous materials and fluids confined in nanopores. My unique expertise is a combination of modeling chemical and mechanical processes in porous materials, with my toolkit spanning various computational techniques that cover both micro and macro scales. I am particularly interested in phenomena where fluids affect the mechanical properties and behavior of solid or soft porous materials. My areas of practical interest are shale gas recovery, CO_2sequestration and lithium-ion batteries.

Research Experience:

St. Petersburg State University:I received fundamental education in theoretical physics at St. Petersburg State University, Russia. My research was in the field of first-order phase transitions. Particularly, I studied how a system consisting of liquid droplets and vapor evolves in time. In my Ph.D. thesis, supervised by Prof. A. P. Grinin and Prof. F. M. Kuni, I developed a novel approach to the kinetic theory of non-isothermal nucleation using methods of thermodynamics, mathematical statistics, and transport phenomena.

Rutgers University: During the following two years I worked as a postdoctoral researcher at the Department of Chemical and Biochemical Engineering at Rutgers University (group of Prof. Alexander Neimark). My main focus was on modeling adsorption in nanoporous materials. I achieved the following results: I developed macroscopic and molecular theories of the deformation of nanoporous materials during adsorption. By means of Classical Density Functional Theory and Monte Carlo simulations I modeled nitrogen adsorption in novel types of nanoporous carbons and developed methods for characterization of these materials. These methods have been further successfully implemented in commercial instruments by Quantachrome Inc. I also modeled adsorption and transport of chemical warfare agents in polyelectrolyte (Nafion) membranes.

Princeton University: At Princeton University I worked with Prof. Jean Prevost (Department of Civil and Environmental Engineering) and Prof. Craig Arnold (Department of Mechanical and Aerospace Engineering). I focused on two main areas: CO_2 sequestration and lithium-ion batteries. More specifically, I developed models for CO_2 behavior during geological storage. It involved the development, implementation, and application of methods of thermodynamics of fluid phases and mechanics for modeling the compositional flow of CO_2-brine mixtures. My other achievement was development of a poromechanical model for lithium-ion batteries separator, contributing to the understanding of mechanical behavior of batteries during working cycles.

Naval Research Laboratory: Since September 2014 I am an NRC Research Associate at Center of Computational Materials Science at NRL. I work on developing atomistic (Electronic Structure DFT) and continuum models for mechanical response of nanoporous and nanostructured materials to sorption of fluids. In addition to theory and simulations, I do some experimental work on materials for lithium-ion batteries.

Teaching: Apart from research activities, I have always enjoyed teaching. I have taught a number of courses to a variety of different audiences: from high-school students to Princeton graduate students.

At the poster session, besides giving an overview of my past and current research, I will discuss my plans in research and teaching as a prospective faculty.

Selected publications (from 30+ total):

[1] **Gor G. Y.*** Adsorption Stress Changes the Elasticity of Liquid Argon Confined in a Nanopore ***Langmuir*** 2014, 30 (45), p. 13564-13569

[2] **Gor G. Y.***, Cannarella J., Prevost J. H., Arnold C. B. A Model for the Behavior of Battery Separators in Compression at Different Strain/Charge Rates ***J. Electrochem. Soc.*** 2014, 161(11), F3065-F3071

[3] **Gor G. Y.***, Paris O., Prass J., Russo P. A., Ribeiro Carrott M. M. L., Neimark A. V. Adsorption of n-Pentane on Mesoporous Silica and Adsorbent Deformation ***Langmuir*** 2013, 29 (27), p. 8601-8608

[4] Landers J., **Gor G. Y.**, Neimark A. V. Density Functional Theory Methods for Characterization of Porous Materials (Review) ***Colloids and Surfaces A*** 2013, 437, p. 3-32.

[5] **Gor G. Y.**, Elliot T. R., Prevost J. H. Effects of Thermal Stresses on Caprock Integrity During CO_2 Storage ***International Journal of Greenhouse Gas Control*** 2013, 12, p. 300-309

[6] **Gor G. Y.**, Thommes M., Cychosz K., Neimark A. V. Quenched Solid Density Functional Theory Method for Characterization of Mesoporous Carbons by Nitrogen Adsorption ***Carbon*** 2012, Vol. 50 (4), p. 1583-1590

[7] **Gor G. Y.**, Neimark A. V. Adsorption-Induced Deformation of Mesoporous Solids ***Langmuir*** 2010, Vol. 26 (16), p. 13021-13027

426006 Genetically Engineered Peptides and Proteins: A Platform for Programming Bio-Inspired Functional Materials and Analytical Assays

Abstract:

In my current position as a Postdoctoral Associate in Research Triangle MRSEC and Biomedical Engineering Department at Duke University (2012 to present) in Prof. Gabriel Lopez's lab and in close collaboration with Prof. Ashutosh Chilkoti, I have designed and constructed a wide range of biopolymers to understand their bulk behavior as a function of their molecular design. I have used these biopolymers in applications ranging from bio-analytics to mechanically robust biomaterials and hydrogels. I used genetic engineering to design complex biopolymers inspired by nature but unique in their existence and characteristics. Using the structure-function relationship of these genetically engineered materials I successfully designed hierarchically structured biomaterials with novel mechanical characteristics. My extensive experience with bio-inspired and genetically engineered materials provides a highly flexible and powerful toolbox that I use to design, modify, and construct virtually any peptide, biopolymer and protein. These materials have utility in applications ranging from theranostics to bioassays to surface engineering and hydrogels. I believe my strong background in chemical engineering, biomolecular engineering, and biopolymer design, in combination with more than 6 years of industrial experience as a chemical engineer and years of teaching and mentoring experience prepared me very well to become a faculty member.

As a new faculty, I will explore a number of diverse but related avenues. Here I present a broad view of a couple of ideas that I will pursue. First, I will use my background in biomolecular, biopolymer, and hydrogel engineering to design tunable bio-inspired, biologically active, hierarchically structured hydrogels. The goal is to use protein and peptide engineering to develop a system in which the components and their assembly process can be chosen on demand and based on the final desired mechanical and functional characteristics of the gel (e.g. toughness, stretchability, bioactivity, bio-responsiveness…). This is especially useful considering that hydrogels are generally assembled based on empirical approaches rather than precise programming of these materials at the building block level. My ultimate goal is to provide the ability to select the physicochemical characteristics and resultant functionality of bio-based hydrogels by tailoring components and conditions that result in such hydrogels with minimum number of trials. I am confident that I can accomplish this through seamless programming of biopolymers and their derivatives.

The second area that I will explore is the use of genetically engineered self-assembled particles (micelles, vesicles, LBL assembled,…) to address the critical issues of *in vivo* diagnostics/monitoring and controlled and sustained drug delivery. In many disease treatment approaches ranging from diabetics to cancers, it is quite challenging to either maintain a sustained release of drug over the course of time or to deliver the active cargo (drug, gene…) on demand. I will address this issue by using protease and peptide biomarkers as a regulatory means for sustained and controlled release of materials from engineered carriers. The combination of my background in stimuli responsive materials, micelle and nano particle engineering, and protease assay development provide me all the needed platforms to design an all-in-one monitoring and release system that works based on continuous feedback from specific biomarkers.

425666 Engineering a Non-Enzymatic Analog of the Glycolysis Pathway

Abstract:

Engineering a Non-Enzymatic Analog of the Glycolysis Pathway

Marat Orazov* and Mark E. Davis

Chemical Engineering, California Institute of Technology

Pasadena, CA 91125

Lactic acid and its esters have been shown to have a high potential as a renewable platform chemical for the production of polymers, green solvents, and fine chemicals. Currently, fermentation is the prevailing route for making L-lactic acid. There is a need for D-lactic acid, so catalytic routes to racemic lactic acid could, after separation, provide economic pathways to either enantiomer of lactic acid.

Glycolysis is the metabolic pathway by which cells convert glucose to pyruvate, capturing the energy associated with the process for further metabolic needs. Phosphorylation reactions that alter the stability of intermediates and minimize the potential of side reactions play a crucial role in natural glycolysis. The other essential reactions involved in the glycolysis pathway are the isomerization of glucose 6-phosphate to fructose 6-phosphate by isomerases, the retro-aldol cleavage of fructose 1,6- bisphosphate to glyceraldehyde 3-phosphate and dihydroxyacetone phosphate by aldolases, and the reaction cascade that, through isomerization, oxidation, and phosphate transfer takes these phosphorylated trioses to the pyruvate, producing energy in the process. The net result of glycolysis is the conversion of glucose to pyruvate, phosphorylation of ADP to ATP, and reduction of NAD^+ to NADH. For the process to keep going, the cells must re-oxidize NADH. Lactic acid fermentation is a strategy certain cells use to sustain glycolysis for energy production in the absence of oxygen by coupling NADH oxidation with pyruvate reduction to lactate.

Synthetic glycolysis is a catalytic strategy that mimics the natural glycolysis and lactic acid fermentation pathways for the conversion of glucose to lactate, without involving phosphorylation. The key steps for a synthetic glycolysis route are: isomerization of glucose to fructose, retro-aldol cleavage of fructose into trioses, glyceraldehyde (GLA) and dihydroxyacetone (DHA), and conversion of trioses into lactates via sequential dehydration to pyruvaldehyde, rehydration, and hydride-shift isomerization. In 2010, the Davis group pioneered the use of Lewis acidic molecular sieve Sn-Beta for the isomerization of glucose to fructose via a 1, 2 intramolecular hydride shift.[1] Since then, a number of groups have extended the applications of the 1, 2 intramolecular hydride shift using similar Lewis acidic zeotypes to other sugars. Notably, the conversion of trioses to alkyl lactates has been shown to be nearly-quantitative when catalyzed by Sn-Beta at moderate temperatures (ca. 100 °C).[2] However, a bottleneck in the overall process still exists, as retro-aldol reactions of hexoses are not thermodynamically favored at low temperatures, and have relatively high activation energies, even with the best reported catalysts. As a result, most investigations of the retro-aldol-based production of lactates have been carried out at elevated temperatures (160 °C and higher), with long reaction times. Under these conditions, ketohexoses like fructose are prone to side-reactions that can lead to poor yields of lactates.

Recently, in our characterization of the active site of Sn-Beta, we showed that alkali-exchanged Sn-Beta can catalyze glucose epimerization to mannose via 1, 2 intramolecular carbon shift[3,4], known as the Bilik reaction. Using such catalysts with fructose as a reagent at moderate temperatures (ca. 100 °C) yielded a complex reaction mixture of sugars, including the branched sugar, hamamelose, ketohexoses (sorbose, psicose, and tagatose), and retro-aldol products (DHA and GLA). Investigation of fructose reactions catalyzed by the traditional Bilik catalysts, i.e. molybdates, yielded similar results. We found that a variety of molybdenum (VI) species (MoO_3, $(NH_4)_6Mo_7O_{24}{\cdot}4H_2O$, $H_3PMo_{12}O_{40}$, and Na_2MoO_4) catalyze retro-aldol reactions of ketohexoses at moderate temperatures in both aqueous and alcoholic solvents. However, under these conditions, the equilibrium of the retro-aldol reaction favors hexoses, and aldol condensation and carbon shift products (ketohexoses and branched hexoses, respectively) begin to form after relatively low concentrations of DHA and GLA are achieved. To overcome this limitation we developed co-catalyst systems where molybdates catalyze retro-aldol reactions of ketohexoses, and microporous stannosilicates promote the formation of lactic acid or alkyl lactates from the established pool of DHA and GLA. With such catalytic systems, we have achieved high yields of lactates (ca. 70% for both methyl- and ethyl-lactate) from fructose at 100 °C.

Overall, our resulting catalytic systems, when used together to convert glucose into lactates, mirror the biological glycolysis pathway. Because we have developed a strategy to decouple the retro-aldol reaction from the active site of the 1, 2 hydride shift utilized in the isomerization reactions, we can easily adjust the ratio of active sites for each reaction, much like a cell can adjust the expression of a particular enzyme in a complicated reaction pathway to meet its metabolic demand.

References:

[1] Moliner, M.; Roman-Leshkov, Y.; Davis, M. E. *Proc. Natl. Acad. Sci. U.S.A.* **2010**, 107, 6164−6168

[2] Osmundsen, C. M.; Holm, M. S.; Dahl, S.; Taarning, E *Proc. R Soc. A*, **2012**, *468*, 2000–2016

[3] Bermejo-Deval, R.; Orazov, M.; Gounder, R.; Hwang, S.; Davis, M. E. *ACS Catal.* **2014**, 4, 2288–2297

[4] Orazov, M.; Bermejo-Deval, R.; Gounder, R.; Hwang, S.; Davis, M. E. *2014 AIChE Annual Meeting*, **2014**, Atlanta, Georgia, Nov 16-21

*Corresponding author. Email: morazov@caltech.edu

425809 Development of Minimally Invasive Tools for Genetic Monitoring of Pancreatic Health

Abstract:

The diagnosis of pancreatic cancer carries an especially grim prognosis, with a median patient survival of less than 6 months, and an average 5-year survival of <5%. One reason for this poor outlook is the fact that pancreatic cancer is often diagnosed very late in the cancer development, resulting in the existence of distant metastases in more than one-half of patients at the time of diagnosis. There is a significant need for effective ways of detecting this cancer, early in its progression, which could lead to the evaluation and identification of better treatments for this deadly disease. Because ~95% of pancreatic cancers are exocrine in nature, pancreatic fluid represents a rich source of potential biomarkers that could lead to earlier diagnoses, and better clinical outcomes. However, the only method that is currently available for sampling pancreatic fluids is endoscopy, which is both time-consuming and highly invasive.

We are engineering a tool that would allow for sampling and genetic screening of pancreatic fluids in a minimally invasive manner. We have identified materials that efficiently capture DNA from raw, untreated pancreatic fluid that has been collected from patients, thereby allowing it to be purified for downstream analysis. We have also performed time-course studies to evaluate fluctuations in pancreatic fluid DNA content upon secretin-stimulated secretion. These samples are now being utilized, in conjunction with next-generation sequencing, to identify resident genetic mutations. Here, I discuss these results and also provide data on our ability to detect rare mutant alleles within these fluid samples.

425816 Engineering Immunity: Design and Development of Customized Nanomaterials with Controlled Immunostimulatory Effects for Biomedical Applications

Abstract:

Engineered nanomaterials (ENMs) are rapidly emerging as a new class of materials that allow tuning of their physicochemical properties for biomedical and pharmaceutical applications, e.g., drug delivery, tissue engineering and vaccines. ENMs could interact with biological entities at the nano-bio interface, which could induce cascades of signaling pathways leading to the activation of immune responses including both innate and adaptive immunity. It has been identified that hydrophobicity, size, and surface charge are the possible parameters affecting the biocompatibility of ENMs. However, the detailed mechanisms by which ENMs could generate immunostimulatory effects at the nano-bio interface are still not fully understood. In light of this, my research will focus on immunoengineering, which will involve developing ENMs with tunable physicochemical properties, investigating the interaction of ENMs with immune systems, and further customizing their applications in vaccine development, tissue engineering, and cancer immunotherapy.

(6ey) Directed Assembly at All Length Scales: The Pathway Towards Future Metamaterials

Bharti, B., North Carolina State University
Findenegg, G., Technische Universität Berlin
Velev, O. D., North Carolina State University

Assembling building block molecules and particles is the key to fabricate functional materials. The equilibrium morphology of the self-assembled state is determined by the building block packing efficiency (entropy) and interparticle interaction (enthalpy). My research focuses on the fundamental understanding of these governing factors and developing new 'assembly principles' for organizing building blocks at all scales; from molecular to particulate. At the *molecular level*, we proposed methods for directing the assembly of amphiphiles on silica nanoparticles and under confined nanopores of SBA-15 materials. We also developed approaches for triggering the release of surface bound amphiphiles on-demand by co-adsorption of displacer molecules. Recently, we discovered a new method for assembling superparamagnetic nanoparticles into magnetically-responsive ultraflexible filaments via 'nanocapillary' lipid bridging. We reported that the filament and their networks can be healed upon mechanical damage. Secondly on the *nanoscale*, we established the fundamentals of colloidal interactions between silica nanoparticles and biologically relevant entities such as globular proteins. We revealed how the adsorbed protein induces a pH-dependent reversible aggregation of the particles. A similar route of modulating colloidal interactions was used to fabricate lignin based environmentally benign antimicrobial nanoparticles. The particles exhibited time-limited activity and were shown to be a potentially safer substitute to the presently used silver nanoparticles. Finally at the *microscale*, we demonstrated that oppositely charged particles can be assembled into rigid permanent chains using dielectrophoresis as a structure directing tool. In addition, we developed novel strategies for assembling anisotropic shaped microparticles into self-reconfigurable and reversibly actuating active structures which can be used as soft-robotic components. The assembly principles developed can be used for engineering a wide range of functional materials providing efficient and environmentally friendly solution to the present day needs of the society.

426187 Hybrid Nanomaterials for Energy Harvesting

Abstract:

While inorganic materials have revolutionized the world of electronic applications, polymers and small molecules serve as the workhorses in consumer goods and pharmaceuticals. As we undergo a scientific revolution in energy, these traditional materials, however, are approaching their performance limits; so to meet the inexorable global demand for energy in a clean and sustainable fashion, new materials must be developed. Hybrid materials are an emerging class of compounds that combine classical inorganic and organic components to yield materials that manifest new functionalities unattainable in traditional composites or other related multicomponent materials, which have additive function only. Effective medium theories are widely used to model several physical properties and modes of energy transport in composites. However, the distinctive feature of a truly hybrid material is that material synergies lead to performance that is greater than the sum of its parts, which can happen when there are strong, non-linear interactions between the constituent components. These interactions can give rise to beneficial energy-transport phenomena, such as excited-state charge transfer in photodetectors and photovoltaics. Unfortunately, we still lack a fundamental understanding of the electronic and thermal properties of these hybrid systems especially those comprising of "hard" and "soft materials that would provide a general and robust framework to guide materials design. Given that hybrid nanomaterials have abundant surface areas stemming from their nanoscale dimensions, the "rules of design" governing the synergistic transport properties of these materials are grounded in the fundamental physics and materials science of nanoscale interfaces. These nanoscale interfaces constitute a fascinating scientific realm straddling the quantum and classical regimes where truly new behavior can be engineered by taking advantage of the unique physical and chemical properties as well as unique transport modes of "hard" and "soft" condensed matter. The goals of my research program are to understand and decouple thermal and charge transport in composite systems (inorganic/organic) and elucidate the role of interfaces which would then provide a platform to design and innovate novel material systems through hybrids targeted towards specific energy-relevant applications involving charge, heat, and mass transport. We will design, develop and characterize novel solution processable nanocomposites in order to gain functional understanding and demonstrate how engineering of nanoscale interfaces can drive novel macroscale behavior. We will present a myriad of approaches that are aimed at understanding the fundamental underlying physics of hybrid materials and devising new applications for the same using novel syntheses.

426276 Catalysis for Energy and Environmental Applications

Sunday, November 8, 2015
Exhibit Hall 1 (Salt Palace Convention Center)
Zhenglong Li, Oak Ridge National Laboratory, Oak Ridge, TN

As petroleum reserves decrease, environmental pollution and the impact of rising atmospheric CO_2 are exacerbated, both sustainable energy supply and environmental protection are needed. Catalysis plays an important role in addressing energy and environmental issues. Since my Ph.D. study, I have been working on developing new technologies and new catalysts for energy conversion and environmental protection.

My Ph.D. research was on fast pyrolysis of biomass and catalytic upgrading of bio-oil to liquid fuels and valuable chemicals. One of the most important contributions is the development of a new electrocatalytic method with a ruthenium electrocatalyst for biomass conversion, converting electrical energy into carbon-neutral drop-in replacement fuels and valuable chemicals. This method provides substantial advantages, including no need of hydrogen gas, operation at mild conditions to reduce coke formation and catalyst deactivation, and a great way to store intermittent renewable electricity as chemical energy in liquid fuels. The other significant contribution is the development of a new screw-conveyor pyrolysis reactor with compact design, enabling flexible application in regional biomass processing depots, showing great potential to solve biomass supply issue.

To deepen and expand my expertise in catalysis, I continued my postdoc research with Prof. Fabio H. Ribeiro and Prof. W. Nicholas Delgass at Purdue University. My research focused on kinetic studies and also involved new catalytic material synthesis (e.g., zeolites and nanomaterials) and catalyst characterizations (*in situ* and *operando*). All the work were performed to establish the fundamental understanding of the catalyst active sites, and the relationship between catalyst structure and catalyst activity. Besides this fundamental research, I am also working on applied catalysis research at Oak Ridge National Lab to develop heterobimetallic zeolites for ethanol conversion to hydrocarbon blend-stock, aiming to overcome the blend wall for bio based fuels. Our technology has been licensed to Vertimass LLC. I have been working on developing new zeolite materials, testing and characterizing new catalysts, scaling up new catalysts, and facilitating technology transfer and development. Furthermore, I am also studying and testing catalysts for automotive emission control.

My future research interests are applying the knowledge of catalyst synthesis, characterization and kinetic studies to solve energy conversion and environmental issues. Specific interests include biomass conversion to fuels and chemicals, natural gas liquefaction and conversion to chemicals, emission control catalysis, electrocatalysis for CO_2 conversion, catalysis for wastewater treatment, etc.

426409 Light-Activatable Nanoconstructs for Mechanism-Based Combination Therapy

Abstract:

My background is in biochemical engineering with extensive experience in cutting-edge photoresponsive nanoconstructs for the delivery of therapeutics. As a postdoctoral fellow my training has broadened into tumor biology and rationally designed combination therapies, with a focus on tackling compelling challenges in translational nanomedicine. My research efforts cover cross-disciplinary boundaries and reflect creativity in developing new nanoplatforms. This multidisciplinary tack has led to influential and innovative work as reflected by my publications, patent and awards. To transition to independence, I seek to build a laboratory bridging biochemical engineering (nanotechnology) and oncology (basic science). It is increasingly apparent that multi-faceted combination treatments will most likely to improve outcomes. However, the selective delivery of multiple regimens to the right place, right time, and in the correct sequence with respect to the mechanistic therapeutic interactions remain major challenges. These unmet needs provide a clear and distinct path for my transition to independence. I will leverage image-guided strategies and engineering to develop a photoresponsive nanoplatform that can co-delivery multiple regimens in a unique manner, where one treatment primes the target for the second modality, and the subsequent evasion pathways are mitigated by a third agent. All agents are rationally selected and released in an appropriate time and sequence to account for mechanistic interactions and to improve outcomes. The principle and the nanoplatform developed here will be adaptable to treating a broad range of diseases. This poster will highlight my previous and current research, mentoring (teaching) philosophy and the distinct future research direction to establish a unique program in image-guided customization of nano-photomedicine.

426591 First-Principals Modeling of Methanol Fuel Cells: Kinetics and Catalyst Design

Abstract:

First-Principals Modeling of Methanol Fuel Cells: Kinetics and Catalyst Design

Glen R. Jenness

The decreasing supplies of petroleum resources coupled with an increase in energy demand has created a need for an alternative power supply. Proton-exchange membrane fuel cells (PEMFCs) are an attractive prospect, due to the high energy density and low emissions. Previous PEMFC technology relied on the oxidation of H_2, but due to the difficulties of storing H_2, this route is not economically favorable. By replacing H_2 with an alternative renewable proton source we can alleviate the cost of storing and transporting H_2, as well as making the PEMFC technology more environmentally friendly. Methanol is an attractive proton source due to low storage costs, good thermal stability, high energy density, and is readily produced from biomass resources. While methanol fuel cells offer a high energy density, the efficiency remains low, limiting the use of methanol fuel cells. As a consequence, there has been an increased effort in recent years to find an efficient catalyst for the oxidation of methanol. Previous catalysts employed a platinum nanoparticle with an oxide support, such as γ-Al_2O_3. However, recent experiments indicate that a platinum nanoparticle supported on a carbon-based support, such as carbon nanotubes (CNT), have led to an increase in the efficiency of methanol fuel cells.

Several challenges still remain to be resolved in order to increase the efficiency of methanol fuel cells. Firstly, the CNT itself can carry an intrinsic reactivity, with the reactivity being determined by the metallic or semi-conducting nature of the CNT, arising from the chirality. However, the effect of the chirality on the reactivity of the CNT support is still debated. I have shown that the choice of nanoparticle support can result in a charge state being induced on the nanoparticle, which in turn affects the bonding/anti-bonding interactions between the reactants/products and the catalyst, with the resulting charge state being related to the band gap of the support material.1,2 As the band gap of the CNT is driven by the chirality, the Pt–CNT interaction will also be driven by the chirality. Additionally, the efficiency of CNT supported catalysts can be increased via the doping of the CNT with nitrogen, which in turn changes the electronic properties of the CNT (i.e. band gap), as well as introduces a Lewis acid center into the support. As the mechanism for the oxidation of methanol on the Pt/CNT catalyst is still in question, the effect of introducing a Lewis acid center is currently unknown. I have also shown that the strength of the Lewis acid center can be related to the atomistically resolved conduction band properties of the Lewis acid center, with the reactivity being related to the strength of the Lewis acidity.3,4 Finally, it should be mentioned that one factor preventing the wide spread use of methanol fuel cells is the cost of the Pt metal, and it would desirable to use a base metal such as nickel or iron, without sacrificing efficiency. Through the use of computational tools (i.e. state-of-the-art ab-initio methods such as density functional theory), I seek to resolve these issues. In particular, a) how the metallic vs. semi-conducting nature of the carbon nanotube affects the methanol oxidation process; b) how the effect of doping the CNT affects the electronic properties of the catalytic system; c) how the introduction of a Lewis acid center into the support affects the mechanism and kinetics of the methanol oxidation process. The investigation of these effects from an atomistic level will allow us to understand the origin of these interactions, which enable a rationale and systematic improvement of the efficiency of methanol fuel cells.

(1) Jenness, G. R. et al. ACS Catal. 2013, 2881–2890.
(2) Hermes, E. D. et al. Mol. Simul. 2014, 1–11.
(3) Jenness, G. R. et al. J. Phys. Chem. C 2014, 118, 12899–12907.
(4) Jenness, G. R. et al. J. Phys. Chem. C 2015, 119, 5938–5945.

426833 From Fluorescence to Magnetic Resonance: Engineering Proteins for Molecular Imaging

Sunday, November 8, 2015
Exhibit Hall 1 (Salt Palace Convention Center)

Arnab Mukherjee, Chemistry and Chemical Engineering, California Institute of Technology, Pasadena, CA; Chemical and Biomolecular Engineering, University of Illinois at Urbana-Champaign, Urbana, IL

Research overview: My research interest broadly focuses on engineering proteins to develop new tools for molecular imaging based on fluorescence and magnetic resonance imaging (MRI).

A. Genetically encoded fluorescent reporters for low-oxygen imaging

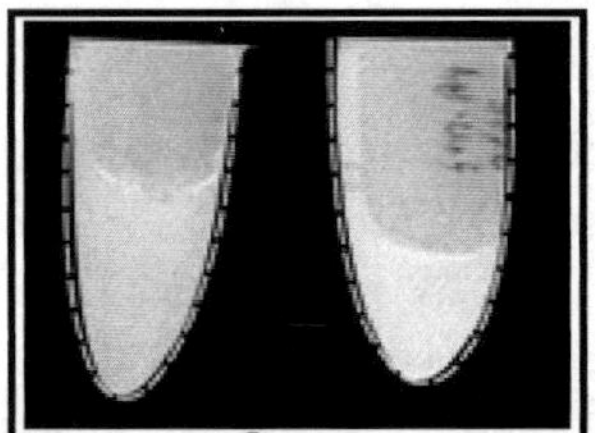

Figure 2: Genome mining was used to identify 2 new LOV- fluorescent proteins

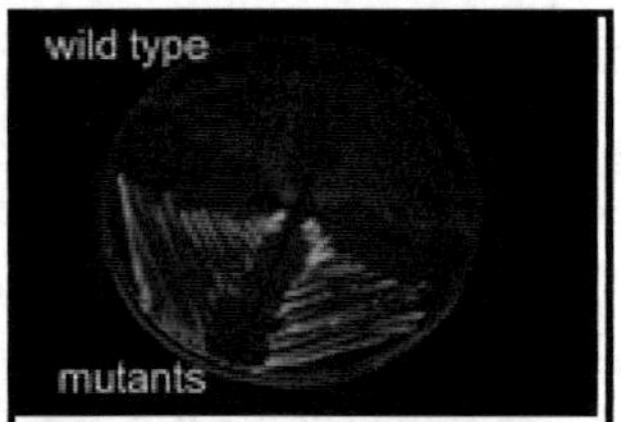

Figure 1: Directed evolution was used to engineer brighter LOV- fluorescent proteins

Low-O_2 regimes are encountered in several clinically and industrially relevant biosystems such as microbial fermentation, rumen microbiota, and biofilms. Hypoxic and anoxic regimes have hitherto pose a long-standing challenge to fluorescence imaging as widely used GFP-based probes depend on O_2 for fluorescence (Mukherjee *et al.*, Curr. Op. Biotech., 2014). To address this challenge, I engineered bacterial & algal photoreceptors to develop viable fluorescent probes for live cell imaging & fluorimetry in low-O_2 environments. Specifically, I used directed evolution to enhance brightness (quantum yield) of a putative O_2-independent fluorescent reporter originally identified in *Pseudomonas putida* (**Fig. 1**) (Mukherjee *et al.,* J. Biol. Eng., 2012). Second, I applied genome mining to identify and engineer algal photosensory proteins to develop two novel fluorescent reporters (**Fig. 2**) (Mukherjee *et al.*, ACS Syn. Bio., 2014). Importantly, I demonstrated that, compared to GFP-based probes, one of the newly discovered algal fluorescent proteins (CreiLOV) is characterized by an overall small size (12 kDa), faster maturation kinetics, and improved performance at extremes of pH and temperature. Finally, I developed a comprehensive biophysical characterization framework to interrogate the maturation of fluorescence in this emerging class of O_2-independent reporter proteins with a view towards elucidating design rules for engineering improved reporters (Mukherjee *et al.*, PLoS ONE., 2013). In this way, my research extends the spectrum of fluorescent imaging to systems that remain intractable using GFP-based reporters.

B. Genetically encoded reporters for magnetic resonance imaging

Figure 3: GS_{ATP} can detect ATP based on enhanced contrast in MR images

Despite the wide prevalence of MRI for clinical imaging, MRI remains a predominantly anatomical (vis-à-vis molecular) imaging modality due to a dearth of MR contrast agents that enable molecular-scale functional imaging. A highly promising albeit under-exploited approach towards addressing this challenge borrows from the GFP paradigm and involves the development of genetically targeted reporters for functional magnetic resonance imaging (Shapiro *et al.*, 2010). To this end, I recently engineered a Mn^{2+} binding bacterial enzyme (glutamine synthetase) to develop the first genetically encoded sensor for noninvasive detection of ATP using MRI (**Fig. 3**). As ATP fluctuations often accompany several neurodegenerative and cerebrovascular disorders, GS_{ATP} is a valuable tool to noninvasively and dynamically monitor ATP levels in cellular and animal models of stress, injury, inflammation, and toxicity.

C. Future research goals

Aim 1. Anaerobic imaging: I propose to leverage LOV-based fluorescent proteins to develop fluorescence complementation-dependent biosensors for imaging Ca^{2+}, ATP, and cAMP dynamics in hypoxic models of cerebrovascular disease, tumor spheroids, and bacterial biofilms

Aim 2. Noninvasive MRI-based in vivo sensors: In this aim, I seek to repurpose a eukaryotic antibacterial Mn^{2+} binding protein to develop a genetically encoded MRI-based sensor for calcium imaging *in vivo*. In addition, I propose to develop a new platform for noninvasive and dynamic imaging of protein interactions in preclinical animal models based on interaction-dependent reconstitution of split metalloproteins

Aim 3. Genetically encoded agents for photodynamic therapy: I propose to engineer cyanobacterial light harvesting proteins as a first-of-its-kind class of genetically targetable therapeutic photosensitizers

426947 New Chemical and Biological Processes for Next Generation Biorefining

Abstract:

Chemical and biological processes for biomass conversion are the key to unlocking the potiential of biomass as a widely available feedstock for sustainable biorefining. To operate these processes economically and sustainably, biomass needs to be refined for the production of not only high-demand low-value commodity chemicals (e.g. ethanol biofuel), but also high value products. New refining technologies need to be developed for feedstock fractionation and conversion, which face challenges associated with the structural complexity and the chemical heterogeneity of biomass.

The low-cost transformation of biomass to platform chemicals (e.g. sugars, organic acids, phenolics) and the subsequent conversion to high value products are instrumental processes in the upgrading biomass feedstocks. To make these processes possible, better understanding in plant biopolymer chemistry and biomass conversion engineering is essential. My research will focus on 1) the use of spectroscopic and microscopic techniques in furthering the understanding of plant cell wall morphology and the key biochemical conversion processes, as well as 2) biomass process engineering including biomimetic enzyme-free catalysis and distributed small-scale biomass processing. A holistic approach that integrates all steps of biorefining will be employed to explore solutions with the greatest potential in commercialization. The research work will also be used to develop curricula and activities that convey renewable concepts in educational programs involving undergraduate students and pre-college students.

426341 Engineering the Flow Properties of Colloidal Materials

Sunday, November 8, 2015
Exhibit Hall 1 (Salt Palace Convention Center)

Lilian C. Hsiao, Department of Chemical Engineering, Massachusetts Institute of Technology, Cambridge, MA

Colloids are micron-sized particles that exhibit Brownian motion when suspended in a dispersed phase. Attractive and repulsive interactions can be induced between these particles, leading to phenomena such as crystallization, phase separation, and glass transition. These colloids can assemble into a broad range of structures that can be tuned for industrial and technological applications. Consequently, there has been a concerted effort to understand the microscopic contribution of particle-particle interactions to the overall structure and dynamics of a system, which ultimately leads to the ability to control material properties. How does one generate, for example, a load-bearing gel structure out of oil droplets or polymeric colloids that can support large yield stresses? What happens when roughness and shape anisotropies are introduced into the system? To address these questions, I generate a scaling law for model colloidal gels in which their non-linear flow properties can be predicted from microstructure. Measurements of the mechanics of composite hydrogels with controlled pore structures are discussed in the context of stress-bearing backbones. Experiments with dense suspensions of rough colloids (often found in consumer applications) demonstrate that the surface roughness reduces rotational freedom and is key to shear thickening and jamming in colloidal suspensions. Given the large interest in assembling targeted structures by tuning particle shape, I also present experiments and simulations that explain the unusual ability of shape-anisotropic particles to rapidly adopt certain kinds of orientational order when assembled in attraction-driven systems. By connecting the underlying physics behind the phenomenon of self-assembly and rheology, I plan to find new possibilities of designing soft materials in which the local microstructure can be modified and improved.

427157 Multiscale Design of Gas-Phase Synthesis of Nanomaterials

Abstract:

Abstract

Multiscale design of aerosol reactors for synthesis of nanomaterials includes continuum, mesoscale, molecular dynamics (MD) and quantum mechanics models spanning 10 – 15 orders of magnitude in length and time, respectively. Quantum mechanics account for the electronic structure of matter determining the interatomic potentials in MD models for accurate estimation of sintering and crystallization rates. Mesoscale models provide the transport properties and coagulation rate of multi-particle structures while continuum models describe the effect of process variables on product particle size and morphology at various process temperatures and residence times (Buesser and Pratsinis, 2012). Multiscale modeling is introduced in the design of scalable processes for gas-phase synthesis of nanoparticles with closely controlled characteristics.

Fundamental particle properties, like structure and crystallinity are determined from first principles as they affect material performance in a score of applications (e.g. sensors, catalysts, biometerials). More specifically, highly parallelized MD simulations are carried out to monitor dynamic properties and physical mechanisms governing the nanoparticle characteristics that cannot be easily obtained experimentally. Sintering rates and crystallinity dynamics that can be incorporated in mesoscale and continuum models are determined *ab initio* for gold nanoparticles.

Discrete Element Modeling (DEM) simulations are used to track the detailed structure and size distribution and agglomerate size of fractal-like particles undergoing Brownian coagulation without coalescence, sintering or surface growth (Goudeli et al., 2015). The DEM-obtained mobility size and number-based geometric standard deviation of the mobility radius are compared to Scanning Mobility Particle Sizer (SMPS) and combined Differential Mobility Analyzer (DMA) and Aerosol Particle Mass (APM) measurements as function of the number of primary particles per agglomerate. Particle morphology and structure are quantified by the fractal dimension, D_f, and mass-mobility exponent, D_{fm}, which are related to the cluster optical and transport properties, respectively, from spherical particles to fractal-like agglomerates until they attain their well-known asymptotic structure.

Easy-to-use expressions are extracted that can be readily interfaced with climate dynamics, meteorological models or computational fluid dynamics describing the reactor operation and particle production. For example, the effect of varying particle structure during the formation of fractal-like crystalline TiO_2 and amorphous SiO_2 is investigated by employing the above DEM-derived D_f evolution in a monodisperse continuum model (Kruis et al., 1993) accounting for concurrent coagulation and sintering. Neglecting the evolving particle structure overestimates the hard-agglomerate diameter by up to 25% for TiO_2 and 30% for SiO_2, especially at high maximum temperatures and cooling rates that hard-aggregates consist of only a few primary particles.

Interfacing the above models that are applicable at multiple length and time scales can facilitate the understanding and scale-up design of aerosol reactors for synthesis of nanoparticles whose properties can be closely controlled during scale-up from laboratory scale to commercial products.

References

Buesser, B., and Pratsinis, S.E. (2012) *Annu. Rev. Chem. Biomol. Eng.*, **3**, 103-127.

Goudeli, E., Eggersdorfer, M.L., and Pratsinis, S.E. (2015) *Langmuir* **31**, 1320-1327.

Kruis, F.E., Kusters, K.A., Pratsinis, S.E., and Scarlett, B. (1993) *Aerosol Sci. Technol.* **19**, 514-526.

427044 Soft Materials Engineering: From Colloids to Biological Interfaces

Abstract:

The scope of soft materials engineering encompasses rationally manipulating the configuration of particles via self-assembly methods in the creation of materials with unique properties to fabricating artificial biological membranes to understand fundamental cellular processes. With such breadth, progress in the field relies on an interdisciplinary approach to apply principles of physics, biology, materials science and chemical engineering to identify and solve current challenges in the field. These include 1) translating fundamental studies to systems with increased technological importance, for example moving from hard sphere colloids to hairy colloids that possess chemical functionality, 2) developing novel experimental methods to probe processes across a wide range of length and time scales, for example the diffusion of a single lipid molecule within a membrane to membrane budding and cell division, and 3) moving beyond microscopic self-assembly to the large-scale application of self-assembly principles to specific particle shapes and chemistries with particular bulk optical, mechanical or thermal properties in mind. I aim to start a research lab that excels at the interface of soft matter engineering- leveraging my diverse background to create a unique group addressing these challenges.

During my doctoral research, I made fundamental progress in the field of colloidal electrokinetics and applied that knowledge to create a new class of acoustic metamaterials. In this work, I constructed and validated a dielectric spectrometer to measure the electrokinetic properties of colloidal suspensions over a wide frequency range. Studies on concentrated colloidal suspensions helped explain the AC electric field-induced disorder-crystalline order transition. A modified AC electric field self-assembly scheme was applied to create ordered thin films of colloidal ellipsoids. The particle anisotropy and order translated to macroscale anisotropy in the materials acoustic properties in the hypersonic frequency regime. This finding is only the beginning of a nascent field in tuning the transport of phonons by rational organization of polymer colloids. While at Delaware, I helped establish two international research collaborations, was a leader in our graduate student group, and participated in the teaching fellows program. My postdoctoral research at ETH Zürich has built upon my soft matter expertise to include thin surfactant films and phospholipid bilayers. I am currently developing a route to millimeter scale, free-standing, planar phospholipid membranes that can be used for a variety of fundamental studies. This technique advances the state-of-the-art in bilayer fabrication and overcomes several limitations to existing technologies, therefore there is a tremendous opportunity for future research- from the fundamental (membrane elasticity and phase behavior) to the applied (protein pore formation and particle endocytosis). While my postdoctoral work will be the focus of my oral presentation at this meeting, this poster will highlight my research experiences and my vision for future research.

Postdoctoral Advisor: Prof. Jan Vermant, ETH Zürich

PhD Advisor: Prof. Eric M. Furst, University of Delaware

Selected Publications:

1. **Beltramo, P. J.,** Schneider, D., Fytas, G. and Furst, E. M., "Anisotropic hypersonic phonon propagation in films of aligned ellipsoids." Phys. Rev. Lett., 113(20), 205503, (2014).
2. **Beltramo, P. J.**, Roa, R., Carrique, F. and Furst, E. M. "Dielectric spectroscopy of concentrated colloidal suspensions." J. Colloid Interface Sci., 408(1), 54-58, (2013).
3. Schneider, D., **Beltramo, P. J.**, Mattarelli, M., Crespy, D., Montagna, M., Pfleiderer, P., Vermant, J., Furst, E. M., and Fytas, G. F. "Elongated Polystyrene Spheres as Building Blocks for Anisotropic Colloidal Crystals: A Particle Vibration Spectroscopy Study." Soft Matter, 9, 9129-9136, (2013).
4. **Beltramo, P. J.** and Furst, E. M. "Predicting the disorder-order transition of dielectrophoretic colloidal assembly with dielectric spectroscopy." Electrophoresis, 34(7), 1000-1007, (2013).
5. **Beltramo, P. J.** and Furst, E. M. "Transition from Dilute to Concentrated Electrokinetic Behavior in the Dielectric Spectra of a Colloidal Suspension." Langmuir, 28(29), 10703- 10712, (2012).
6. **Beltramo, P. J.** and Furst, E. M. "Dielectric spectroscopy of bidisperse colloidal suspensions." J. Colloid Interface Sci., 380(1), 34-41, (2012).

(6fj) Hierarchical, Nature-Inspired Nanomaterials for Electrochemical Energy Conversion/Storage Devices

Trogadas, P., University College London

Abstract

Several studies on polymer electrolyte fuel cell (PEFC) durability have shown that the components of the central membrane electrode assembly (MEA), especially the polymer electrolyte membrane (PEM) and electrocatalyst, deteriorate during long term operation. PEM chemical degradation as well as electrocatalyst dissolution are reported to be the major contributors to PEFC lifetime limitations.

The chemical degradation of the PEM occurs in a two step process: (i) formation of reactive oxygen species (ROS) such as hydroxyl (OH*) and hydroperoxyl (HOO*) radicals (OH* are far more reactive); and (ii) reaction of ROS with the PEM leading to chain scission. Additionally, loss of electrochemically active area is observed under long term operation due to nanoparticle growth by Ostwald ripening where large Pt particles grow at the expense of small ones via interparticle transport of single atoms.

In order to mitigate those degradation mechanisms, the efficacy of rare earth metal oxide nanoparticles on reducing the membrane degradation rate as a function of their microstructure, redox properties and location within the fuel cell is presented. On the catalyst dissolution front, the investigation of the effect of uniform particle size distribution on catalyst stability is presented via theoretical and in-situ experimental studies.

To further improve the design of highly stable hierarchical electrocatalysts, in-depth research on the effect of nano-architecture on reaction/transport kinetics is necessary. Inspiration can be derived from nature as it is full of hierarchical designs that are intrinsically scaling. Developing fundamental understanding of how such desired properties of biological systems are related to their architecture and self-assembly dynamics can guide the development of novel hierarchical catalytic nanomaterials and nature-inspired electrochemical devices.

Short Bio

Dr. Panagiotis Trogadas is a Senior Postdoctoral Fellow at the Centre of Nature Inspired Engineering (UCL) and Teaching Fellow at the Chemical Engineering Department (UCL). He completed his undergraduate studies in Chemical Engineering at National Technical University of Athens, Greece and obtained his doctorate degree from Illinois Institute of Technology, Chicago at the end of 2009. He worked as a Postdoctoral Fellow at Georgia Institute of Technology (Prof. Tom Fuller) and Technical University Berlin (Prof. Peter Strasser) from 2010 until 2014. He is recipient of a grant funding proposal (Nov. 2014) from the German National Science Foundation (DFG).

428086 Engineered Natural Biomaterials for Understanding the Interplay Between Cells and Their Environment

Abstract:

Dynamic biomaterial tools are needed to address modern medicine challenges ranging from the repair and replacement of tissues and organs to the treatment of diseases such as fibrosis and cancer. In particular, the ability to engineer biomaterial milieus that re-create the heterogeneous microenvironments of physiological and pathological conditions is critically important. My PhD work in Prof. Brendan Harley's lab at the University of Illinois focused on the design of spatially-patterned collagen biomaterials to engineer orthopedic interfaces and direct cell behavior using approaches with broad applicability to a variety of regenerative medicine challenges. During my postdoctoral work in Prof. Jason Burdick's lab at the University of Pennsylvania I have developed a suite of mechanically dynamic hyaluronic acid hydrogel substrates useful for a range of fundamental mechanobiology studies, with a particular focus on understanding hepatic stellate cell behavior during liver fibrosis. As a principal investigator I will leverage my experience in chemical engineering, biomaterial fabrication and characterization, and mechanobiology to design regenerative templates for tissues including skin and skeletal muscle. On a more fundamental level, I will use these materials to investigate how cells interrogate their environment during growth factor signaling and in disease states such as liver fibrosis and cancer.

Selected Publications (out of 16 first author published/submitted/in preparation):

7) **S.R. Caliari**, J.A. Burdick, 'A practical guide to hydrogels for cell culture,' In preparation, 2015.

6) **S.R. Caliari**, M. Perepelyuk, B.D. Cosgrove, R.L. Mauck, R.G. Wells, J.A. Burdick, 'Visible light-mediated stiffening hydrogels for probing the dynamics of myofibroblast differentiation in fibrosis,' In preparation, 2015.

5) **S.R. Caliari**, D.W. Weisgerber, W.K. Grier, Z. Mahmassani, M.D. Boppart, B.A.C. Harley, 'Collagen scaffolds incorporating coincident gradations of instructive structural and biochemical cues for osteotendinous junction engineering,' *Adv Healthc Mater*, 4(6):831-7, 2015.

4) **S.R. Caliari**, E.A. Gonnerman, W.K. Grier, D.W. Weisgerber, J.M. Banks, A.J. Alsop, J. Lee, R.C. Bailey, B.A.C. Harley, 'Collagen scaffold arrays for combinatorial screening of biophysical and biochemical regulators of cell activity,' *Adv Healthc Mater*, 4(1):58-64, 2015.

3) **S.R. Caliari**, B.A.C. Harley, 'Composite growth factor supplementation strategies to enhance tenocyte bioactivity in aligned collagen-GAG scaffolds,' *Tissue Eng Part A*, 19(9-10):1100-12, 2013.

2) **S.R. Caliari**, M.A. Ramirez, B.A.C. Harley, 'The development of collagen-GAG scaffold-membrane composites for tendon tissue engineering,' *Biomaterials*, 32(34):8990-8, 2011.

1) **S.R. Caliari**, B.A.C. Harley, 'The effect of anisotropic collagen-GAG scaffolds and growth factor supplementation on tendon cell recruitment, alignment, and metabolic activity,' *Biomaterials*, 32(23):5530-40, 2011.

428095 Exploring Structure-Function Correlations of Nanomaterials in Energy Conversion and Storage

Sunday, November 8, 2015
Exhibit Hall 1 (Salt Palace Convention Center)

Weiqing Zheng, Catalysis Center for Energy Innovation, Department of Chemical and Biomolecular Engineering, University of Delaware, Newark, DE

"And yet it is only by studying function that we can understand function, so that the kinetic aspect must be allowed to retain its place and assigned its part in the unravelling of the great mystery" – C. N. Hinshelwood, 1947

Catalysis is the science and technology of influencing the rates of chemical reactions, which are at the core of chemical industry. Practical success of catalysis needs the interdisciplinary intersection of chemistry, chemical engineering, and material science. My research interests lie on understanding the mechanism of chemical reactions and the role of catalysts under reaction conditions in order to transform qualitative observations and knowledge into predictive science.

In heterogeneous catalysis, under reaction conditions, active centers produce products. The presence of active centers depends on the local geometry of a catalyst particle, which can be varied by tuning the interactions between catalyst particle and the support. As a first of its kind demonstration of this concept, in my Ph.D. work, the ruthenium-carbon interactions were varied by depositing Ru nanoparticles on carbons with different local geometry and functionalities. Using atomic level TEM analysis, we found that the local disorder of the support induces local disorder to the Ru particle[1] and influences catalysis as confirmed through the direct observation of the reaction intermediate (adsorbed dinitrogen) using *in situ* XPS over Ru/C catalyst operating at 400 °C and 0.5 mbar ammonia.[2]

Identification of active centers requires interdisciplinary characterization techniques. For instance, the high H_2 production from ammonia over Mo_2C was due to the prevalence of highly energetic sites (twin-boundaries, stacking faults, steps and defects) which were confirmed with pre- and post TEM and EELS analysis.[3] Similarly, in my postdoctoral work on biomass conversion, by combining a series of characterizations of XPS, *ex/in situ* EXAFS and TEM along with reactivity studies, we have discovered that the high yield of 2,5-dimethylfuran from 5-hydroxymethylfurfural benefits from dual functionality catalytic centers of Ru and RuO_2.[4,5],

The knowledge gained from realizing the active centers can lead to optimized catalysts. In a first instance, we have modified iron catalysts by alloying Co or Ni leading to enhanced activity and stability for ammonia decomposition.[6]Furthermore, we have developed a novel route to produce nanosized early transition metal carbide particles embedded in carbon, which show an improvement of ammonia cracking activity and a promotion effect in PEM fuel cells.

Reference

[1] W. Zheng, et al, *ChemSusChem*, **2010**, *3*, 226-230.

[2] W. Zheng, R. Schloegl, in preparation.

[3] W. Zheng, R. Schloegl, et al, *J. Am. Chem. Soc*, **2013**, *135*, 3458-3464.

[4] J. Jae, W. Zheng, R. Lobo, D. Vlachos, *ChemSusChem*, **2013**, *6*, 1158-1162.

[5] J. Jae, W. Zheng, A. Karim, R. Lobo, D. Vlachos, *ChemCatChem*, **2014**, *6*, 848-856.

[6] J. Zhang, J. Mueller, W. Zheng, et al, *Nano Lett*, **2008**, *8*, 2738-2743.

(6fo) Engineering Biomimetic Self-Assembled Materials

Leon, L. F., University of Chicago

My research interests involve the design, characterization, and application of novel self-assembling materials. In particular, I am interested in biomimetic systems in which carefully tuned molecular interactions create functional and responsive materials for applications ranging from nanoscale electronics to drug delivery. Electrostatic interactions play a fundamental role in both creating spatially patterned interfaces and also creating separate polymer rich phases (complex coacervation) for both equilibrium and non-equilibrium processes, and yet are still largely misunderstood. Using rational design of peptides and polymers I intend to explore the role that these interactions play in compartmentalized delivery systems and as a dynamic templates to create hierarchical hybrid materials composed of organic and inorganic constructs. My goal is to thoroughly understand the thermodynamic and kinetic processes driving the self-assembly of these materials and take them from the bench to their functional application.

My postdoctoral work with Professor Matthew Tirrell (The Institute for Molecular Engineering, University of Chicago &Argonne National Laboratory) involves the design, characterization, and application of polyelectrolyte based materials. We have explored the fundamentals of polyelectrolyte complexation using oppositely charged polypeptides and found methods of tuning the phase behavior of the resultant complex from liquid to solid by manipulating polymer chirality. Additionally, we can stabilize these complexes at the nanoscale by forming micelles with either liquid or solid cores depending on their secondary structure, which provides unique self-assembly and degradation properties. After carefully characterizing model systems, we designed polyelectrolyte complex micelles that contain therapeutically relevant charged molecules such as nucleic acids and peptides, specifically for the treatment of atherosclerotic lesions and cancer. The modular nature of these assemblies enables the addition of a targeting ligand to increase the efficacy of delivering nucleic acids, creating polyelectrolyte complex micelles with a targeting ligand outside the corona of the micelle. Our rationally designed micelles have successfully delivered miRNA inhibitors *in vitro* and are currently being tested *in vivo.*

My graduate work with Professor Raymond Tu (Chemical Engineering Department, The City College of New York) involved the interfacial templating of inorganic materials using self-assembled peptides. In nature, biological molecules form interfaces that assemble patterns of chemical functionality with exceptional precision. The role of dynamics during the assembly of biological molecules appears to be important for mineralization processes. Our work applied model sheet-forming peptides at interfaces to explore the dynamics of assembly in order to template mineral growth. Thermodynamic analysis of structure formation with increasing surface pressure allowed us to understand the nature of self-assembly with iterative changes in the peptide sequence. Additionally, we investigated the dynamics of the self-assembled state, where the organic phase switches between short- and long-range order as a function of surface pressure. This model system allowed us to explore the influence of electrostatic interactions on self-assembly and additionally the influence of short- and long-range order on the nucleation and growth of inorganic material. This is in contrast to a system that starts with a well-ordered preformed template that defines the epitaxial growth of the mineral phase. Versions of our model peptides are modified to include histidine in order to nucleate Au nanocrystals in both the short and long range ordered organic matrix showing that the phase behavior of the peptide influences the crystallinity and shape of the resultant nanocrystals.

428604 Understanding Solar-Fuel Systems from the Nanoscale to the Device Level

Sunday, November 8, 2015
Exhibit Hall 1 (Salt Palace Convention Center)

Miguel Modestino, School of Engineering, École polytechnique fédérale de Lausanne, Lausanne, Switzerland

The world's development and the implementation of a sustainable lifestyle depend on significant advances in the alternative energy sector. These advances will require basic scientific developments in energy generation, capture and storage as well as large scale economic policy drivers to motivate the necessary changes in our current energy infrastructure. Developing viable energy conversion technologies can significantly help trigger these changes. I have actively participated in the search for solutions to energy challenges with deep fundamental and technological implications in the field of artificial photosynthesis. Artificial photosynthesis devices are complex energy conversion systems that need to absorb sunlight, create and transport charges, and electrochemically generate energy rich molecules such as hydrogen that can be used as a fuel. Given the complexity of the process, integrated systems need to be designed to efficiently carry-out all of these processes in parallel. My research has tackled aspects ranging from the materials nanostructure level to the overall device design structure, with particular emphasis on the role of hybrid polymeric materials for solar-fuel generation. Here I will highlight research insights from the basic understanding of functional nanoparticle self-assembly in polymer membranes, to the structural and transport characteristics of ion-conducting polymers at inorganic interphases for electrochemical water splitting. Also, broader aspects on the implementation and device design of cost-effective solar-fuels technologies will be discussed.

Electrochemical energy conversion devices, such as the solar-fuel generators described above, rely strongly on active multifunctional materials in order to perform various electrochemical processes involved in their operation. Polymer composites can play a key role in the functioning of energy conversion devices, as they act as a versatile platform for the achievement of properties unattainable by single component material systems. Additionally, combining the properties of polymeric and inorganic materials, multifunctionality can be achieved with the potential to impact a wide range of energy applications. **My proposed group's research will focus in underpinning structure-property relationships of polymer composites in order to enable deployable energy conversion devices.** We will depart from classical research schemes, and we will not only focus on fundamental aspects of particular material systems, but we will close the loop and design active materials envisioned to enable a particular technology or application. This would allow us to not limit ourselves to existing material problems of current devices, but to create disruptive technologies where the materials and device design go hand-in-hand and achieve unconventional functionalities. **Within my scientific program, I will implement a research model that will iteratively circle around 3 pillars: (a) development of energy conversion devices and processes based on**

material properties, (b) development of multifunctional polymer composite materials to enable device functionality and (c) strong structural and physicochemical characterization of materials and devices.

Specifically, we will first design materials using basic physical models to understand the drivers for performance and potential for a particular application, as well as devise the required morphologies to achieve the desired functionality. Basic continuum models to describe transport properties (i.e. mass, heat and electrical), chemical processes, and the mechanical behavior of materials will inform my group about particular materials morphologies or device designs that satisfy the desired requirements for a given application. Then, we will devote the bulk of our research efforts to understand the self-assembly aspects of composites that lead to the required morphologies and functionalities. To achieve this goal, a wide range of polymer processing techniques and structural characterization tools will be used to obtain and assess composites' internal structure. Methodologies to obtain materials that self-assemble into the desired morphology will be employed, as well as processing steps to kinetically trap morphologies that are not thermodynamically equilibrated. Characterization tools that will be instrumental into assessing the internal microstructure of materials include basic electron microscopy techniques such as scanning or transmission electron microscopy, as well as X-ray diffraction or scattering tools that can probe large volumes of materials. Specifically, my group will leverage my expertise in synchrotron materials characterization techniques such as small- and wide-angle X-ray scattering, SAXS and WAXS respectively, together with scattering modeling tools to elucidate the internal structure of active materials as they change during operation. Lastly, understanding the effects of composites' nanostructures on their properties (e.g. electrical, chemical and physical) will allow us to achieve optimized hybrid materials that can be implemented into devices and will also point out directions for improvements in their design.

Initially, this research model going from materials design to device implementation will be employed (i) to reinvent hybrid membrane-electrode assemblies used in water–splitting devices and (ii) to develop solid-state photocatalytic hybrid materials. In the future, and as my research program grows, I will leverage the knowledge, protocols and the infrastructure created in the initial projects to divert and expand my research program to other energy conversion systems. I anticipate that the initial proposed projects will provide the basic foundations and understanding of new hybrid materials that will enable a wide range of applications. Energy applications that the family of hybrid materials described above can tackle involve: chemically resistant transparent conductors for solar-fuel applications, supercapacitor electrodes, and materials used in flow batteries.

428971 Multi-Physical / Multi-Scale Modeling for Nanotechnology Convergence Systems

Abstract:

During the past decades, along with remarkable scientific and engineering achievements in many areas, the world has faced immense challenges. Increasing need for higher efficiency and reduction in system size motivates the investigation of novel nanoscale systems and materials. Various enhancements have been reported by tuning atomistic properties, chemical structures, and functionalities of advanced materials leading to eye-opening progress. The tremendous improvements in nanotechnology provide the necessary understanding of nanoscale physics to control and design the system properties, while at the same time focusing on the critical issue of integration with macroscopic system scales resulting in converging technology and phenomena at different time and length scales.

My research interest focuses on this challenging problem by introducing a novel holistic multi-scale integration in nanotechnology by hybridizing atomistic / molecular and mesoscale / continuum models, and developing unique methodologies for hierarchical integration of the complete system at different scales, and expanding to include multi-phenomena / physics. Information transfer from one scale to neighboring scales allows one to eventually predict macroscopic behavior or device scale functions from first principles in top-down / bottom-up / middle-out approaches. These approaches along with system optimization based on a clear understanding of nanoscale phenomena will lead to a novel multidisciplinary analysis paradigm.

In this poster session, I present case studies, which have been examined in the area of information technology and energy technology & sustainability for multi-scale modeling via a hierarchy of scale specific models coupled with optimization for the process system integration. Nano-tribology in the data storage system was chosen as a benchmark example utilizing atomistic / molecular simulations for the nanoscale film of functionalized oligomers / polymers with hybridization of novel graphene technologies and extended to mesoscale / continuum theories for the whole system integration [1,2]. In energy & sustainability area examples, we analyzed polymer electrolyte membrane fuel cell including membrane electrode assembly (polymer electrolyte membrane & catalyst layers) via atomistic / molecular simulations and transport phenomena in the gas diffusion layer via lattice Boltzmann method [3]. In addition, I will introduce and discuss the details of my future research topics and plan in this poster session.

Reference:

[1] P.S. Chung, S.H. Vemuri, S. Park, and M.S. Jhon, "Molecular rheological analysis on binary blends of perfluoropolyether lubricants," J. Appl. Phys., 115, 17B738 (2014).

[2] P.S. Chung, D.S. So, L.T. Biegler, and M.S. Jhon, "Nanotechnology convergence and modeling paradigm of sustainable energy system using polymer electrolyte membrane fuel cell as a benchmark example," J. Nanopart. Res., 14 (8), 853 (2012).

[3] P.S. Chung, R. Smith, S.H. Vemuri, Y.I. Jhon, K. Tak, I. Moon, L.T. Biegler, and M.S. Jhon "Multiscale/ multi-physical modeling in head/disk interface of magnetic data storage," J. Appl. Phys., 111 (7), 07B712 (2012).

428342 Composite Nanoparticles for Energy Generation & Storage Applications

Abstract:

Advances in synthesis techniques have enabled a revolution in the design of colloidal dispersions containing particles with uniform size, shape and composition. In turn, studying these uniform particle populations has provided a sophisticated understanding of how the intrinsic characteristics of a nanoparticle determine the strength and range of inter-particle interactions and consequently the extrinsic properties of the dispersion. The intimate link between the intrinsic and extrinsic properties of a colloidal dispersion provides a powerful design paradigm for the development of new materials for a wide range of engineering problems. I will present two examples of the use of novel nanoparticle formulations to address specific challenges in the development of renewable energy. In both examples, understanding both the intrinsic and extrinsic relationships of the complex formulation have important implication for nanoparticle design improvements with regards to performance. By focusing on characterization techniques that can separate out the intrinsic properties of the nanoparticle from the extrinsic properties of the dispersion, we show that it is possible to establish structure-function relationships that can establish a basis for design. In the first example, aqueous dispersions of composite conjugated polymer/fullerene nanoparticles are investigated for organic solar cell applications. Similar to their thin-film counterparts, the performance of these composites is a function of the distribution of polymer and fullerene within the particle volume. Contrast Variation Small Angle Neutron Scattering (CV-SANS) is used to investigate how the internal structure of these composite nanoparticles changes as a function of the production process. These structural features are related to their electronic performance using conductive probe Atomic Force Microscopy (AFM). In the second example, I will show ongoing work in the design and synthesis of nanoparticles that possess a mixed ionic and electronic conducting adsorbed polymer layer. We hypothesize that this polymer layer will allow for electrical charge percolation in concentrated suspensions without the onset of mechanical percolation or prohibitively high viscosities. We use neutron scattering and electron microscopy to interrogate the properties of the polymer layer and relate these to dispersion's viscosity and dielectric properties. These new materials will have relevance to flow capacitor and battery applications.

429138 Next-Generation Bioengineering and Biodesign

Abstract:

Biological materials have been used by human being for thousands of years, from enhancing the leisure experience with fermented foods, to manufacturing essential therapeutics to cure diseases. In our perspective, we observe living cell as a design tool, which has the potential to be engineered to synthesize and build anything. This means that we are not only interested in the cellular or secreted specific products, but also consider the cell itself as a functional building block. The faculty candidate's research strategy is bidirectional—enhancing the relationship with bio-related industry by advanced technology, and educating the civilians to learn and to use biotechnology through biodesign. The ultimate goal is to bridge the interface between bioengineering and biodesign for novel, functional, and interactive products which benefit both industry and consumer. In the poster, the faculty candidate would like to present two recent projects within this vision: 1) on-line monitoring of product in mammalian cell culture; 2) novel microbe-human interface for stimuli-responsive transformational bio-hybrid materials.

428977 First-Principles-Based Multiscale Modeling of Functional Nanomaterials

Abstract:

The past several decades have witnessed major advances in the fields of electrochemistry, heterogeneous catalysis and materials science due to the development of experimental and theoretical tools capable of elucidating the atomic-scale structures and processes underlying macroscopic phenomena. This atomic-scale knowledge has proven invaluable in the development of functional molecules and materials for myriads of diverse technological applications, and is indispensible in obtaining a fundamental understanding of the physical processes these technologies harness. Nevertheless an atomic level understanding for all but the simplest model systems remains elusive due to the complexity of most real world systems.

First-principles-based multiscale modeling provides a powerful tool for overcoming this difficulty by using derived methods to extend the powerful lens on the atomic-scale offered by modern *ab initio* computational methods to achieve descriptions of phenomena, whose description requires time and length scales well in excess of the practical limitations of *ab intio* methods. My research involves the development of these derived methods (e.g. reactive force fields, kinetic Monte Carlo) and their application to elucidate complex systems and phenomena in electrochemistry, catalysis and materials science. In doing so we aim at extending our fundamental understanding of nature to increasingly complex systems and enabling the rational design of novel functional molecules and materials.

429517 Accelerating the Speed and Scale of Metabolic Engineering in Challenging Biological Contexts

Abstract:

Biological systems are fantastic platforms for solving myriad societal problems, as they contain a diverse set of highly specific catalysts, operate at mild reaction conditions, self-replicate, and can undergo evolution to improve function. In order to harness the full potential of living systems, metabolic engineers require detailed models of biological processes to enable rational design as well as high-throughput approaches for generating improved systems where uncertainties remain. In this poster, I will describe my efforts to fill these critical gaps between metabolic engineering practice and desired biological function in difficult-to-engineer systems.

My graduate research focused on expediting the development cycle of engineered yeast, which is a workhorse organism for the production of diverse chemicals, yet was slow and time-consuming to engineer. In particular, tuning the rate of biocatalyst synthesis, biocatalyst properties, and the genetic background of the yeast host are each critical to optimizing a bioprocess, yet yeast engineers have been traditionally faced with a choice between working with a handful of suboptimal genetic parts and a time-consuming, iterative process for improving them. To expedite the optimization of parts related to biocatalyst synthesis, I identified phenomena limiting biocatalyst production and in response developed computational techniques which enable metabolic engineers to quickly refactor these genetic parts and generate novel ones tailored toward their desired application, thus improving biocatalyst production in a single design cycle. Because the design rules governing biocatalyst properties and host performance remain undiscovered, I further developed experimental techniques for high-throughput optimization of these qualities in yeast. In each case, my innovative approaches expedited the design cycle by several orders of magnitude over the current state-of-the-art and opened exciting vistas in strain development and biological discovery.

As an independent researcher, I aim to broaden the scope of metabolic engineering to include the microbial communities that live in the human gut, as this complex ecosystem plays critical roles in the extraction of nutrients, production of vitamins, and the progression of disease, yet whose engineering is currently regarded as a "grand challenge" in human health. As this organ is, in its simplest sense, a packed bed reactor, chemical engineers are ideally poised to make significant contributions in this area. In my postdoctoral work, I am currently laying the foundation for engineering this system through the determination of design rules which collectively ensure the stability, robust performance, and safety of engineered probiotics (i.e. the catalysts) in this complex environment. Looking ahead, these design principles will enable my future lab to engineer "smart" probiotics which monitor the gut environment for signatures of disease and subsequently synthesize and deliver therapeutics in close proximity to affected areas. Additionally, integration with approaches developed during my graduate work will enable the construction of highly robust therapeutic strains which adjust their genetic content for improved gut function in response to changing environmental conditions. Third, these engineered strains will enable the elucidation and control of the three-dimensional structure of gut microbiota, enabling further insight into gut disease progression through modeling the currently unknown spatial reaction networks in this environment. More broadly, the paradigms my lab develops for engineering of the gut community will be applied to other microbial ecosystems in order to address continuing challenges in health, energy, and the environment.

429809 Accelerating Discovery of Advanced Materials through Simulation

Abstract:

Discovery of materials with desired properties and development of new processes involving these materials are essential to our efforts to meet a variety of health, environmental and energy challenges. Simulations accelerate materials discovery and process development by deciphering how the materials properties depend on molecular structure, providing molecular-level principles to guide experimental designs, and screening large databases to search for optimal structures. My PhD and postdoctoral research exemplify how simulations can be used to discover materials and understand processes.

My PhD research at the University of Washington showed how simulations could be used to drive the development of zwitterionic anti-biofouling materials. We investigated (1) why zwitterionic materials resist biofouling differently from polyethylene glycol, and (2) hydration, protein interaction, ion interaction, and self-association of zwitterionic materials using quantum mechanical calculations, atomistic molecular dynamics simulations, free energy perturbation and well-tempered metadynamics simulations. This work elucidated the mechanisms by which zwitterionic materials resist biofouling and helped explain how properties of zwitterionic materials depend on their molecular structure. We also developed a systematic approach to computationally design new zwitterionic anti-biofouling materials. One zwitterionic material designed using this approach was verified by experiments to be able to resist biofouling in blood.

My postdoctoral research at North Carolina State University shows how simulations can be used to help understand and predict toxicity of nanoparticles. When nanoparticles are immersed in the human body, a corona of proteins forms around them. The composition of the protein corona determines how the nanoparticles interact with the biological environment. One key step in the formation of the corona is the competitive adsorption of multiple protein species on the nanoparticle surface. To better understand this process, we investigated the adsorption of a variety of proteins and their mixtures on gold nanoparticles with different radii. We: (1) developed a model that can describe protein-protein and protein-nanoparticle interactions by coarse-graining results of atomistic simulations, and (2) investigated the competitive adsorption of multiple types of proteins on gold nanoparticles with different radii using discontinuous molecular dynamics simulations. This work helps us to understand how the corona composition depends on the nanoparticle size and the physicochemical properties of proteins.

My future research group will focus on discovering advanced materials and developing processes that can help us face critical health, energy and environment challenges. We propose to computationally design: (1) amyloid-forming peptides for catalysts, (2) peptide nucleic acid-peptide conjugates for DNA sensors; and (3) zwitterionic moieties for Li^+ batteries and solar cells.

429896 Towards the Next Generation of Magnetic Resonance Spectroscopy: Harnessing Light and Spin

Sunday, November 8, 2015
Exhibit Hall 1 (Salt Palace Convention Center)

Jonathan King, Materials Sciences Division, Lawrence Berkeley National Laboratory, Berkeley, CA; Department of Chemistry, University of California, Berkeley, Berkeley, CA

Nuclear magnetic resonance spectroscopy (NMR) and imaging (MRI) are two of the most powerful analytical techniques available to engineers, scientists, and medical professionals, enabling a wide range of chemical and structural analysis. However, they require costly infrastructure and are plagued by low sensitivity resulting from small nuclear spin magnetization. My laboratory will harness the highly polarized spin states of photons and electrons to control nuclear magnetization, resulting in greatly enhanced signals from magnetic resonance techniques. We will build upon my previous work with nitrogen vacancy (NV^-) centers in diamond [1,2], where I have demonstrated a 170,000-fold increase in NMR signal from the ^{13}C nuclei in the diamond by coupling the electron and nuclear spin states with microwave irradiation. Future work involves the transfer of this nuclear magnetization to arbitrary samples of interest. This will require development of high-surface area diamond structures, such as diamond nanoparticles, as well as fundamental understanding of transfer mechanisms such as nuclear spin cross relaxation and cross polarization at the diamond surface. These signal enhancements will be transformative, reducing the timescale of medical imaging, allowing the analysis of dilute biomolecules and microscopic samples, and generating technology for portable, low-cost magnetic resonance instruments.

1. **J. P. King**, P. J. Coles, J. A. Reimer, "Optical polarization of ^{13}C nuclei in diamond through nitrogen vacancy centers." *Physical Review B*, 81, 073201 (2010).
2. **J. P. King**, K. Jeong, C. C. Vassiliou, C. S. Shin, R. H. Page, C. E. Avalos, H-. J. Wang, and A. Pines. "Room-Temperature in situ Nuclear Spin Hyperpolarization from Optically-Pumped Nitrogen Vacancy Centers in Diamond." *arXiv:1501.02897 [quant-ph]* (2015).

429909 Systems and Synthetic Biology of Photosynthetic Organisms for Biorenewable Chemicals

Sunday, November 8, 2015
Exhibit Hall 1 (Salt Palace Convention Center)

Rajib Saha, Department of Biology, Washington University, St. Louis, MO

Background: In the FY 2016 Budget, the Department of Energy (DOE) requests $10.7 billion including $2.72 billion for Energy Efficiency and Renewable Energy (EERE), and $645 million for renewable energy generation technologies. In addition, National Science Foundation (NSF) recently set a goal of pertaining crops the ability to fix nitrogen instead of using expensive, chemically synthesized, and artificial fertilizers. To this end, NSF started a novel research program in order to revolutionize current farming methods and awarded $12 million to four different groups from the US and the UK. Photosynthetic organisms such as cyanobacteria and plants derive energy from light and carbon from atmospheric carbon dioxide, and, thus, lie in the core of achieving these goals. However, the feasibility of utilizing these organisms hinges on the development of efficient platforms and genetic toolkits by systems and synthetic biology approaches.

Motivation: The scope and breadth of genome-scale metabolic reconstructions have continued to expand over the last decade. However, only a limited number of efforts exist on photosynthetic metabolism reconstruction. In addition, less well studied are the similarities and differences in their metabolism and diurnal behavior. The unique demands of living on light as a primary nutrient make the use of many established gene expression control systems difficult or impossible to apply in cyanobacteria. Hence, developing an effective genetic toolkit to regulate gene expressions would facilitate biotechnological applications in cyanobacteria.

My Research: My postdoctoral work at the Pakrasi Lab (Washington University in St. Louis) is aimed at developing an efficient system for regulating gene expression and also, applying bioinformatics approaches to explore cyanobacterial physiology. My doctoral work at the Maranas Lab (The Pennsylvania State University) involved reconstruction and analysis of genome-scale metabolic models of photosynthetic organisms. This poster will focus on three broad topics of my research that can widely be applied to develop platforms for producing biorenewables:

- **Reconstruction of genome-scale metabolic models:** Developed high-quality genome-scale models for *Cyanothece* sp. ATCC 51142 (*i*Cyt773), *Synechocystis* sp. PCC 6803 (*i*Syn731), and *Zea mays* (a global model *i*RS1543 and a leaf-specific model).
- **Development of engineered riboswitch:** Combined a theophylline-responsive riboswitch with the T7 control system to demonstrate better regulation of gene expression than that of the existing systems.
- **Application of bioinformatics approaches to analyze cyanobacterial physiology:** Analyzed microarray data to shed light on the global coordination of cellular processes over light/dark cycle in *Synechocystis* sp. PCC 6803 and also studied RNAseq data to find regulatory RNAs in the *nif* cluster of *Cyanothece* sp. ATCC 51142.

429996 Multifunctional Electrocatalysts for Waste Utilization

Abstract:

Multifunctional Electrocatalysts for Waste Utilization

Sujan Shrestha[a], Elizabeth J. Biddinger[a], William E. Mustain[b]

[a]Department of Chemical Engineering, City College of New York, 140th Street and Convent Avenue, New York, New York 10031

[b]Department of Chemical Engineering, University of Connecticut, Storrs, CT 06269

The applications of electrocatalysts range from sensors to energy conversion devices. Electrocatalysts can be fine-tuned by applying potential and current. The energy required for catalysis reactions can be provided in the form of electrons; therefore, many complex reactions can be performed at room-temperature. To catalyze intricate electrochemical reactions involving molecules with one or more carbons, the need for multifunctional electrocatalysts is realized. However, frameworks for designing and synthesizing multifunctional electrocatalysts are lacking. The research projects that I will establish will address the design and synthesis of multifunctional eletrocatalysts for catalysis of carbon-based molecules. The targeted applications of these research projects will be utilization of carbon-based wastes from various sources using energy from clean and renewable sources.

In my graduate work, I explored nitrogen-doped carbons supports to increase the stability and activity of Pt for oxygen reduction reaction. Although carbon is the most popular support for Pt, it does not provide strong metal-support interaction to intrinsically enhance activity and stability of Pt catalysts. By doping carbon with nitrogen, electronic properties of carbon can be changed so that the surface has stronger interaction with Pt. My work highlights how microstructure affects the electronic character of the carbon and, in turn, its role as a support of Pt for oxygen

At my present appointment, I am studying electrochemically induced phase transformation in ionic liquids (ILs), which are salts with melting point below 100 °C. Among the many applications of this study, is the recovery of fission platinoids (Pd, Rh, Ru) through electrodeposition in ILs. I am interested in controlling the structure and morphology of these metals by understanding how they undergo electrochemical phase transformation in ILs. The work I have carried out, in this regard, shows how interfacial interactions can be exploited to control the electrochemical phase transformationin ILs.

These past research experiences in electrocatalysis and electrochemical phase transformation have equipped me with necessary tools to execute my future research projects, which I will discuss in my poster presentation.

430195 Engineering Faster Reactions: Catalysis and Transport from Energy to Pharmaceutics

Abstract:

By identifying and systematically characterizing rate controlling phenomena in complex reacting systems, it allows for the development of new approaches to a diverse set of problems. Applying this fundamental approach to reaction engineering, several complex systems are simplified to their underlying rate controlling phenomena, revealing key insight into their mechanisms and potential future optimizations. Four distinct systems are considered: 1) biomass fast pyrolysis, 2) surface barriers in hierarchical zeolites, 3) microreactor design for multiphase catalytic reactors, and 4) amino acid activation for automated peptide synthesis.

1. High speed imaging techniques on a micron-scale are combined with extensive chemical analyses to understand the fluid dynamics of ***biomass fast pyrolysis*** including the revelation of the reactive liquid intermediate, identification of the aerosol ejection method for nonvolatile transport, and the presence of the Leidenfrost effect with controllable heat flux with tunable surface properties.
2. ***Surface barriers*** are shown to dominate transport in extremely small ***zeolites,*** with diffusivities diminished by three orders of magnitude in hierarchical materials. Several dynamic experimental techniques (ZLC, frequency response) and computational methods (kinetic Monte Carlo), are utilized to systematically characterize the barrier limitation in zeolites, resulting in the conclusion that the vast majority of the surface is dominated by a structural restrictions that act as effective pore blockages.
3. By incorporating small amounts of catalysts into ***micropacked beds***, systematic reaction analysis (kinetic, transport) is performed for gas/liquid/solid systems. Automation of all reactor components and in line analytics allow for rapid and comprehensive system mechanism and kinetic analysis and catalyst screening.
4. ***Amino acid activation*** becomes rate relevant during automated peptide synthesis over Merrifield resins. Mechanistic analysis and reaction kinetics are assessed using a rapid flow-based microfluidic platform to characterize the reaction and optimal conversion conditions.

430265 Separation and Catalysis Using Nanoporous Materials: A Computational Approach

Abstract:

Theory and molecular modeling have seen tremendous growth in recent years; capitalizing on theoretical and algorithmic developments and advancements in computer hardware, many previously intractable models for complex systems can now be solved numerically to sufficiently high levels of fidelity that their results can augment experiments. Applications of computational methods have begun to penetrate into all fields of sciences. In this poster presentation, I will highlight applications in the field of separation and catalysis using crystalline nanoporous materials, primarily zeolites, and show how molecular modeling can be used both to gain a microscopic understanding into phenomena difficult to probe experimentally and to predict accurate data at a scale beyond the reach of the traditional trial-and-error approach. First, I will describe the development of a new, transferable force field for all-silica zeolites, TraPPE-zeo, which is applicable to any type of sorbate molecule and framework structure. With the improved intermolecular potentials as well as better methodologies, it was then possible to study several complex systems involving articulated molecules and liquid mixtures, such as the multi-component adsorption of aqueous alcohol, polyol, glucose, and furfural solutions. These studies demonstrate the important effects of hydrogen-bonding, which render both the bulk and adsorbed phases highly non-ideal and are for certain molecules so strong that adsorption becomes entropically rather than enthalpically driven. In another application, the unusual adsorption and diffusion behaviors in a hybrid microporous/mesoporous zeolite are investigated [1], and our simulations allow us to unravel the distinct features exhibited by such hierarchical materials. More recently, using first-principles calculations, factors contributing to the reactivity and selectivity of zeolites as solid acid catalysts were carefully examined [2]. A better understanding achieved from this study on the different effects, including the chain lengths and architectures of reactant molecules and the different degrees of framework confinement of the reactant/product as well as transition states, affords the potential of rational catalyst design and optimization. Finally, such an effort is described, in which a large number of zeolites were screened with the purpose of finding better candidates for two energy-related applications, ethanol/water separation and hydrocarbon iso-dewaxing. Predictions from this work were subsequently verified by experiments and together this demonstrates the possibility of using theory and computation in a predictive mode to expedite the pace of materials discovery for complicated, realistic, and industrially relevant applications.

[1] "Understanding unique diffusion behavior in hierarchical zeolites," to be presented at 2015 AIChE Annual Meeting, Symposium on 'Diffusion, Transport and Dynamics in Adsorption Systems,' Salt Lake City, UT.

[2] "Chain-length and framework effects in alkane cracking on solid acid catalysts," to be presented at 2015 AIChE Annual Meeting, Symposium on 'Catalysis with Microporous and Mesoporous Materials,' Salt Lake City, UT.

430324 Metal-Organic Framework (MOF) Thin Films for Functional Materials Enabled By Atomic Layer Deposition

Abstract:

Metal-organic frameworks are crystalline porous materials that consist of metal ions/clusters and organic linkers. MOFs typically exhibit high surface area, good adsorption capacity and opportunity for post-synthetic modification. While thousands of MOFs have been reported in the past decade, the insoluble MOF powders typically obtained through conventional solvothermal methods may not be the ideal configuration for applications like gas filters, chemical sensors and smart membranes etc.[1] Growing MOF thin films onto various functional substrates, however, could expand the material set of MOF-based composites and enable new applications.

Atomic layer deposition (ALD) enables precise control of thin film coatings and surface functionality on a variety of substrates at low processing temperature. Using a lab-designed ALD reactor system, I have successfully grown conformal Cu-BTC MOF thin films on to Al_2O_3-coated polypropylene (PP) fibers using both solvothermal and layer-by-layer methods.[1,2] Compared to previous reports, my approach significantly enhances MOF mass loading and greatly improves the growth uniformity of MOF thin films. MOF functionality is fully maintained after integration, as indicated by high BET surface area and large adsorption capacity for hazardous gases such as NH_3 and H_2S. I have also demonstrated the wide applicability of this process scheme to other MOFs and substrate materials.

Though an extensive investigation for the effect of different ALD metal oxide nucleation layers, I found an ultra-fast reaction path for Cu-BTC formation at room temperature using ZnO. I achieved a space-time-yield up to 2.9×10^4 $kg{\cdot}m^{-3}{\cdot}d^{-1}$ for producing Cu-BTC powder, which is more than 10× larger than previously reported. Same synthesis strategy has also been applied to MOF thin films and MOF patterns using ALD ZnO seed layers. The results indicate my method dramatically improves Cu-BTC production rates and widely expand the material set of MOF-functionalized composites.

In addition to the research progress highlighted above, I have also applied ALD to the synthesis of ZnO nanostructures, designed and built an automated reactor that realizes fast layer-by-layer deposition of MOF thin films (>8x faster than manual processes), and investigated UiO-66-NH_2 growth on different ALD surfaces and its photocatalytic activity towards organic dye degradation. In the future, I will actively apply my research findings to application-oriented device configurations, and continue the effort of scaling up processes for MOF and MOF-based composite materials.

1. Zhao, J. *et al.* Conformal and highly adsorptive metal–organic framework thin films via layer-by-layer growth on ALD-coated fiber mats. *J. Mater. Chem. A* **3,**1458–1464 (2015).

2. Zhao, J. *et al.* Highly Adsorptive, MOF-Functionalized Nonwoven Fiber Mats for Hazardous Gas Capture Enabled by Atomic Layer Deposition. *Adv. Mater. Interfaces* **1,** (2014). doi: 10.1002/admi.201400040

430531 Rational Way of Designing Microfluidic Devices for Energy and Bioengineering Applications

Abstract:

My research interests are in developing novel microfluidic platforms that aid in solving the current challenges in energy and health industry. During my PhD, I worked on developing models based on physics and graph theory for droplet microfluidic systems. I also worked on developing technology that helps in designing microfluidic devices based on applications. The projects proposed here are an outgrowth of my research that I performed during my PhD and as co-founder at SysEng LLC.

Microfluidic technology has shown great promise in providing novel solutions for biochemical analysis. The greatest impact of microfluidic technology would be in developing robust lab-on-chip devices with varied capabilities - *e.g.*, producing perfectly controlled emulsions, point-of-care diagnostics and rapid screening of cells. While the promise is exciting, at present the functionalities of these devices focus on basic tasks- such as droplet production, storing, merging and sorting -but not on a complete design solution. There is a need to develop modeling and designing tools to rationally develop microfluidic devices that have multiple functionalities with flexibility and robustness. These tools will enable microfluidic technology to transition to practice and result in an impact similar to what was witnessed in the silicon revolution. Students working in this area should be trained in interdisciplinary fields consisting of biochemical analysis, optimization, process control and fluid mechanics to conceptualize and build designs that are commercially viable. I got an NSF STTR grant funded for this technology. Therefore, I plan to work on both fundamental research and also on applied engineering that impacts society in immediate future.

Microfluidic systems are also used by nature for optimal transport of fluids. So, I envision that my research on microfluidics will not only help in developing lab-on-chip devices for biochemical analysis but more broadly could help in understanding natural systems. This would translate to bio-mimicry and bio-robotics; I present some of my initial interests in these areas. These projects present my long-term research interests that will allow my group to use its knowledge of microfluidics and systems engineering to make valuable contributions to health and energy needs of the world.

430759 Towards Accurate and Fast Discovery of Compound Materials As Catalysts: Lessons Learned from Oxides

Abstract:

Non-metal, compound catalytic materials are used in mature, well-established industrial applications and but also lie at the forefront of state of the art technologies. These materials are typically discovered through costly trial and error. Density functional theory (DFT) calculations can elucidate key mechanistic details of catalytic processes and structure-property relationships of both known and potential catalytic materials. For metal catalysts, work using these principles have produced a number of successful examples of high-throughput discovery of novel, inexpensive, and active materials. Unfortunately, our ability and understanding on non-metal compound catalysts pales in comparison. Transition metal oxides (TMOs) constitute a large family of compound, catalytic materials. In contrast to metals, oxides can exist in multiple oxidation states and crystal structures, hindering our ability to make comprehensive models for predicting their activity. Standard implementations of DFT also fail to capture the behavior of correlated *d*-electrons present in TMOs. These failures limit our ability to capture accurate energetic data, which is required for a kinetic and thermodynamic analysis of catalytic processes.

Currently in my PhD with my advisor Prof. John Kitchin, we have addressed both of these difficulties. The hallmark method of our work is the coupling of structural perturbation calculations and identified structure motifs across phase space. We first demonstrate how these calculations can be used to discover subtle electronic structure relationships that describe catalytic effect of changing composition and crystal structure. Using well known structure-property relationships of metals as a foundation, we have gained insight to the critical features of the electronic structure that are responsible for adsorption across the transition metal series of oxides.[1] Similarly, we identify the strain and ligand effect present in mixed metal oxides, using structural perturbations to relate calculated electronic and adsorption features to known chemical properties.[2] This work establishes one of the first models that can predict the activity of mixed metal oxides from known chemical data. Our most recent work explores the potential activity of polymorph oxides, where we realize the unique chemical properties offered by polymorphic structures and relate these properties to the both the crystal and electronic structure.

Secondly, we use structural perturbations coupled with a self-consistently calculated Hubbard U to construct a DFT+U(V) method for more accurate predictions of relative stability of oxide materials.[3] We find that our method is both computationally faster and more accurate than the current state-of-the-art electronic structure methods. We also show how using the linear response U leads to more accurate predictions of the activity of late TMOs for the oxygen evolution reaction (OER), which is the current bottleneck towards the electrochemical production of the alternative fuel H_2.[4] We are currently extending our method to other transition metal compounds (sulfides, nitrides, etc) to ascertain their need for DFT+U.

My future work aims to create predictive models that relate the structure and composition of an oxide to its activity. These models will be validated with a wide variety of known compounds and will take advantage of the large databases of crystal structures already constructed in the materials science community (materialsproject.org, OQPMD, etc). This work will include development of algorithms of the construction of the surface facets as well as atomistic-thermodynamic models to predict the structure and surface segregation of mixed metal oxides. Using structure-property relationships determined in my PhD as a foundation, machine learning techniques will be applied to pinpoint chemical properties key for determining their activity/stability. Combined with the aforementioned established materials databases and structure-property relationship insight, high-throughput evaluations of catalytic activity can be realized. Finally, the methods and ideas established in my PhD will also be applied to other compounds such as nitrides, phosphides, carbides, etc.

Relevant Publications

(1) **Xu, Z.**, & Kitchin, J. R. (2014). *Catal. Commun.*, *52*, 60–64.

(2) **Xu, Z.**, & Kitchin, J. R. (2015). *J. Chem. Phys.*, *142*(10), 104703

(3) **Xu, Z.**, Joshi, Y. V, Raman, S., & Kitchin, J. R. (2015). *J. Chem. Phys.*, *142*(14), 144701

(4) **Xu, Z.**, Rossmeisl, J., & Kitchin, J. R. (2015). *J. Phys. Chem. C.*, *119*(9), 4827–4833

430453 Advanced Biological Imaging Probes and Sensors Using the Intrinsic Optical Signals of Single-Walled Carbon Nanotubes

Abstract:

The intrinsic band-gap photoluminescence (fluorescence) of single-walled carbon nanotubes (SWCNTs) exhibits exceptional photostability, narrow bandwidth, near-infrared (nIR) tissue-penetrating emission, and microenvironmental sensitivity, which potentiates their usage in a variety of biomedical imaging applications. Amphiphilic bio-polymers, such as single-stranded DNA, were found capable of singly-dispersing SWCNTs, enabling photoluminescence in aqueous solutions. A DNA sequence-dependent scheme for SWCNT species (chirality) separation has been observed with DNA-SWCNT binding strengths that were highly correlated. Using all-atom replica exchange molecular dynamics simulations (REMD), equilibrium structures confirmed that a variety of novel DNA secondary structures were forming when confined to the cylindrical surface of the SWCNT in a DNA-sequence and SWCNT-chirality-dependent fashion.

In cellular applications, DNA-encapsulated SWCNTs were found to enter primary mammalian endothelial cells by means of an energy-dependent endocytosis mechanism and remain in the endosomal pathway. A novel spectral imaging approach was able to spatially resolve over a dozen chiralities of single SWCNTs in live cells, in excised mouse tissue, and in live zebrafish. Additionally, in the endosomes of live cells, the SWCNT's ability to sense perturbations in its microenvironment, via modulations in emission intensity and peak center wavelength, was exploited as a non-destructive and long-term monitoring nIR optical reporter. Again using REMD simulation, there was observed correspondence between the optical signals of the SWCNT and the nanostructured analytes and solvent molecules.

430636 Corona Phase Molecular Recognition of Fibrinogen

Abstract:

Corona Phase Molecular Recognition (CoPhMoRe) is a method whereby a heteropolymer is adsorbed onto a nanoparticle surface, templating it for the recognition of a specific target analyte. Recently demonstrated for small molecules [1,2], CoPhMoRe applicability has not been extended yet to macromolecules such as proteins. In this work a CoPhMoRe screen of single walled carbon nanotubes (SWCNT) wrapped in a library of phospholipid-poly ethylene glycol derivatives and oligonucleotides against a panel of human blood proteins reveals a phase highly selective to fibrinogen. This corona phase has one of the highest relative coverage of the nanotube surface, of more than 85%, and upon the interaction with fibrinogen, an 80% decrease in fluorescent emission intensity is observed. The three nodules of fibrinogen are shown to bind sequentially in a three step mechanism, starting with the outer D-regions followed by the central E-region, with dissociation constants that increase with the SWCNT diameter, ranging from 3.5 nM for the (6,5) chirality to 31.4 nM for the (11,3) chirality. Our results open a new direction in the fundamental search of novel synthetic non-biological antibodies for the recognition of bio-macromolecules, proteins, and peptides, for biological applications.

References

[1] Zhang J. *et. al.*, *Nature Nanotechnology*, 8, 959–968, 2013

[2] Kruss S. *et. al.*, *J. Am. Chem. Soc.*, 136 (2), 713–724, 2014

430788 Engineering Nanoparticles As Theranostic Probe and Understanding Their Interaction with the Lysosome-Autophagy System

Abstract:

Engineered nanoparticles have shown great promise for the diagnosis and treatment of a broad range of diseases. My research will investigate the engineering design principles for nanoparticle-mediated early diagnosis and efficient therapy. A particular interest will be on designing lanthanide based theranostics probe – a class of nanoprobe extremely promising for deep tissue optical imaging, magnetic resonance imaging (MRI), X-ray luminescence optical imaging, and therapy.

When nanoparticles are internalized into cells, like other nano-sized material perceived by the cell as foreign or toxic (such as viruses), they may stimulate the activation of cellular clearance mechanisms. The main clearance pathway that mediates degradation of toxic and aberrant internalized material is autophagy. My other major research interest is focused on understanding the interaction of nanoparticles with the cellular and subcellular organelles particularly on the lysosome-autophagosome system. I would also like to predictively determine deposition, translocation of the environmental particulate matters and inhaled drugs in the respiratory track.

430839 Design of New Materials and Understand Emergent Behavior Using Computational Methods

Abstract:

Naga Rajesh Tummala

Design of New Materials and Understand Emergent Behavior using Computational Methods

My research experiences have provided me with the opportunity to learn and implement computational techniques at multiple length and time scales. I learned to investigate physics behind many challenging research problems (nanotube dispersions, packing of fullerene derivatives, nano-particle transport in porous media, fracture and electronic disorder in polymer:fullerene mixtures), and to describe them fundamentally at molecular level. I have also acquired knowledge of methods to identify both physical interactions involved in various phenomena and the techniques needed to relate those interactions to macroscopic properties. With this knowledge, I am confident that my group will successfully pursue following research topics.

- Understanding ion/charge transfer properties in polymers
  - Correlating building blocks to ion/charge transport in crystalline and amorphous regions
  - Effect of polymer-particle (nano-particles of various shapes and adducts) interactions on morphology and charge transport

Potential Impact: Design rules for polymers for improved energy storage.

- Aggregation and crystal formation of organic molecules and metal nano-particles
  - At interfaces (fluid/fluid, air/fluid, solid/fluid) and surface effects
  - Understanding kinetics of formation of metal nano-particles in various solvent conditions

Potential Impact: Solvent affects and anisotropic effects on the dynamics of aggregation and better predictive models for self-assembly.

- Interactions and dynamics in porous media and at interfaces
  - Understand how electrostatics determine the (de)wetting phenomena at various interfaces
  - Interactions and dynamics of hydrocarbons/polymers/surfactants in porous media
  - Dynamics of nano-particles/soft particles in random and structured porous media.

Potential Impact: To provide molecular level understanding of the surface-molecule interactions and their effect on the dynamics of molecules and particles.

My research interests also include interfacial phenomenon of colloids, surfaces, polymers, nano-particles, and charge transfer at these interfaces. I am eager to collaborate on other research ideas apart from the independent proposals.

PhD: 2005 – 2010, University of Oklahoma, Advisor: Prof. Alberto Striolo, present: UCL, London

Post-Doctoral Fellow: 2010 – 2011, University of Oklahoma, Advisor: Prof. Dimitrios Papavassiliou

Post-Doctoral Fellow: 2011 – 2014, Georgia Institute of Technology, Advisor: Prof. Jean-Luc Bredas

Research Scientist: 2014-Present, Georgia Institute of Technology, Advisor: Prof. Jean-Luc Bredas, present: KAUST, Thuwal, Saudi Arabia

Google Scholar: https://scholar.google.com/citations?hl=en&user=Gou93qcAAAAJ

431034 Understanding and Predicting the Activity of Zeolite Catalysts

Sunday, November 8, 2015
Exhibit Hall 1 (Salt Palace Convention Center)
Florian Göltl, Chemistry, UW Madison, Madison, WI

Zeolites are the most important industrial catalysts today. Their main application in catalysis is within crude oil refinement, but more recently they have been shown to be very effective systems in the reforming of biomass and in cleaning car exhaust gases. The activity for a given process is determined by the framework type and by the preparation of the catalyst. Similar to quartz, zeolites are mainly composed of Si and O. Exchanging Si^{IV} by e.g. Al^{III} creates a local charge imbalance, which in turn is compensated by a counter ion. These ions act as catalytically active centers. The activity of the catalyst will depend on the distribution of these active sites, which in turn is determined by the distribution of Al framework atoms.

The goal in my research is to characterize this underlying distribution, to show how it can be used to understand the catalytic behavior of zeolite catalysts, and, in a next step, to computationally predict this distribution and design optimal catalysts.

One of the largest challenges in characterizing zeolite catalysts is the mismatch between the calculated spectra, which are typically modeled statically and correspond to experimental conditions of 0K, and experimental measurements at ambient conditions. Therefore the goal is to develop and apply theoretical, ensemble based methods that are comparable to experimental conditions. This includes the use of molecular dynamics simulations, distribution based techniques and Monte-Carlo simulations to calculate e.g. IR, NMR- and UV-vis spectra as well as studying the impact of water under the given conditions on the catalyst.

Another key point is to understand the impact of this distribution on the catalytic activity. This can be as simple as comparing catalytic activity measurements to the distribution of active sites in the material, but can also be highly complicated when the positioning of the active sites influences their chemical activity or changes the influence of the framework on the reaction. These confinement effects have been a major direction within recent research in the field and will also play a key role in my future research.

With the increase in computational power approaches to predict possible new catalysts have become more and more important. The mid- to long-term goal in my research is to develop a screening based approach to design zeolite catalysts. Based upon the ideas described in the previous paragraph it is therefore necessary to identify descriptors for chemical influence of the active sites and the zeolite framework. The second step is to screen various zeolite structures for these parameters, which in the end will lead to potential candidates for the reactions in question. This includes understanding the errors of theory by comparing with experiment to arrive at a quantitatively accurate approach.

431113 Model Colloid System for the Direct Observation of Interfacial Sorption Kinetics

Abstract:

Particle adsorption to an interface can be a complicated process, motivating detailed measurements of the relevant factors that govern adsorption coverage and rates. My research uses micron-sized particles as an experimental model system to investigate the effect of particle characteristics on adsorption kinetics through direct observations at the individual particle level. The particle-interface interaction strength can be tuned to be comparable with nanoscale systems and experimental conditions can be carefully controlled through modifications to the particles and/or interface. For instance, the effect of shape on adsorption kinetics is investigated by tracking colloidal dimers near a solid interface. Kinetics are measured by tracking transitions between an entropically trapped adsorbed state and a desorbed states driven by Brownian motion. The observed kinetics can be directly compared to theoretical predictions based on the measured interaction potential and local diffusivity. Multiple light microscopy methods are employed for 3-dimensional particle tracking, including total internal reflection microscopy (TIRM) for positions near the surface and off-focus tracking for positions in the bulk. This experiment approach can be extended to study complex adsorption processes in detail, including effects such as crowding, deformability, and applied fields.

431152 Smart Membranes with Hygro-Responsive Surfaces for Efficient Separation of Liquid Mixtures

Abstract:

There is a critical need for new energy-efficient solutions for separating liquid mixtures. Among the variety of current separation technologies, membrane-based operations are attractive because they are relatively energy-efficient, and are applicable to a wide range of industrial effluents. Our research explores the systematic design of membranes, as well as, the development of smart methodologies that enable separation of a wide variety of both immiscible and miscible liquid mixtures.

In this work we have developed novel membranes with hygro-responsive surfaces, which are both superhydrophilic and superoleophobic in air and underwater environments. Our membrane can separate several liters of oil-water mixtures, including surfactant-stabilized emulsions, solely using gravity, in a single-unit operation, with > 99.9% separation efficiency, by using the difference in capillary forces acting on the two phases. Our separation methodology is solely gravity-driven and consequently is expected to be highly energy-efficient. We demonstrate the separation of several liters of oil-water mixtures using a scaled-up apparatus. We also demonstrate continuous separation of oil-water emulsions for over 100 hours without a decrease in flux.

In addition to the separation of immiscible liquids such as oil and water, separation of miscible liquids is also important in various industries, including oil refining, biofuel production and food industry. Although distillation is one of the most widely used separation methods for separating miscible components from a liquid mixture, it has a high energy-cost, being a thermally driven process. Further, it is unsuitable for the separation of components with similar boiling points and azeotropes.

Liquid-liquid extraction, also known as solvent extraction, is typically used to separate azeotropes or components with overlapping boiling points when simple distillation cannot be used. For an efficient extraction of components, maximizing contact between the feed and the extractant is critical. Emulsions, especially those stabilized by surfactants, provide a large interfacial area and greatly enhance the mass transfer in extraction, the subsequent separation of emulsions can be energy-intensive and less economical. Consequently, there is a great need and a significant opportunity to develop a new energy-efficient separation methodology with enhanced mass transfer during extraction and facile separation of the surfactant-stabilized emulsion.

Here we have developed a new energy-efficient methodology that combines liquid-liquid extraction using surfactant-stabilized emulsions, and solely gravity-driven separation of these emulsions into a single unit operation, using membranes with hygro-responsive surfaces. We have demonstrated that our separation methodology is useful for a wide variety of separations, including the separations of azeotropes of alcohol-hydrocarbons, miscible dyes and sulfur compounds from oils. We anticipate that our separation methodology has numerous applications, including the production of diesel with ultra-low sulfur content, separation and purification of bio-fuels, removal of dyes from jet fuels and separation of azeotropes.

(6gn) Responsive Hydrogels for 4D Cell Culture and Controlled Drug Delivery

Tibbitt, M. W., University of Colorado
Langer, R., Massachusetts Institute of Technology

Responsive hydrogels for 4D cell culture and controlled drug delivery

Mark W. Tibbitt and Robert Langer, Massachusetts Institute of Technology

Interactions between mammalian cells and the surrounding extracellular matrix (ECM) are critical in the regulation of stem cell differentiation, cell migration, tissue morphogenesis, and disease progression. Cell decisions are often made by integrating these complex interactions, which are dynamic in 4D: three-dimensional space and time. Synthetic polymer-based hydrogels have emerged as an attractive cell culture platform to investigate the influence of these cell-matrix interactions on cell function in a systematic manner. Hydrogels are appealing mimics of the ECM on account of their high water content, tissue-like elasticity, and facile transport of nutrients, waste, and soluble factors. Furthermore, poly(ethylene glycol) (PEG) hydrogels can be formed under mild, cytocompatible conditions and are easily modified to present various biochemical and mechanical signals to cells during culture.

While cell-laden PEG hydrogel constructs have been employed for regenerative medicine and fundamental biological applications, there is still a lack of materials that allow the user to recapitulate the spatially heterogeneous and dynamic environment that cells interact with in vivo. My Ph.D. research has focused on the development, characterization, and application of photoresponsive PEG-based hydrogels that afford the user control of the mechanical and biochemical properties of the scaffold in both 3D space and time. This class of materials facilitates experiments that probe the relationship between spatially heterogeneous and dynamic signals from the ECM and cell function in a well-defined environment to study the fundamentals of the cell-material interactions, stem cell differentiation and plasticity, as well as the cellular response to local changes in soluble factor concentration.

Specifically, we have developed photoresponsive materials for the dynamic modulation of the ECM during 2D and 3D culture. These materials have been used to investigate stem cell response to dynamic changes in elasticity and the plasticity of stem cells during microenvironmental mechanotransduction. Importantly, we uncovered fundamental information about mechanical dosing and memory in mesenchymal stem cells that is regulated by YAP/TAZ signaling. Further, we have utilized photoresponsive hydrogels to develop photodegradable microspheres for the targeted delivery of soluble factors to study tissue morphogenesis and chemotaxis. Thin films of photodegradable hydrogel have been employed for the selective capture and release of cells for diagnostic application.

In my Postdoctoral research, we have developed responsive hydrogels for controlled delivery of therapeutics to intervene during disease progression. Specifically, we have developed a class of shear-thinning and self-healing hydrogels based on specific and reversible polymer-nanoparticle (PNP) interactions. These materials are facile to fabricate, scalable, and amenable to a broad range of biomedical and industrial applications. As the PNP hydrogels can be formed from biocompatible materials, they have been leveraged for local injection and dual-stage therapeutic release as well as cell carriers. Additional efforts are exploring the use of PNP gels for local cell recruitment within the body.

(6go) Recovery of Folded Heterologous Proteins in the Extracellular Space from Bacterial Culture

Metcalf, K. J., University of California, Berkeley
Valdivia, E., University of California, Berkeley
Azam, A., University of California Berkeley
Rosales, S., University of California, Berkeley
Finnerty, C., University of California, Berkeley
Bevington, J.,
Yao, B.,
Reid, M.,
Tullman-Ercek, D., University of California, Berkeley

Heterologous protein production in bacteria is a batch process, where the cells are lysed and the protein of interest is purified from the cellular milieu. Protein secretion into the extracellular space simplifies protein purification and enables continuous processing. We use synthetic biology and protein engineering techniques to optimize a bacterial strain for production and secretion of heterologous proteins of interest (POIs). We start by engineering the apparatus of the type III secretion system (T3SS) of *Salmonella enterica*, a heterogeneous multi-protein structure that spans the periplasm, inner and outer membranes, and projects out from the cell, resembling a hypodermic needle.

To increase secreted protein titers, we must control the expression of the secretion apparatus. Critically, expression of the genes that code for the apparatus is tightly regulated and very sensitive, and native expression of these genes is low and only occurs in a fraction of the culture. We achieve a greater than 10-fold increase in secreted protein titer, exceeding 10 mg/L, using transcriptional control. We control the expression of the positive transcription factor, HilA, the native regulator of the apparatus genes. By synthetically controlling expression of the apparatus genes, we increase expression of the apparatus genes and activate nearly all the cells in the culture for secretion. This approach is generalized for all proteins tested, as it does not manipulate the POI.

Additionally, we characterize the folding of the secreted protein product. Recent structural data shows that the protein is unfolded and linearized as it is secreted through the "needle", which is 2 nm in diameter and 50 nm long. Spontaneous refolding in the extracellular space is desired for recovery of correctly folded proteins. We use enzymatic activity as a proxy for protein folding, and demonstrate enzyme activity in the extracellular space, indicating that secreted proteins indeed refold after secretion. Genetic and chemical methods are used to probe the folded state of diverse models proteins. Standard functional assays indicate that these proteins refold into functional forms after secretion. In addition, our data suggests that post-translational modifications, such as metal cofactor biding, multimerization, and disulfide bond formation, spontaneously occur post-secretion. Together, recovery of folded heterologous protein at high titer in the extracellular space exhibits a strength of the use of a protein secretion chassis.

431362 Understanding and Controlling the Mechanical Properties of Polymeric Networks

Abstract:

The mechanical properties of polymer networks are vitally important to the performance of materials, their dynamic response under force activation and their longtime stability. Emerging applications of polymeric networks in energy, biomedical engineering, and defense technologies demand further advancement in understanding the structure-mechanics relationship, as well as development of methods that allow predicative control over mechanical properties of materials.

In this poster, I will present my graduate research on understanding and controlling the mechanical properties of model associating polymeric networks. First, I investigate the network relaxation hierarchy over a large window of length scales and time scales. By combining various experimental techniques such as mechanical spectroscopy, forced Rayleigh scattering and nuclear magnetic resonance, I identify two major deviations from the ideal Maxwellian behavior, which originate from the presence of dynamic inhomogeneity and the difference between the Rouse time of network strands and the junction dissociation time. Next, I demonstrate a new approach to control the mechanics of associating networks by manipulating the molecular architecture of network constituents. Specifically, topological entanglement is engineered into the network to drastically slow down the relaxation of polymer chains at long times. This strategy effectively improves the materials' toughness, extensibility, creep resistance and stability in solutions. The entanglement effect and mechanical properties are further controlled by introducing short grafts or branches on the polymer architecture. Experiment results show qualitative agreement with predictions from sticky Rouse and sticky reptation theories. Finally, I will discuss new synthetic approaches to prepare hybrid macromolecules mimicking the biological functions and mechanical properties of natural mucins and proteoglycans.

In the future, I would like to carry research towards (1) unraveling the complex structure-property relationship in complex natural biomaterials, and (2) developing new platform technologies to prepare multifunctional materials interfacing with living matters.

(6gq) Accelerated Computational Discovery of Materials for Production, Storage, and Efficient Use of Energy

Gomez Gualdron, D. A., Northwestern University

One of the greatest challenges that mankind is facing is to continue the development of technologies that improve our quality of life without depleting our energy resources. At the center of addressing this challenge is the need to develop new materials that allow us to produce, store, and efficiently use energy. The advent of nanotechnology has given us unprecedented access to the atomistic structure of matter and enabled the rational design of materials by precisely controlling their nanostructure. My primary research interest is to develop high-throughput screening strategies to use with quantum and classical mechanics molecular simulations and other computational methods to (help) design materials that can play pivotal roles to address energy-related challenges in fields such as transportation and chemical industry.

At its core, my research aims to elucidate the links between properties, structure, performance, and processing of materials, with the goal of using this knowledge to accelerate the discovery of new materials. In this poster, I will provide an overview of relevant research carried out during my doctoral and postdoctoral work. During my doctoral studies, I used molecular simulations (density functional theory (DFT) and reactive molecular dynamics (RMD) to elucidate how the processing or catalyzed synthesis of single-walled carbon nanotubes (SWCNTs) could be modified to produce desired structures of these materials selectively. Selective synthesis is perhaps the "holy grail" of the nanotube synthesis field, and it would eliminate current bottlenecks for the exploitation of nanotube-based technologies arising from the well-established structural dependence of nanotube properties, and costly and energy-intensive nanotube separations methods. I investigated the feasibility of using the catalyst nanoparticle structure to act as a template to the nascent nanotube, and thus provided useful synthesis guidelines to experts in nanotube synthesis.

During my postdoctoral work, I have sought to design metal-organic frameworks (MOFs) for applications such as natural gas storage, hydrogen storage, carbon capture, and selective catalysis. MOFs are novel, remarkable, crystalline materials whose nanopore structure can be finely tuned for a given application by appropriate selection of essentially limitless possible combinations of inorganic "nodes" and organic "linkers." Thus the challenge is to efficiently identify the best possible material and elucidate the link between structure and performance, as well as performance boundaries. In addressing these challenges, my work has involved i) developing computational methods to automatedly generate thousands of MOFs structures for subsequent high throughput screening of their properties, and ii) developing simulation strategies that can facilitate the efficient screening of these materials.

In the first case, I have helped developed an automated algorithm used to generate thousands of MOFs using a topologically guided strategy. The goal was to establish limits (through data mining, analysis and visualization) and topologically-dependence performance of this class of materials for natural gas and hydrogen storage applications, and identify the best materials within these boundaries. Through close collaboration with synthetic chemists, some of the best MOFs identified via simulation have been synthesized, and their promising adsorption characteristics have been experimentally confirmed.

In the second case, I have contributed to the development of efficient MOF screening strategies based on genetic algorithms, which has been used as an alternative to the "brute force" screening methods. As a proof-of-concept, this strategy has been used to identify high-performance MOFs for pre-combustion carbon capture from a database of over hundred thousand materials, and it has demonstrated a reduction of computational expenses used during screening of several orders of magnitude. Also, I have developed a simulation strategy to investigate the design of hybrid nanoparticle at MOFs (NP@MOF) catalysts for regioselective oxidation of n-butane to n-butanol. Mimicking the working mechanism of enzymes, the nanopores of the MOF are expected to provide control on how the molecule "attacks" the surface of the nanoparticle, thus potentially engendering selective catalysis which is highly desirable for energy-efficient chemical synthesis (separation processes typically account for 50-60%of the energy cost to operate a chemical plant). A major challenge in modeling reactions at the NP/MOF interface includes the large number of atoms typically needed to construct a chemically meaningful model of the system. In principle, this problem could be solved using QM/MM methods, but unknowns about the exact structure of the NP/MOF interface structure, among other factors, also hinder the efficient construction of appropriate simulation models. To overcome this problem, I implemented a conceptually simple approach that can be used to efficiently explore how changes in the NP composition affect the energetics of a given reaction under sterically constrained conditions, as would occur at a NP/MOF interface. In this approach, two size- and shape-tunable rings composed of helium atoms placed on top of the catalyst surface mimic the steric constraints that would arise from a given MOF pore. This model can be implemented with standard DFT programs, and it has been used to investigate the oxidation of butane to butanol on palladium and palladium-alloy catalysts.

Building on my expertise in materials science, computational methods, catalysis, and data mining, my future research plans include the efficient computational design and discovery of catalytic interfaces and MOFs for storage, production, and efficient use of energy, as well as addressing the emerging challenge of polymorphism control on MOFs.

Doctoral Advisor: Professor Perla B. Balbuena, Texas A&M University, College Station TX

Postdoctoral Advisor: Professor Randall Q. Snurr, Northwestern University, Evanston IL

431629 Applied Synthetic Biology for Engineering Metabolism and Synthetic Microbial Communities

Sunday, November 8, 2015
Exhibit Hall 1 (Salt Palace Convention Center)

Charles Rutter, Chemical and Biomolecular Engineering, Univeristy of Illinois - Urbana Champaign, Champaign, IL

Applied Synthetic Biology for Engineering Metabolism and Synthetic Microbial Communities

Charles David Rutter, University of Illinois, Urbana-Champaign

2nd Year Postdoctoral Fellow

Research Interests:

Synthetic Biology is a multi-disciplinary field focused of design and construction of biological devices for a desired application. Work is this field has led to development of tools that allow manipulation of microbial species beyond simple overexpression or knockout of protein function. Particularly, it has demonstrated the applicability of synthetic genetic constructs in both rational and combinatorial approaches involving manipulation of microbes at the genetic, transcript, and protein levels to improve performance. As such, these approaches are well suited for application to complex systems such as improved engineering of metabolism by addition of metabolite sensor elements and development of optimized synthetic microbial communities for enhanced lignocellulosic biomass degradation or complex biotransformations.

Postdoctoral Project: "Metabolic Engineering and Systems Biology of Oleaginous Yeasts"

Under supervision of Christopher V. Rao, Chemical and Biomolecular Engineering, Univeristy of Illinois, Urbana-Champaign

PhD Dissertation: "Engineering *E. coli* Toward Consolidated Bioprocess of Cellulose"

Under supervision of Rachel R. Chen, Chemical and Biomolecular Engineering, Georgia Institute of Technology

Undergraduate Thesis: "Characterization of Cellulosic Sugars by the Microalgae *C. reinhardtii*"

Under supervision of John Morgan, Chemical Engineering, Purdue University

Research Experience:

My research experience falls within the fields of synthetic biology, systems biology, and metabolic engineering. I have had the opportunity to work with all types of industrially relevant microorganisms (bacteria, yeast, and algae). As a PhD student I gained expertise in the relevant processing and metabolism of cellulosic substrates, approaches for improvement of protein secretion in bacteria as well as relevant metabolic engineering strategies for conversion of cellulose into a variety of valuable products. As a postdoctoral researcher I have had the opportunity to extend my knowledge to include the metabolism of traditional and oleaginous yeasts as well as their mechanisms for regulation at the genetic, transcript, and posttranscriptional level. This has left me with extensive knowledge in the fields of metabolic engineering, synthetic biology, molecular biology, microbial genetics, and fermentation technologies.

Teaching Experience:

In addition to my time spent as a researcher, I have gained substantial teaching experience. As an undergraduate at Purdue University I spent 3 semesters as a TA for the undergraduate math modelling and statistical analysis course. As a member of the Zeta Chapter of Omega Chi Epsilon I organized and led supplement instruction sessions for core courses in the undergraduate curriculum including momentum transfer, thermodynamics, and reaction kinetics across 5 semesters. I also worked as a tutor through the Purdue Athletic department to provide one-on-one and group tutoring for student-athletes in courses such as Spanish, chemistry, calculus, and differential equations. As a graduate student I had the opportunity to TA the Senior Operations Laboratory, which involved instruction of senior Chemical Engineering students in the lab as well as the Senior Capstone Design course in which I worked with professors and other TAs so create a problem statement and evaluate the designs submitted by students. In additional to these more formal teaching assignments, I have served as a mentor for younger students in the lab as both a graduate student and a postdoctoral research helping researchers to understand molecular and synthetic biology principles, analytical techniques, and experimental design.

Future Direction:

As a faculty I would like to continue my research in metabolic engineering by employing the synthetic biology techniques I have learned as well as those which I plan to develop. Firstly I would like to develop systems that are capable of sensing the metabolic and redox state of microbial cell factories to differentially regulate metabolism post-transcriptionally to remove inhibitory compounds and improve flux through rate limiting steps to improve product formation. This will require development of a variety of biosensors, which will be capable of sensing metabolites and reacting to regulate transcript levels dynamically.

Additionally I would like to continue to develop systems for consolidated bioprocessing of lignocellulosic substrates using synthetic microbial communities to leverage the natural capabilities of different organisms to create a cost-effective system for utilization of the most abundant biomass on the planet. This will require development of synthetic biology tools to manipulate various types of microbes including bacteria, cyanobacteria, fungi, and algae to cooperatively degrade substrates and application of creative metabolic engineering approaches to maximize product formation

Selected Publications:

C. Rutter, Z. Mao, and R. Chen. 2013. Periplasmic expression of a *Saccharophagus* cellodextrinase enables *E. coli* to ferment cellodextrin. *Applied microbiology and biotechnology* (2013): 1-10.

C. Rutter, R. Chen. 2014. Improved cellobiose utilization in *E. coli* by including both hydrolysis and phosphorolysis mechanisms. *Biotechnology Letters.* 36:301-307

HD Shin, SH Yoon, J Wu, **C Rutter**, SW Kim, R Chen. 2012. High-yield production of meso-2,3-butanediol from cellodextrin by engineering *E. coli* biocatalysts. *Bioresource Biotechnology.* 118:367-73

C Rutter, C.V. Rao, S. Zhang. 2015. Engineering *Yarrowia lipolytica* for Production of Medium-Chain Fatty Acids. In Preparation.

430650 Interfacial Interactions and the Design of Smart Materials

Abstract:

Interactions at interfaces (solid/solid, solid/liquid, liquid/vapor) mediate many processes, including colloidal/self-assembly, charge transfer, and separation. One type of system in which these interactions are important is stimuli-responsive materials. Stimuli-responsive materials are typically comprised of various components and phases of matter, therefore having many types of interfaces. Thus, the study of appropriate interfacial phenomena can greatly aid in the design of smart materials. First, the engineering of systems responsive to magnetic field and light is presented. These systems are complex multiphase dispersions fabricated by exploiting interactions between anisotropic particles in solution and at air-liquid interfaces. We examine the magnitude of material response to an applied field as a function of system composition and rheological properties. Next, the study of interactions at solution-particle interfaces is presented in context of its importance to carbon nanotube separations. The effect of a molecular dispersant on the nature of the effective particle facilitates nanotube separation by diameter, electronic type, and chirality in solution-based purification processes. In our studies, chirality and length sorted single-wall carbon nanotubes dispersed in ionic surfactant serve as model colloids for refining analytical ultracentrifugation as a tool to characterize colloidal systems, and for understanding nanotube-surfactant interactions. From these works, it is clear that studying interactions at particle-complex solution interfaces, particle-particle interfaces, as well as liquid-vapor interfaces at different environmental conditions and for non-spherical particles is not only fundamentally and industrially relevant but is also crucial for the intelligent design of smart materials.

(6gr) Modeling Biomass and Its Conversion to Fuels and Specialty Chemicals

Rabideau, B. D., RWTH Aachen University

My research uses modeling and simulation to develop a significantly deeper understanding of the nature of lignocellulosic biomass and the processes by which it is converted into target fuels and chemicals. There are still a number of major challenges affecting the profitability of second and third generation biorefineries and each project will work to address one of these different challenges. The use of modeling and simulation is uniquely suited to these problems, with length scales at the molecular-level. Developing uniqe insights at these scales will enable us to make informed decisions in the implementation of these processes for converting biomass into desired products and hopefully lead to breakthroughs in the full valorization of lignocellulose that would help bring this greener and more sustainable future to fruition.

431586 Dual Conduction Polymers for Energy Conversion and Storage

Sunday, November 8, 2015
Exhibit Hall 1 (Salt Palace Convention Center)

Bhooshan C. Popere, Chemical Engineering & Materials, University of California Santa Barbara, Santa Barbara, CA and Rachel Segalman, Departments of Materials and Chemical Engineering, UCSB, Santa Barbara, CA

Organic materials that transport electrons, ions or both are promising candidates for alternative sources of energy. The vast potential of available molecular design principles enables a deeper understanding of how structure at the molecular and mesoscopic length scales affects the macroscopic properties of these functional materials. Despite the recent meteoric progress in the development of high efficiency materials for electron (charge) transport, in general, the fundamental mesoscopic-level understanding of such materials remains somewhat elusive. Drawing on my understanding from physical organic chemistry, polymer physics, self-assembly and transport phenomena in condensed matter, my research has been primarily focused on developing molecular strategies to gain a deeper insight into the structure-property relationships of such functional materials.

During my doctoral research at UMass Amherst (with Prof. Sankaran 'Thai' Thayumanavan), I developed a novel class of pi-conjugated polymeric materials, based on the BODIPY core, inspired by optoelectronic applications. These materials possess panchromatic absorption in the visible and near-IR region of the electromagnetic spectrum making them strong light absorbers in photovoltaic devices. Rational molecular design guided by density functional theory calculations enabled us to develop materials with independently tunable frontier molecular orbital energy levels, while retaining the panchromatic absorption and high charge carrier mobility – a highly desirable feature not commonly observed for pi-conjugated polymers. As a postdoc with Prof. Rachel Segalman at UCSB, I am investigating a new set of materials based on polymerized ionic liquids (PILs)viaDielectric Relaxation Spectroscopy with the aim of understanding the fundamental dielectric properties of these systems. An attractive feature of such polymeric materials is their ability to form electrical double layers upon polarization at the interface with an electrode. The large electric fields due to the small electrical double layer (EDL) thickness results in unusually large areal capacitance values in comparison with conventional organic materials. This makes PILs attractive candidates as high k dielectric materials for low-voltage operable field effect transistors (FETs). Our preliminary experiments with using these PILs as gate dielectrics in organic FETs indeed suggest a dependence of operational threshold voltage on the specific capacitance determined at low frequencies. Specifically, for conventional organic semiconducting polymers like P3HT and PBTTT, the threshold voltages with our PILs as gate dielectrics are less than 1V, indicating a true electrolyte gating mechanism.

In the future, my research will focus on understanding the molecular and mesoscopic underpinnings of the next generation of charge transporting materials. Dual-conduction polymeric systems that simultaneously transport ions and electrons are of great interest both, from a fundamental perspective, and as promising materials for energy storage and conversion, electromechanical actuators and biomimetic systems, amongst many others. However, understanding the physics of charge transport through pi-conjugated polymers in the electrochemical milieu is non-trivial. My approach will be based on rational molecular design, guided by theory and simulations to systematically decouple the dynamics of charge transport in these mixed conducting systems.

431718 Accelerating Materials Discovery with Data Science

Abstract:

From genomics to high-energy physics, data science is transforming many branches of science and engineering. My research group will use tools from data science and molecular modeling to transform how we discover new materials to tackle challenges in energy, environment, and human health.

My postdoctoral research has addressed energy and environmental challenges, focusing on an emerging class of adsorbent materials called metal-organic frameworks (MOFs). As part of multi-institution team within the Nanoporous Materials Genome Center (NMGC), we created a database of "computation-ready, experimental" (CoRE) MOFs, which contains the atomic coordinates of thousands of experimentally synthesized MOFs. As an initial demonstration of the utility of the database, we carried out high-throughput grand canonical Monte Carlo simulations to find best performing MOFs for methane storage and delivery, a challenging chemical isomer separation, and pre-combustion carbon capture. Through this work, we were able to quickly identify promising MOFs for target applications and provide molecular-level insights into the performance limit of MOFs for these applications. To augment the efficiency of high-throughput computational screening, we developed a genetic algorithm to identify high-performing MOFs for precombustion CO_2/H_2separation. High-performing MOFs identified through the process were then synthesized and tested by experimental collaborators.

Materials discovery will be the central theme of my research group, where we will use tools from data science and molecular modeling to rapidly generate, identify, and disseminate promising materials for their applications in energy, environment, and human health. The initial foci will be on the discovery of new polymer electrolytes for energy storage systems, the discovery of new MOFs for carbon capture, and the discovery of new materials for drug delivery system. At the same time, my group will use a combination of simulation, theory, and experiments to investigate the fundamental questions related to how these novel materials are synthesized. The initial focus of this research will be on the crystal formation of high-performing MOFs for carbon capture.

431790 The Synthesis, Design, and Tunable Functionalization of Nano-Engineered Materials

Abstract:

The goal of my research is to synthesize materials with precise control over composition and functionality to establish and advance new principles, techniques, and methods in nanotechnology. In this work I investigated the fundamental properties of nano-engineered materials and focused on developing model systems to design efficient and cost-effective materials for various applications. My efforts will be focused into three research directions, all of which cut across the traditional boundaries of science disciplines – materials science, chemical engineering, and nanotechnology. In particular, I have demonstrated the design and development of nano-engineered materials for various applications, including: 1) Nanoscale polarization manipulation and switching of ferroelectric perovskites for nano-device applications; 2) Processing-structure-property relationships that govern organic-inorganic nano-interfaces for organic transistors and solar cells; and 3) Materials synthesis of transition metal oxides for heterogeneous catalytic systems.

431854 Optoelectronic and Electronic Interfaces to the Brain

Abstract:

Nanotechnology-based breakthroughs in Biology and Energy

My research will focus on nanotechnology-based approaches to understand and control fundamental electronic and mechanical properties of biological materials for advancing both biomedical and energy sciences. I am particularly interested in interfacing biology with functional electronics to yield breakthroughs in drug development, to enable new avenues of powering biomedical devices, and to probe biological changes in a truly non-invasive way. For my doctoral research, I developed a novel method for understanding charge transport at nanoscale metal-oxide interfaces with the use of scanning probe techniques and extended that knowledge to understanding and controlling polarization switching in ferroelectric devices. Ferroelectrics are highly efficient mechanical couplers and are ideal candidates for interfacing with biology for a number of applications. As a postdoctoral fellow, I developed new techniques to measure tight junction adhesion forces at piconewton resolution. With innovative electrode designs, I developed a novel bioimpedance system to track changes in blood-brain barrier dynamics – in-real time – while modulating ion transport gateways. As a Faculty Research Associate, I am interfacing optoelectronic devices with the peripheral nervous system for optogenetic control of nerve signals. With my background and skills, I am uniquely positioned to launch an interdisciplinary, collaborative Nanotechnology in Biology and Energy program that incorporates the expertise of both scientists and engineers.

431975 Unraveling the Chemistry of Energy Systems

Sunday, November 8, 2015
Exhibit Hall 1 (Salt Palace Convention Center)

Nicole Labbe, Chemical Sciences & Engineering Division, Argonne National Laboratory, Argonne, IL

The need for reliable energy sources is a constantly evolving problem with unique challenges. For the foreseeable future, much of our energy demands will still be met with hydrocarbon fuels, though the landscape of types of fuels and fuel-sources will continue to evolve. This evolution in fuel composition is currently being witnessed in the energy sector with the emergence of petroleum-derived fuels from shale oil and tar sands that have greater naphthenic content and fuels obtained from biomass with their higher content of oxygenated molecules. Each of these fuel sources/types poses challenges to infrastructure and eventual end-use applications, particularly in the transportation sector. A second challenge is that in recent years, legislature has been passed mandating significant improvement of transportation fuel efficiency, reduction of carbon-emissions, tighter pollution regulations, and increased use of bio-renewable energy sources. To meet these new regulations, new technologies are continuously being developed to increase fuel efficiency and reduce emissions, such as HCCI engines as an alternative to traditional spark ignition engines and membrane separation technology to sequester greenhouse gas emissions. Under this evolving energy landscape, it is imperative to have a detailed understanding of the complex chemistry and thermodynamics of these energy applications to accelerate the development of new green energy technologies.

Chemical kinetic modeling allows for the application of theory to interpret experimental data and elucidation of the chemical pathways essential to understanding harmful emissions formation. Kinetics models offer the practical engineer tools to improve fuel efficiency and reduce emissions, and to aid in the design and development of the next-generation of energy technologies. The development of chemical models used to model energy applications has evolved over the past several decades, both in fuel complexity and in the breadth of temperature and pressure conditions to be considered. Even simple fuels such as ethanol require models with hundreds of distinct molecules, which undergo thousands of unique elementary reactions. The complexity of these models are expanded exponentially when considering "real" fuels (e.g. diesel), which are blends of many high molecular weight species. Further complicating the problem is that energy systems operate over a wide range of conditions, requiring accurate kinetics over an incredible range of temperatures (300-2500 K) and pressures (< 0.1-100 atm). As a result of this complexity, most models to date have relied on kinetic estimations and manual tuning of kinetic rate constants to reproduce benchmark experimental data. The resultant models are often not reliable or are highly uncertain beyond the scope of the targeted benchmark experimental data, and cannot be universally applied to predict the chemistry of all energy applications.

My research will focus on the development of robust chemical kinetic models using state of the art theoretical methods to accurately unravel chemistry relevant to practical energy problems. This work will facilitate the development of new technologies to increase fuel efficiency, decrease harmful emissions, and reduce dependence on non-renewable energy sources. In my doctoral work at the University of Massachusetts Amherst, I gained extensive experience working with low-pressure flame experiments, analyzing molecular beam mass spectrometry speciation data, and building chemical kinetic models for real fuels such as hypergolic rocket fuels to solve real-world problems. I expanded my knowledge of kinetics during my postdoctoral appointment at Argonne National Laboratory where I focused on the fundamental chemistry of fuels, exploring the effects of pressure on rate constants and the effects of chemically-activated radical chemistry by combining high-level theoretical calculations of potential energy surfaces with the development of kinetic models emphasizing uncertainty reduction. These two experiences make me uniquely qualified to develop predictive kinetic models for energy applications that balance the need for immediate results, providing qualitative information on performance trends to identify targets for further research and development, with the long-term need for accurate quantitative information for specific fuels to stimulate breakthroughs in sustainable energy.

Acknowledgements

This material is based upon work supported by the U.S. Department of Energy, Office of Science, Office of Basic Energy Sciences, Division of Division of Chemical Sciences, Geosciences, and Biosciences, under contract number DE-AC02-06CH11357.

(6gz) 3D, Self-Assembled, Membrane-Electrode Assemblies for Advanced Electrochemical Devices

St. John, S., University of Tennessee

3D, Self-Assembled, Membrane-Electrode Assemblies for Advanced Electrochemical Devices

Samuel St. John, University of Tennessee

sstjohn1@utk.edu

2nd Year Postdoctoral Fellow

Research Interests

I am interested in self-assembled anion exchange membranes, meso-structured electrodes, and alkaline electrocatalysis for electrochemical devices (e.g., fuel cells and batteries). In general, the area-specific resistance, reactant crossover, and cell performance are closely related to ion transport across the membrane/catalyst layer interface and within the membrane electrode assembly. Understanding electrocatalysis and transport within structured membrane/electrode assemblies and controlling that transport is the key to realizing high-current-density devices. The research questions I explore will enable rationally designed architectures in membranes, electrodes, and their interfaces. Science teaches us that nature uses hierarchical approaches to overcome the challenge of assembly across many length-scales. Such a foundational approach underpins the primary methods I use in my research, which include: self-assembly of phase-segregated block copolymer membranes; layer-by-layer assembly of blended membrane/electrode interfaces; and chemical-vapor impregnation of catalysts onto fully formed electrode supports. This research is well-funded by the NSF CBET division and the DOE BES program and will result in an outstanding record of publications, educate many PhD students, and advance the reputation of my future department.

Proposal Writing Experience

Funded (Written by S. St. John):

- Graduate School Distinguished Dissertation Fellowship, University of Cincinnati

Funded (Contributed to by S. St. John as a graduate student):

- NSF DMR-1410118 Investigator: Angelopoulos, Anastasios: NSF SusChEM: Sustainable Synthesis of Bismuth-Based Core-Shell Nanoparticles for Alternative Energy Applications

Contributed to 13 additional unfunded NSF, DOE, and ACS grant applications

Postdoctoral Project

Alkaline Electrocatalysis in Anion Exchange Fuel Cells and Batteries

Advisor: Dr. Thomas A. Zawodzinski, Jr., University of Tennessee

PhD Dissertation

Hierarchical Electrocatalyst Structure Control to Study Cathodic and Anodic Overpotential in Proton Exchange Membrane Fuel Cells

Advisor: Dr. Anastasios P. Angelopoulos, University of Cincinnati

Research Experience

My doctoral work focused on oxygen reduction and ethanol oxidation electrocatalysts, which are critical to the commercialization of proton exchange membrane fuel cells. By developing a nanoparticle preparation method that allowed me to study the structure of the electrocatalyst surface during use, I revealed an unexpected relationship between catalyst geometry and activity that depended upon the proportion of stepped surface active sites. These promising results led me to investigate the ligand-mediated nanoparticle growth mechanism, equilibrated particle size, and the strength of the metal-ligand interaction using a suite of spectroscopic techniques. I plan to apply similar methods to investigate the interaction of ligands with metal electrocatalysts and the binding energy of active intermediates (e.g., H and OH) during electrode polarization, the understanding of which is central to my current electrocatalysis research.

My present collaborative research focuses on catalysis and transport in anion exchange membrane fuel cells (AEMFCs). These devices are receiving renewed interest because of recent advances in anion exchange polymer membrane conductivity and stability that offer the possibility to dramatically reduce the noble metal (Pt) and specialty polymer (Nafion) content. Identifying suitable alternative hydrogen oxidation reaction (HOR) catalysts in alkaline pH is foundational for the technological promise of AEMFCs to be fulfilled. My colleagues and I have just prepared a paper demonstrating the critical understanding of alkaline HOR needed to develop low-cost AEMFC anodes. Additionally, I am involved in understanding and manipulating the structure of a new class of membranes with exceptional hydroxide conductivity, how this transport correlates with aqueous domain size, and how directly manipulating the membrane-electrode interface influences transport across said interface.

Teaching Experience

I have a robust portfolio of teaching experience. I taught Material and Energy Balances at the University of Cincinnati and can provide a detailed teaching evaluation and student feedback. Additionally, I have experience mentoring both graduate and undergraduate students as a post-doctoral scholar, graduate student, and employee at The Procter &

Gamble Company. It is my desire to use my students' own experiences to help them develop an intuition for chemical engineering concepts, the mastery of which will be important for making connections between curriculum and contemporary scientific pursuits in the world at large. My time in academia and industry have taught me that mastery of chemical engineering concepts in the context of contemporary challenges is the key to being a successful engineer. My primary teaching interests include undergraduate courses in all of the transport phenomena, material and energy balances, as well as reactor engineering. I would be happy teaching professional development and/or toolkit (e.g., engineering data analysis with Matlab) courses for graduate students in addition to graduate transport phenomena, reaction engineering, and electrochemical engineering.

Future Direction

My research direction seeks a fundamental understanding of material properties and how they to electrocatalysis and transport in membrane-electrode assemblies (MEAs). More broadly, however, ion-conducting membranes are central the to performance of electrochemical devices and play a large role in technologies related to the water-energy nexus, e.g., electrodialysis for water purification. I envision that structured membrane synthesis will lead to the development and commercialization of thin, durable, and stable structures for these applications. Block copolymerization is a flexible chemistry for the development or ordered, self-assembled membranes with tailored material properties. By precisely understanding the polymerization and membrane self-assembly conditions, I hypothesize that it is possible to design membranes and membrane interfaces for a variety of specific applications. This is one example of how fundamental understanding of structure-property relationships in electrochemical devices will lead to high impact science across a number of important technologies.

Selected Publications

- **S St. John**, RW Atkinson, III, RR Unocic, AB Papandrew, and TA Zawodzinski. Ruthenium- alloy electrocatalysts with tunable hydrogen oxidation kinetics in alkaline electrolyte. *J. Phys. Chem. C*, Accepted, 2015

- **S St. John**, P Boolchand, and AP Angelopoulos. Improved electrocatalytic ethanol oxidation activity in acidic and alkaline electrolytes using size-controlled Pt-Sn nanoparticles. *Langmuir*, 29:16150-16159, 2013.

- **S St. John** and AP Angelopoulos. In-situ adatom analysis of optimum surface atom coordination for Pt nanoparticle oxygen reduction electrocatalysts. *Electrochim. Acta*, 112:258-268, 2013.

- **S St. John**, Z Nan, N Hu, DW Schaefer, and AP Angelopoulos. A nanoscale-modified LaMer model for particle synthesis from inorganic tin-platinum complexes. *J. Mater. Chem. A*, 1(31):8903-8916, 2013.

- **S St. John**, N Hu, DW Schaefer, and AP Angelopoulos. Time-resolved, in situ, small- and wide-angle x-ray scattering to monitor Pt nanoparticle structure evolution stabilized by adsorbed $SnCl3^{-}$ ligands during synthesis. *J. Phys. Chem. C*, 117(15):7924-7933, 2013.

- P Turner, L Petzold, A Shiflet, I Vakalis, K Jordan and **S St. John**. Undergraduate computational science and engineering education. *SIAM Review*, 53(3):561, 2011.

- **S St. John**, I Dutta, and AP Angelopoulos. Enhanced electrocatalytic oxygen reduction reaction activity through electrostatic assembly of Pt nanoparticles onto porous carbon supports from SnCl2-stabilized suspensions. *Langmuir*, 27(10):5781-5791, 2011.

- **S St. John**, D Lee*, I Dutta, and AP Angelopoulos. Acceleration of electrocatalytic activity on Pt–Sn nanoparticles. *J. Electrochem. Soc.*, 157(8):B1245–B1250, 2010.

- **S St. John**, I Dutta, and AP Angelopoulos. Synthesis and characterization of electrocatalytically active platinum atom clusters and monodisperse single crystals. *J. Phys. Chem. C*, 114(32):13515–13525, 2010.

432349 Transition Metal-Oxides for Sustainable Energy Conversion and Storage: The Computational Catalysis Perspective

Abstract:

My primary research focus will be on the fundamental understanding of the **catalysis and surface phenomena in transition metal-oxides** (TMO) and their application towards **sustainable energy conversion and storage**. While TMOs are frequently used as photosensitive semiconductors (TiO_2), in recent years, they have become a central component of electro-chemical processes. They are commercially used as the cathode material in batteries[1] and have shown promising potential as heterogeneous catalysts[2] and electrode materials for pseudocapacitators[3] and solid-oxide fuel cells.[4]Since electrochemical processes occur at interfaces, it is crucial to develop a fundamental understanding of TMO surfaces.

A good illustration of this topic in my latest research on ultra-thin layers of TMOs, which can be stabilized when interfaced with precious metal supports such as Au(111) or Pt(111). Additionally, the gold supported Co/Ni/Mn-based catalysts have also been experimentally proven to exhibit higher oxygen evolution reaction (OER) activities than other metal supported oxide catalysts.[5–7]However, the explanation of synergistic effect of contact with gold support remains elusive and detailed theoretical models of the catalysis at the active site are need to fully explain this phenomenon.

TMOs are also known to exist in a wide variety of surface structures, which exhibit distinct physical properties: surface relaxation and structural defects, localization of electronic states, modifications from the electrolyte environment and variations in voltage bias, and site-specific interactions. All such phenomena have only been weakly explored and are therefore poorly understood. Conventional *ab-initio* methods are generally not accurate or efficient enough to be predictive of the unique physical properties of TMOs surfaces. Therefore, higher accuracy theoretical methods with acceptable efficiency are urgently needed, in order for theory-guided optimization of TMO surfaces.

From the above perspective, the main objectives of my research will be a) to discover new materials with enhanced surface properties for applications in energy conversion and storage accelerated by b) classification of important surface structures of TMOs and mapping their energetics with versatile descriptors and further improved by c) development of more accurate but scalable state-of-the-art *ab-initio*simulation methods applicable to TMOs.

(1) Tarascon, J.-M.; Armand, M. *Nature* **2001**, *414*(6861), 359–367.

(2) Luo, J.; Im, J.-H.; Mayer, M. T.; Schreier, M.; Nazeeruddin, M. K.; Park, N.-G.; Tilley, S. D.; Fan, H. J.; Grätzel, M. *Science* **2014**, *345*(6204), 1593–1596.

(3) Augustyn, V.; Simon, P.; Dunn, B. *Energy Environ. Sci.* **2014**, *7*(5), 1597–1614.

(4) Fergus, J.; Hui, R.; Li, X.; Wilkinson, D. P.; Zhang, J. *Solid Oxide Fuel Cells: Materials Properties and Performance*; CRC Press, 2008.

(5) Yeo, B. S.; Bell, A. T. *J Am Chem Soc* **2011**, *133*(14), 5587–5593.

(6) Yeo, B. S.; Bell, A. T. *J. Phys. Chem. C* **2012**, *116*(15), 8394–8400.

(7) Kuo, C.-H.; Li, W.; Pahalagedara, L.; El-Sawy, A. M.; Kriz, D.; Genz, N.; Guild, C.; Ressler, T.; Suib, S. L.; He, J. *Angew. Chem.* **2014**, n/a – n/a.

432365 Rational Design of Redox Materials and Catalysts for Conversion and Storage of Renewable Energy

Sunday, November 8, 2015
Exhibit Hall 1 (Salt Palace Convention Center)
Ronald Michalsky, Department of Mechanical and Process Engineering, ETH Zurich, Zurich, Switzerland

This poster presents current and previous research on materials and technologies for renewable energy conversion and energy storage in the form of chemicals and chemical fuels. Generally, the presented work employs a dual approach, based on quantum-mechanical computations to understand and tailor the activity of a material for a certain application and lab-scale materials synthesis, reactor construction and experimental testing to verify trends, demonstrate concepts, and determine limitations. In addition to the presented work, I will discuss research plans aiming at an advanced understanding and development of chemical looping, heterogeneous catalysis and electrocatalysis for conversion and storage of renewable energy.

Specifically, the poster focuses on three examples for the production of solar-derived chemicals: First, metal oxide redox materials are developed for solar-driven production of syngas, the precursor for renewable liquid hydrocarbons. The performance of such metal oxides is limited by their oxygen conductivity and thermochemical stability, while the solar-to-fuel energy conversion efficiency of solar syngas production is limited by irreversible heat losses from the reactor. The poster introduces design principles for metal oxide redox materials and shows how to employ these for the development of advanced materials for a continuous and isothermal splitting of CO_2 and H_2O, which circumvents irreversible heat losses from heating and cooling of the redox material.

The second example focuses on rational design of electro- and heterogeneous catalysts for the production of synthetic hydrocarbons. The catalytic activity of a material is often limited by a strong correlation between the adsorption energies of reaction intermediates among the transition metals. The presented work demonstrates how the introduction of non-metals into metal catalysts confers to the material desirable activities due to systematic departures from adsorption energy scaling relations. The conclusions are shown to extend beyond atomic probe adsorbates to molecular fragments, making these concepts generally useful for the theory-based design of catalytic materials.

Finally, materials research for a solar-driven low-pressure synthesis of ammonia (NH_3) is presented. Ammonia is one of the most important synthetic chemicals in today's economy. It is used as synthetic fertilizer, responsible for growing the food for approximately half of the world's population, and has attracted attention as hydrogen carrier and as a direct substitute for fossil fuels in internal combustion engines. As an alternative to the industrial fossil fuel-driven synthesis of NH_3 at some of the most extreme conditions of the chemical industry, the poster presents work towards a sustainable production of NH_3 from N_2 and H_2O at ambient pressure via chemical looping of reducible metal nitride and metal hydride catalysts.

431657 Programming Macromolecular Assemblies with Controlled Architecture and Size Towards Packaging and Delivery of Peptide-Based Therapeutics

Abstract:

My research professorship in biomedical engineering at Duke University (2014-present) and postdoctoral work (Harvard University, 2011-2014, 2 years 3 months) are connected by soft matter systems for materials science and engineering, specifically: (i) using high-throughput drop microfluidics for building complex polymeric micro-structures and (ii) programming the phase behavior and self-assembly of genetically engineered polypeptides into hierarchical architectures. My research focuses on developing an in vitro platform to resolve sequence-property relationships of proteins within the context of liquid-liquid phase behavior for encoding the stimulus-induced phase behavior of protein mixtures. I demonstrate the reversible formation of a variety of protein-rich architectures, ranging from uniform nano-, meso-, and micro-scale puncta (small, distinct particulates) to orthogonal, multilayered granules. This work focuses on proteins possessing sequences of low-complexity (SLC), as they are prototypical building blocks for the reconstitution of cell-free protein granule assays because of their compositional and structural similarity to the native proteins involved in cellular assemblies. My current interest is using this approach to engineer protein-based hydrogels for packaging of protein therapeutics and studying multilayer protein adsorption. In addition to this work, I have mentored and supervised undergraduates, graduates and postdocs on a number of multidisciplinary projects. Thus, I believe that my training and expertise position me advantageously for transition into a tenure track faculty position.

As a junior faculty member, my first aim is to take advantage of programmable self-assembly of macromolecules to systematically investigate and optimize engineered hierarchical nano- and micro-structures. I will initially study the structure-property relationship by investigating how subtle polypeptide sequence mutations direct phase behavior and surface energies to obtain quantitative structure-property correlation rules. Programming the hierarchical assembly of polymeric components provides scaling capabilities (e.g. production rate, size) not easily offered by existing droplet microfluidic approaches. I will demonstrate this scaling capability by the biomanufacturing of functional, hierarchical macromolecular assemblies in the meso- and nano-scale range. This unique suite of materials will consist of phase separated protein assemblies and synthetic polymeric structures, with architectures including core-shell, blended alloys, and Janus-like arrangements. Finally, I will engineer functionality into these hierarchical assemblies to form tunable materials with potential applications in drug delivery, drug discovery, and bioanalytics.

My second aim addresses the challenges arising from extended controlled delivery of peptide-based therapeutics for disease treatment. This exciting class of pharmaceuticals is increasingly used for the treatment of a variety of diseases; however, their main drawbacks include short half-life, poor stability, proteolytic susceptibility and poorly controlled pharmacokinetic profiles. To address this challenge, I will program the self-assembly of ELP fusion peptide therapeutics to build smart-release, monodisperse protein nano- and microparticles containing hierarchical structure. My first case study will be to provide sustained and tunable release of glucagon-like peptide-1 (first in vitro and subsequently from sub-cutaneous depots) for treatment of type 2 diabetes. I will investigate controlled release of intact long-circulating GLP1-ELP fusions from the particle depot by (i) including protease recognition sites within the crosslinked matrix for controlled release, (ii) by engineering self-assembled multicomponent core-shell peptide microparticles for tunable release profiles and (iii) engineering stimulus-induced closed-loop delivery of peptide therapeutics. My hypothesis is that programming smart particle assemblies of GLP1 protein fusions with intrinsically disordered ELPs will enable (1) unprecedented low-dosage, extended release of GLP-1 fusions for in-vivo glucose control and (2) a clinically translatable protein-based platform adaptable for a host of pharmacokinetic profiles.

(6hd) Expanding the Genome Engineering Toolkit: Increasing Signal to Noise

Sandoval, N. R., University of Delaware

Sequencing and synthesis of DNA have plummeted in price (5¢/Mb and 23¢/base, respectively), making genome engineering and analysis quicker and cheaper than ever before. Synthetic biology technologies are advancing rapidly, making it easier to manipulate cellular machinery on the genome and regulatory level. My research objective is to use directed evolution to engineer microbes for desired functions, to develop advanced synthetic biology tools, and to apply high throughput methods of analysis. A powerful strategy is to generate an initial genetic diversity (from the gene to genome level) and then select or screen for the desired trait. In the past, both the genetic diversity created and the biological mechanisms of the resultant phenotype would be black boxes, but now it is possible with next generation techniques to construct libraries with specific mutations and subsequently characterize whole genomes and transcriptomes for individuals as well as populations.

The question now turns to what/how/why mutations are to be made and analyzed for a biotechnologically relevant outcome. That is, the aims of my efforts are to expand the relevant search space to explore as many functional variations as possible. This means rethinking what the search space comprises in order to boost signal and decrease noise: reduce the irrelevant and redundant, expand the functional search space in the initial population, and develop methods to deconvolute signal from noise in the post-screen analysis. Non-model organisms and the metagenome are rich sources of enzymatic capability that are largely unexplored due to the difficulties in genome manipulation and/or difficulties in heterologous expression in a genetically tractable organism. I aim to develop a research group that expands the genome engineering toolkit to include high-throughput, quantitative methods for the greater exploration of functional capabilities in non-model organisms. Furthermore, I will apply methods on the DNA, RNA, protein, and metabolite levels to elucidate biologically relevant mechanisms.

Research Experience:

Postdoctoral Research Fellow, University of Delaware

Mentor: Prof. Eleftherios T. Papoutsakis

Award: NIH Ruth L. Kirschstein Postdoctoral National Research Service Award (NRSA)

Projects: Enhanced production of butanol from waste glycerol with *C. pasteurianum*; Heterologous sigma factor expression for metagenomic library screening using flow cytometry; Engineering Synthetic Methylotrophy in *E. coli*

PhD Dissertation, University of Colorado Boulder

Advisor: Prof. Ryan T. Gill

Award: NSF Graduate Research Fellow

Thesis: Genome Engineering to Improve Acetate and Cellulosic Hydrolysate Tolerance in E. coli for Improved Cellulosic Biofuel Production

Teaching Experience:

Lecturer – Colorado Mesa University – University of Colorado Boulder Mechanical Engineering Partnership Program

Courses: Thermodynamics, Heat Transfer, Dynamics, Measurements Lab

Graduate Student Teaching Fellow – University of Colorado Boulder, Chemical and Biological Engineering

Course: Material and Energy Balances

(6he) Design Principles and Performance Metrics for Realizing Cost-Effective Electrochemical Technologies for Energy Storage

Su, L., Massachusetts Institute of Technology

The widespread integration of locally-available intermittent renewable electricity generators, such as solar photovoltaic and wind turbines, necessitates advances in the science and engineering of energy storage devices for cost-effective deployment. Redox flow batteries (RFBs), sometime referred to as regenerative fuel cells, have shown a favorable balance of cost, safety, and performance for meeting this storage need. RFBs are rechargeable electrochemical devices that utilize the reversible redox reactions of two soluble electroactive species for energy storage. Aqueous RFBs have been the subject of the vast majority of research effort to date. Shifting from aqueous to non-aqueous electrolytes may promise a higher cell voltage due to the extended electrochemically stable window (~ 4 V) and an enriched selection of redox materials due to the broader variety of organic solvents. A key RFB component is the separator which is used in keep the positive and negative electrolytes from mixing while selectively enabling working ion transfer between the two fluids to maintain electroneutrality. The ideal membrane should have negligible electronic conductivity, high ionic conductivity, good chemical stability and mechanical strength, and low active species crossover. However, to date, no membrane has been identified that simultaneously fulfills all of these requirements, especially high ionic conductivity and low crossover rate. In fact, most aqueous and non-aqueous RFBs adopt ion-selective Nafion® membranes, originally designed for chlor-alkali production or size-selective separators designed for lithium-ion batteries. To date, few membranes have been specifically designed for RFB operation and performance requirements for this application have yet to be established. Therefore, these effort seeks to define the performance metrics of separators for both aqueous and non-aqueous RFBs using stringent cost targets established by the US Department of Energy. The metrics are based on a detailed analysis of the electrochemical and transport phenomena within RFBs and the impact of rechargeable operation. Building on these benchmarks, design principles for high-performance separators are determined by a combination of experimental and computational analysis. Of particular interest is the behavior of state-of-the-art aqueous separators under non-aqueous conditions. The results from these studies provide a framework for understanding and optimizing membranes and electrolytes for not only RFBs but also a range of electrochemical technologies and processes of societal and industrial relevance (e.g., fuel cells, electrolyzers).

432747 Non-Equilibrium Self-Assembly and Structures

Abstract:

My research focuses on non-equilibrium self-assemblies (NESAs) that require energy supply to maintain their order or, as cells, to "survive". Non-equilibrium assemblies are energy (external field) driven systems that reside in steady states or long-lived metastable structures, some of which can be highly dissipative and are directed by production of the entropy within the system. DNA transcription machinery, self-propelling micelles, liquid crystals under external fields, cytoskeletal fibers, and molecular motors are all examples of various aspects NESAs. The ability to control the ordered non-equilibrium state paves the way for new types of materials that can be reassembled when needed. This shifting ability between different configurations offers variety of applications in several branches: they can be implemented in energy storage purposes due to the high energy stored in their metastable conditions, they can be utilized as sensors responding to outside stimuli by adapting their mode of organization, and in biological applications such as drug delivery. Theoretical studies are promising for these systems due to the presence of time varying different length scales that changes from molecular and nanoscale to mesoscopic and macroscopic ones. In my research, these complex assemblies are studied using the tools of non- equilibrium thermodynamics and molecular simulation techniques. The NESA structures are influenced by the intra- and intermolecular forces imposed by their building blocks, external forces such as magnetic and electromagnetic fields, capillary interactions, and also their solvent media or solid interfaces in the surrounding environment. Therefore, the proposed mesoscopic model should capture the relation between microscopic structure and macroscopic behavior. These dissipative structures have many applications in biological, electronics, catalysis, material science, and chemical worlds with multi-tasking and self-repair properties that can be predicted by the analytical tool of the molecular simulation. Until now, NESA in artificial systems has been achieved by modifying the interactions between the constituent objects/parts. In the proposed methodology, I will look at the problem from a fundamentally different perspective – namely, I will keep the interactions between the components intact but instead change the agitation mode to achieve different structures. For this purpose, I aim to control the particle self- assembly (SA) in non-equilibrium regime by applying local anisotropic vibrations (for the solvent or particles or both), which effectively translating into anisotropic temperature that can be considered through hydrodynamic effects. The hydrodynamic effects along with dynamic anisotropy bias the particles toward the desired configuration. Also, the role of the external fields such as static and alternative magnetic fields in SA of magnetic fluids to develop a complete theoretical approach for dynamic SA far from equilibrium is investigated. Molecular dynamics simulation has employed to test the above hypothesis during my postdoctoral time.

Sample proposed Project: Reaction Networks in Non-Equilibrium Self- Assembled Structures

In biology and chemistry complex biochemical networks displaced far from equilibrium coupled with transport phenomena maintain the NESA structures. These out-of-equilibrium chemical reactions provide a strong tool for controlling NESA systems through a signaling network. The network is a mathematical toolbox composed of several nodes that are responsible for the chemical reactions that control the NESA behavior. By implementing a step-by-step approach the most critical nodes of the network for controlling the NESA systems are identified. Also, the nodes that have most synergistic effects on the network functioning are investigated. These networks have important influences on switching behavior (by imposing molecular stimuli), motility, and memory storage. The challenge here is to combine reaction network with dynamic (motion) behavior of the system that provide the necessary information about the evolution of the particles. As a start we will consider bioinspired simple chemical oscillators that generate waves inside NESA such as a pH-driven self-assembly coupled to a chemical oscillator that cycles the pH of the solution. This switching behavior causes the surface molecules to protonate/deprotonate and self-assemble/disassemble as their attractive forces are turned on and off. Also, pH change can cause coil-globule transition and dramatic volume changes in polymers and biology. A good example of pH responsive materials can be found in polyelectrolytes namely polyacids and polybases. Polyacids are protonated at low pH and is charge neutral. In this situation polymer adopts a hydrophobic collapsed conformation. By increasing the pH (above the pKa) of the acidic monomer, the polymer becomes ionized and deprotonated. Moreover, the negative charges on the polymer chain makes counter ions and associated water molecules to increase the total volume of the polymer. This volume change can be switched "on" and "off" by cycling the pH above and below the pKa of the polyacid. Polybases also protonate at low pH and deprotonate at high pH. However, they swell and ionize at low pH due to protonation of the basic groups. Examples of this behavior is observed in triblock copolymers comprising poly(methacrylic acid) and poly[(2-diethylamino)ethyl methacrylate] as polyacid and polybase pairs. The pH intensive endblocks used in each case includes poly(methyl methacrylate). The advantage of using polyacid and polybase pairs is the device fabricated from these two building blocks is bipolar. Under the same pH condition one polymer is swollen while the other is collapsed. This phenomenon can be used in artificial muscles that need contraction and extraction. Another application is in cell biology where protein-protein interactions (PPI) inside cytoskeleton affect the NESA structure through a network of chemical reactions. It has been well established that in metastatic cancer cells not only a single protein but also entire network of interacting proteins—specifically, those regulating actin filament assembly and disassembly— are up- and/or downregulated resulting in deregulated cell motility and reaction network patterns. The NESA of the cytoskeleton causes constant destruction, self-repair, and/or construction in the structure is the key factor for cellular motions. In addition, cell interactions with surrounding extracellular matrix also influence the NESA. What is not known is how small changes in functioning of the complex intracellular PPI networks translate into global changes inoverall NESA patterns. The situation becomes even more complex when one starts to consider the combined effects of intracellular protein networks and cell interaction with extracellular matrix, surrounding cells and/or soluble factors. Non-equilibrium thermodynamics has been promising in modeling biological systems such as transport through membranes and protein- folding/unfolding. Hence, it has the ultimate tools for complex cellular systems that have both PPI and membrane transport. Considering that the cells are composed of cytoskeleton dynamic events that span a large spectrum of time and length scales, the mechanism and stochastic character of cell movements are difficult to understand based solely on experimental observations. To tackle this complex problem, I propose an advanced theoretical and molecular simulation approach. The physical insight in a complex nonlinear dissipative biological system can be coherently expressed in a mathematical framework in non-equilibrium thermodynamics. To minimize the simulation effort but still develop an efficient technique, I propose a thermodynamically guided simulation/theoretical approach, which will use experimental and simulation data of network PPIs between actin regulators cell surface/membrane proteins and soluble peptide. Specifically, we will implement a non-equilibrium thermodynamic approach that utilizes the experimental/simulation data of the functional protein interaction networks as additional potential wells (intra- and intermolecular interactions) to guide the coarse-graining approach in the simulation technique. We will focus on the protein interaction networks that control polymerization and turnover of actin and thus act as a driving force for cell movements and is a key feature, which is upregulated in metastatic cancer. We will develop a model/technique that allows determining the most important nodes/their combinations of the network for controlling the actin polymerization and consequently NESA and dynamic patterns specific to metastatic cancer cells. Non-equilibrium molecular dynamic simulations will also consider the external interactions with matrix and soluble factors. The proposed research will be the first attempt, to the best of our knowledge, to combine the network module relevant/specific to cancer metastasis with non-equilibrium molecular dynamics approach. Our hope is that understanding of how intracellular interacting protein networks and NESA of the cytoskeleton control overall cell motility patterns in the context of complex microenvironments will enable managing of the cancer cell motility and ultimately would provide rationale for stopping cancer progression. Our theoretical approach would be validated experimentally in collaboration with scientists in experimental research groups while supported by the National Institutes of Health (NIH) and National Science Foundation (NSF).

Outlook and Conclusions:

The goal of this proposal is to develop the theory of NESA based on the experimental data and molecular simulation observations. The projects that I proposed are for the first 7 years of my academic career. However, the methodology that I suggested can be extended to other branches of research that is suitable for the next 20 years. These areas include: electrorheological fluids, cell agglomeration in lysosome-rich cells by light scattering ability at near-IR wavelength that create a long range force for the cells to attract each other at distances up to 200 μm, and light-switch aggregation of magnetic colloids, nanoparticle self-assembly, micellar surfactants or surface chemical reactors, design of catalysts by assembly of metal colloids dispersed in block copolymers, self-replicating micelles, and reactions in vesicle membranes.

(6hh) Materials and Systems Engineering for Healthcare and Energy Applications – from Discovery to Design

Singh, M. R., Lawrence Berkeley National Laboratory

My research is focused on developing mathematical, computational, and experimental tools for i) discovery of crystalline forms of pharmaceutical drugs and energy-conversion materials, and ii) identification and control of systems for the synthesis of pharmaceutical drugs and renewable fuels.

Research Area 1: Discovery and Growth of Crystalline Materials

Description: The crystal structure and morphology are two critical determinants of the physicochemical properties such as solubility, hardness, cleavage, optoelectronic properties, electrical and heat conductivities, and piezoelectric effect of crystalline materials. The polymorphs (or crystal structure) are determined in the early stage of crystallization process whereas the morphologies depend on growth, dissolution, and milling conditions. Figure 1 shows my research scheme to probe processes at different length scales using a combination of experimental and mathematical tools to control crystal structure, shape, and size by tuning various environmental conditions. The proposed research is fundamental to crystal growth in different areas of applications such as pharmaceuticals, semiconductors, catalysts, and specialized materials.

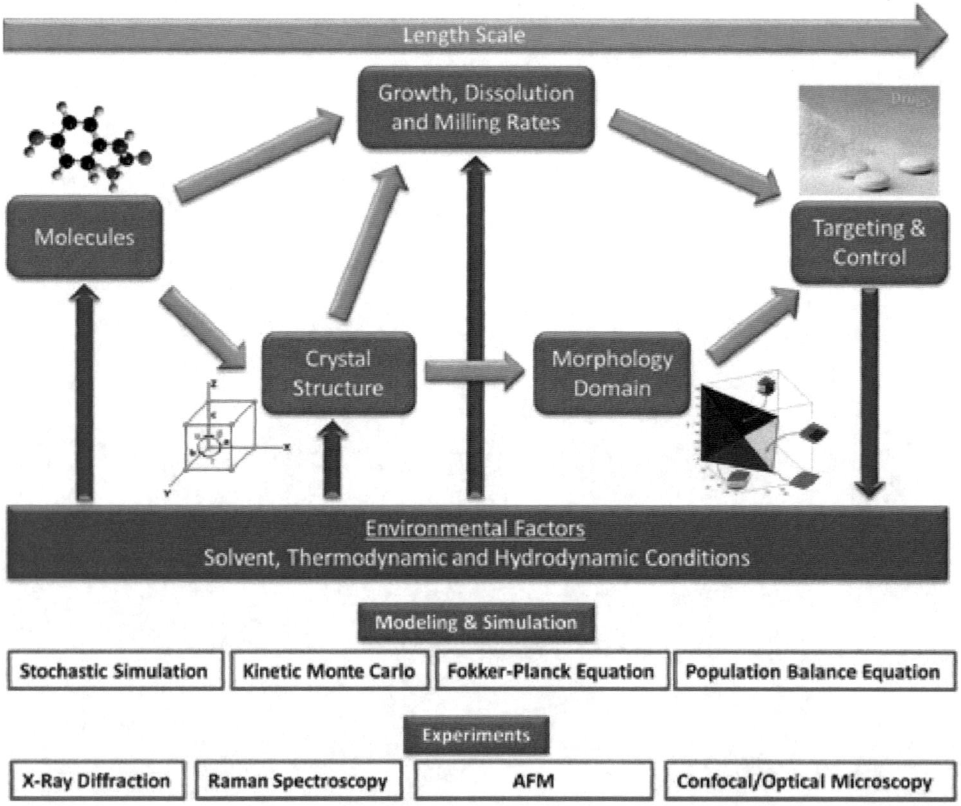

Figure 1: Research Plan for Discovery and Growth of Crystalline Materials

Selected Publications:

1. Meenesh R. Singh, Jayanta Chakraborty, Nandkishor Nere, Hsien-Hsin Tung, Shailendra Bordawekar and Doraiswami Ramkrishna, ?Image-Analysis-Based Method for Measurement of 3D Crystal Morphology and Polymorph Identification using Confocal Microscopy,? Crystal Growth & Design, 12 (7), 3735-3748, 2012

2. Meenesh R. Singh and Doraiswami Ramkrishna, ?A Comprehensive Approach to Predicting Crystal Morphology Distributions with Population Balances,? Crystal Growth & Design, 13 (4), 1397 ? 1411, 2013.

3. Meenesh R. Singh and Doraiswami Ramkrishna, ?Dispersions in Crystal Nucleation and Growth Rates: Implications of Fluctuation in Supersaturation,? Chemical Engineering Science, 107 (7), 102-113, 2014.

4. Doraiswami Ramkrishna and Meenesh R. Singh, ?Population Balance Modeling. Current Status and Future Prospects,? Annual Review of Chemical and Biomolecular Engineering, 5 (1), 2014

5. Meenesh R. Singh, Nandkishor Nere, Hsien-Hsin Tung, Samrat Mukherjee, Shailendra Bordawekar, and Doraiswami Ramkrishna, ?Measurement of Polar Plots of Crystal Dissolution Rates using Hot-Stage Microscopy. Some Further Insights on Dissolution Morphologies,? Crystal Growth & Design, 14 (11), 5647 ? 5661, 2014.

Research Area 2: Production of Renewable Fuels and Fertilizers

Description: Harnessing solar energy to produce transportation fuels is a promising route to support future energy demands. The solar fuel generators such as photoelectrochemical cells (PECs) and PV-Electrolyzers involve a two-step conversion of sunlight to fuels using a photovoltaic (PV) cell directly (wireless) or indirectly (wired) connected to an electrochemical cell. My research in this area is focused on identifying materials and systems to develop an efficient, robust, cheap and durable solar-fuel generator for water-splitting, carbon dioxide reduction and nitrogen fixation.

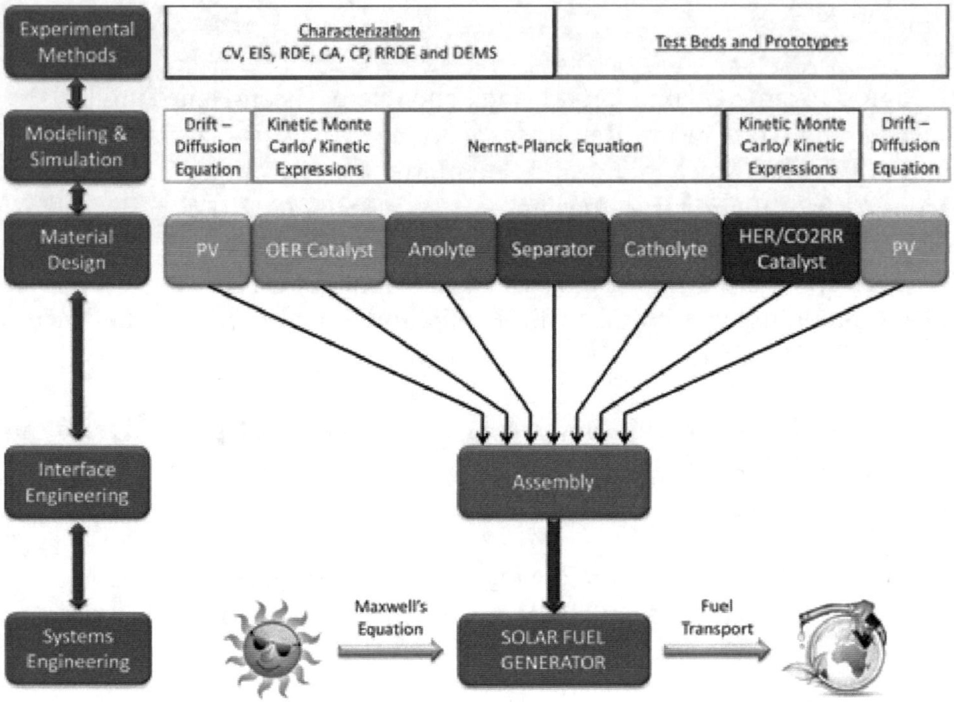

Figure 2: Research Plan for the Production of Renewable Fuels and Fertilizers

Selected Publications:

1. Meenesh R. Singh, John C. Stevens, and Adam Z. Weber, ?Design of Membrane-Encapsulated Wireless Photoelectrochemical Cells for Hydrogen Production?, *Journal of The Electrochemical Society*, 161 (8), 2014

2. Jian Jin, Karl Walczak, Meenesh R. Singh, Chris Karp, Nathan S. Lewis, and Chengxiang Xiang, ?Experimental and Modeling/Simulation Evaluation of the Efficiency and Operational Performance of an Integrated, Membrane-Free, neutral pH solar-Driven Water-Splitting System,? Energy & Environmental Science, 7 (10), 3371-3380, 2014.

3. Christopher M. Evans, Meenesh R. Singh, Nathaniel A. Lynd, and Rachel A. Segalman, ?Improving the Gas Barrier Properties of Nafion via Thermal Annealing: Evidence for Diffusion through Hydrophilic Channels and Matrix,? Macromolecules, 2015.

4. Meenesh R. Singh, Ezra L. Clark, and Alexis T. Bell, ?Effects of Electrolyte, Catalyst, and Membrane Composition and Operating Conditions on the Performance of Solar-Driven Electrochemical Reduction of Carbon Dioxide,? *under review*

5. Meenesh R. Singh, and Alexis T. Bell, ?Thermodynamic and Achievable Efficiencies of Solar-Driven Electrochemical Conversion of Water and Carbon Dioxide to Transportation Fuels,? *submitted*

432879 Pathway Engineering in Yeast: Overcoming Challenges in Design and Optimization By Scaling and Parallelizing Elements of the Design-Build-Test-Learn Cycle

Abstract:

To achieve the goal of rapid and reliable pathway optimization in yeast, strategies for a fast and standardized design-build-test-learn cycle must be developed. Herein, we describe an approach to yeast engineering by large-scale assembly of combinatorial pathway variants. To enable this approach, a semi-automated platform for yeast parts characterization, semi-automated assembly of pathway variants, and computational modeling for part and pathway performance were developed.

These developments were applied to understanding the behavior and bottlenecks of a six-gene pathway in *Saccharomyces cerevisiae*. In the process, over a thousand individual yeast transcription units were characterized for expression and nearly two hundred pathway variants constructed in the first combinatorial library. With these advances in the parallelization and scale of pathway engineering, effects and interactions that were not previously possible to characterize may be observed. Specifically, eukaryotic promoters and terminators both influence gene expression, thus understanding the interactions between them are essential for controlling pathway protein levels. As a result of the large-scale parts characterization, we have found that genome-integrated transcription unit expression of GFP can be modeled with a simple second-order expression. Furthermore, by constructing many pathway variants in parallel, enzymes that must be overexpressed for pathway function are easily identifiable, and interdependencies in pathway gene expression may be observed.

With these advances, we demonstrate a design-build-test-learn cycle that can be generally applied to any multi-gene system in yeast.

432911 Data-Centric Optimization: Methods and Applications

Abstract:

As an engineering community we have been working towards developing detailed mathematical simulation models which accurately capture the behavior of complex systems. This trend is reinforced by our drive to couple multiscale information which ranges from the atomistic and molecular level to continuum, process and even plant and enterprise levels. At the same time, abundance of data is transforming traditional process systems engineering operations in this "Big Data" era, complementing existing mechanistic models or even enabling the description and analysis of problems for which a model is not available. However, as we move towards this paradigm we create a gap between the current state-of-the-art high-fidelity models, and the models which are used in design, optimization and control applications.

My past and future research interests integrate theory from various fields, such as statistics, machine learning, operations research, mathematics and engineering, to solve significant problems which are characterized by abundance in experimental and/or computational data; lack of closed-form numerically stable models; and integration of data and models across scales. The common denominator in the presented work is development of global optimization methods for systems without mathematical closed-form equations and consequently, without derivative information [1-5]. The main application areas pursued are (a) design and optimization of pharmaceutical manufacturing systems [6-9], (b) design and optimization of adsorption systems for energy applications [5,10], (c) integrated refinery and chemicals planning.

References:

1. F. Boukouvala, M.G. Ierapetritou. Feasibility analysis of black-box processes using an adaptive sampling kriging-based method. *Computers & Chemical Engineering*, 36, 2012.

2. F. Boukouvala, Y. Gao, F.J. Muzzio, M.G. Ierapetritou, Reduced order Discrete Element Method (DEM) modeling, *Chem. Eng. Science*, 95, 2013.

3. F. Boukouvala, M.G. Ierapetritou, Derivative-free optimization for expensive constrained problems using a novel expected improvement objective function, *AICHE Journal*, 60(7), 2014.

4. Boukouvala, F. , Misener, R., Floudas, C.A.: Global Optimization Advances in Mixed-Integer Nonlinear Programming, MINLP and Constrained Derivative-free Optimization, CDFO. European Journal of Operational Research. Under Review (2015).

5. Boukouvala, F., Hasan, M.M.F., Floudas, C.A.: Global Optimization of General Constrained Grey-Box Models: New Method and its Application to Constrained PDEs. Submitted to Journal of Global Optimization (2015).

6. F. Boukouvala, F.J. Muzzio, M.G. Ierapetritou. Design Space of pharmaceutical processes using data-driven based methods. *Journal of Pharmaceutical Innovation*, 5(3), 2010.

7. F. Boukouvala, V. Niotis, R. Ramachandran, A. Vanarase, F.J. Muzzio, M.G. Ierapetritou. Integrated approach for dynamic flowsheet modeling and sensitivity analysis of continuous pharmaceutical manufacturing. *Computers and Chemical Engineering*, 42, 2012.

8. F. Boukouvala, A. Chaudhuri, R. Zhou, F. Muzzio, M.G. Ierapetritou, R. Ramachandran, Dynamic flowsheet simulation of pharmaceutical tablet production through wet granulation, *Journal of Pharmaceutical Innovation*, 8(1), 2013.

9. F. Boukouvala, F.J. Muzzio, M.G. Ierapetritou, Surrogate-based optimization of expensive flowsheet models for continuous pharmaceutical manufacturing, *J. of Pharm. Innovation*, 8(2), 2013.

10. M.M.F. Hasan, F. Boukouvala, E. First, C.A. Floudas, Nationwide, Regional and Statewide CO2 Capture, Utilization, and Sequestration Supply Chain Network Optimization. *Ind. & Eng. Chem. Research*, 53(18), 2014.

432403 Nanoscale Simulation and Design for Molecular Sensors and Reaction Engineering

Abstract:

Shrinking sensors to the nanoscale introduces novel selectivity mechanisms and enables the ultimate sensitivity limit, single-molecule detection. Single-walled carbon nanotubes, with a bright fluorescence signal and no photobleaching, are a platform for implantable near-IR sensors capable of selectively detecting a range of small-molecules including the radical signaling molecule nitric oxide, the hormone estradiol, and sugars such as glucose. Selectivity is achieved by engineering an adsorbed phase of polymers, DNA, or surfactants at the nanotube/solution interface. Understanding these sensors requires a range of modeling and simulation tools and presents a unique opportunity to learn how these phases interact with small molecules. I will discuss methods and limits to integrating data from many noisy stochastic sensors, show how we can use these sensors to monitor nitric oxide inside cells with unprecedented spatiotemporal resolution, and describe what is needed to engineer a selective adsorbed phase.

432619 Optimization of Protein/Peptide Electrostatic Properties for Bioengineering Applications

Abstract:

If current technological trends continue, significant advancement in the area of biomolecular engineering is needed, since the ability to manipulate macromolecules, e.g. proteins, for novel purposes is central to many problems in bioengineering and medicine. Proteins are complex biomolecules with intricate three-dimensional structures stabilized through networks of noncovalent interactions, including hydrogen bonds, Van der Waals forces, and electrostatics. In particular, electrostatic interactions contribute significantly to protein structure, dynamics, and consequently function (1), especially the pH-dependent aspects of function such as association, stability, and catalysis (2). Much of my previous work has been focused on developing tools for analysing the role of electrostatics in protein function (3-7), and how to manipulate the electrostatic character of proteins for improved function (8-9).

Future areas of interest are centered around the development and application of methods for optimizing electrostatic interaction networks, with specific focus on engineering pH-dependent properties of proteins/peptides. Global optimization, including data driven methods, will be an essential component for tackling the significant combinatorial complexity of the protein design problem. Such methods could be applied to the design of more stable therapeutic antibodies, or the design of novel enzymes for the production of biofuels. Another area of great interest is the development of peptide-based technologies, such as antimicrobial/cell penetrating peptides, pH-sensitive peptide materials, or peptide catalysts, which would be produced using solid-phase peptide synthesis.

1. Gorham, R., Kieslich, C. A., and Morikis, D. (2011) Electrostatic clustering and free energy calculations provide a foundation for protein design and optimization. Ann. Biomed Eng, 39(4): 1252-63.

2. Kieslich, C.A.*, Tamamis, P*, Gorham, R.D., López de Victoria, A., Sausman, N.U., Archontis, G., Morikis, D. (2011) Exploring protein-protein and protein-ligand interactions in the immune system using molecular dynamics and continuum electrostatics. Curr. Phys. Chem., 20(2):324-343.

3. Kieslich, C.A., Yang, J., Gunopulos, D., and Morikis, D. (2011) Automated computational protocol for alanine scans and clustering of electrostatic potentials: application to C3d.CR2 association. Biotech. Prog., 27(2): 316-325.

4. Kieslich, C.A., Gorham, R.D., and Morikis, D. (2011) Is the rigid-body assumption reasonable? Insights into the effects of dynamics on the electrostatic analysis of barnase-barstar. J. Non-Cryst. Solids, 357(2): 707-716.

5. Kieslich, C.A. and Morikis, D. (2012) The two sides of complement C3d: Evolution of electrostatics in a link between innate and adaptive immunity. PLoS Comp. Bio., 8(12): e1002840.

6. Kieslich, C.A., Shin, D., López de Victoria, A., González-Rivera, G., Morikis, D. (2013) A predictive model for HIV-1 co-receptor selectivity and disease progression. AIDS Res. Hum. Retroviruses, 29:1386-1394.

7. Kieslich, C.A., Goodman, G., Vazquez, H., López de Victoria, A., and Morikis, D. (2011) The effect of electrostatics on Factor H function and related pathologies. J. Mol. Graph Mod., 29(8): 1047-55.

8. Pyaram, K., Kieslich, C. A., Yadav, V. N., Morikis, D., and Sahu, A. (2010) Influence of electrostatics on the complement regulatory functions of Kaposica, the complement inhibitor of Kaposi.s sarcoma-associated herpesvirus. J.Immunol., 184(4): 1956-1967.

9. Liu, Y., Kieslich, C.A., Morikis, D., Liao, J. (2014) Engineering pre-SUMO4 as efficient substrate of SENP2, Protein Eng. Des. & Sel., 27, 117-126.

432946 Integrating Experimental and Computational Approaches to Discover and Design (Therapeutic) Proteins

Abstract:

Proteins are an essential component of all life, fulfilling remarkably diverse roles ranging from structure to catalysis to locomotion and beyond. Their broad diversity of function means that they have potential applications in numerous human endeavors. Over the years, numerous experimental and computational techniques have been developed to design proteins, with each approach having advantages and disadvantages, but they are rarely utilized together. With the recent expansion of high throughput screens and the corresponding growth of "big data," it is an opportune time to combine these complementary approaches.

I will lead a research laboratory that integrates both computational and experimental protein methods, with a specialization in therapeutics. The unifying theme of this research will be the computational analysis of large, experimentally-obtained datasets to extract the biologically-relevant information and generate hypotheses that are validated in the laboratory. As the title suggests, the research will be broadly divided into two categories: the *design* of therapeutic proteins (e.g. antibodies, fibronectin type III domains, anticalins, etc.) and the *discovery* of therapeutically-relevant proteins (e.g. characterizing the complete repertoire of antigens bound by an individual's antibodies). By integrating computational analysis with experimental testing, my research will facilitate the rapid development of disease diagnostics and treatments.

432996 Molecular Simulations of Chemical Reactions

Abstract:

A chemical reaction requires the activation of reactants and successful passage through transition states to products. Reactants and products are long lived stable (or metastable) states, whereas transition states are activated complexes that are rarely and briefly visited. This separation of timescales inherent in chemical reactions is characteristic of rare events generally, including nucleation of 1st order phase transitions, molecular isomerizations such as protein folding, and transport phenomena in solids and glasses.

Molecular simulation of rare events are complicated by the long waiting times for a transition to spontaneously occur. In my PhD work with Baron Peters and Joan-Emma Shea at the University of California, Santa Barbara, we developed new, simple methods for simulating rare events at the molecular level. We applied these methods to study aqueous ion-pair dissociation, vacancy diffusion in crystals, and the association of biomolecules in water.

One of my key frustrations as a PhD student was only being able to model systems at the classical level. In my postdoctoral work with Ed Maginn at the University of Notre Dame, we are integrating quantum calculations into classical simulations of condensed phases. To model CO2 reactivity in a task specific ionic liquid, we compute the change in free energy as CO2 binds to an anion via DFT and the solubility of CO2 in the ionic liquid via reaction ensemble Monte Carlo.

Applied Pharmaceutical Process System Engineering

Ravendra Singh
Assistant Research Professor, NSF Engineering Research Center for Structured Organic Particulate Systems (C-SOPS), Department of Chemical and Biochemical Engineering Rutgers University, 98 Brett Road, Piscataway, NJ 08854, USA
Email: *ravendra.singh@rutgers.edu*, ***Phone:*** *(001)848-445-4944;* ***Fax:*** *(001)732-445-2581*

Abstract

Process system engineering (PSE) is a very advanced field and has been very successful in solving complex problems of chemical based product manufacturing. However, the particulate systems are still poorly understood and there is different level of complexities involved, therefore new techniques and tools need to be developed for pharmaceutical manufacturing through extensive fundamental and applied research. Author's main research area is applied pharmaceutical process system engineering. This area of research and education are highly encouraged and supported by pharmaceutical companies, academic institutions, regulatory authority (FDA) and national science foundation (NSF). For example, recently national science foundation of USA has established a 100 million dollar research center for structured organic particulate systems (C-SOPS) at Rutgers University New Jersey. Furthermore, more than 40 companies are supporting this center economically and providing scientific inputs so that the research outcomes can be directly applied to the companies.

In this work, a scientific foundation for the optimal design, optimization and control of the continuous process involving structured organic particulate systems has been developed. The mathematical models for continuous manufacturing of active pharmaceutical ingredients and final product tablet, has been developed for virtual experimentation and applied for plant design, optimization and control. Advanced modelling techniques such as PBM, DEM and CFD have been applied for pharmaceutical process modeling. The developed continuous tablet manufacturing plant and process model has been adapted by several pharmaceutical companies. A flexible monitoring and control system incorporating different PAT techniques and tools and control strategies for continuous pharmaceutical manufacturing process has been designed and implemented to Rutgers plant. Different control strategies such as PID, PID coupled with dead time compensator, and MPC has been compared and evaluated practically for pharmaceutical process control in terms of set point tracking and disturbances rejection ability, robustness, and stability/instability. A new hybrid MPC-PID control strategy has been developed and implemented into the pharmaceutical tablet manufacturing plant and its performance has been found better than the other control strategies. For the first time, the closed-loop operation of the tablet manufacturing plant using NIR sensor has been demonstrated, which is a significant advancement in pharmaceutical manufacturing as endorsed by the Food and Drug Administration (FDA) and other regulatory authorities as part of the Quality by Design (QbD) paradigm. For the first time NIR has been applied for real time monitoring of powder bulk density. Feedforward control system has been integrated with feedback control system to take proactive mitigation actions on raw materials and process variability. A systematic generic framework including the methods and tools through which the cost and profit can be estimated and the Dynamic Real Time Optimization (DRTO) can be integrated with hybrid control system has been also developed. The integrated DRTO provides the optimal operational set points for the plant control system in real time optimizing an objective function. DRTO integrated with hybrid control strategy ensures the maximum possible profit irrespective of the market demand fluctuations.

Presently, I am working as an assistant research professor at Department of Chemical and Biochemical Engineering of Rutgers University and currently managing and leading ten projects funded by NSF, industries and regulator with total grants exceeding 18 million USD. I was involved in proposal writing of these projects and I am Co-PI of some of these projects. Prior to this, I worked in a large scale European pharmaceutical project (F^3 Factory project), which involves 25 partners from industry and Academia; where my responsibility was to manage and lead the DTU's parts of the F^3 Factory project and also to coordinate with the other partners associated with it. My doctoral research work, done at Department of Chemical and Biochemical Engineering, Technical University of Denmark, has been awarded for ***prestigious "EFCE Excellence Award 2010", given in Recognition of an Outstanding PhD Thesis, from European Federation of Chemical Engineering***. I have exemplary academic records, previous teaching experiences, and an established trail of valued research accreditations'. I have published several peer reviewed scientific international Journal articles, popular reputed magazine articles, book chapters and presented in more than 70 international conferences. I am actively serving as a conference session chair, guest editor of Journal, manuscripts reviewer and master and PhD projects supervisor.

The objective of this presentation is two-fold. First to give the overview of my research and teaching experience, and second to demonstrate the design, optimization, monitoring and control of continuous tablet manufacturing plant.

Teaching and demonstrations: Experience on independent course material preparation, independent class room teaching, and conducting exams. Also prepared and given course modules for external attendees such as FDA and industry employees. Given invited crash courses to industries. Conducted several workshops and in each workshop more than 50 people from industry and academia has participated. Conducted several hands-on training sessions. Served as a teaching assistant for several courses and was involved in course planning, preparing lectures and tutorials (fully responsible for the tutorial classes and assignments/exercises). Supervised several bachelor, master and PhD students. Presented research works in more than 70 well reputed international conferences.

Research: Main research area is applied process system engineering. Expertise in the area of continuous pharmaceutical manufacturing. Founder of pharmaceutical process control. Contributed significantly in the field of process monitoring and control, process modelling and simulation, optimization, PAT, QbD, control hardware/software integration, new methodology and software development.

Grants: Extensive experience in funding proposals writing. Wrote several grants proposals for national science foundation (NSF), FDA, and pharmaceutical industries. 10 projects with total grants more than 18 million USD got funded in which I have contributed significantly. Received DAAD fellowship from Germany and DTU grant to complete master and PhD projects.

Publications: Scientific articles (43); Book chapters (3); International conference presentations (70); Invited industrial/academic lectures (8); Workshops (5); Thesis (3); Reports (>100).

Activities in AIChE 2015 conference

Session chair/co-chair

1. **Chair of session #399, "Dynamic Simulation and Optimization II".**
   Tuesday 3:15-5:45 PM, Salon E (Salt Lake Marriott Downtown at City Creek).
2. **Chair of session 721#, "Advances in Information Management".**
   Thursday 3:15-5:45 PM, Salon G (Salt Lake Marriott Downtown at City Creek).
3. **Co-chair of session #347, "Dynamic Simulation and Optimization I".**
   Tuesday 12:30-3:00 PM, Salon D (Salt Lake Marriott Downtown at City Creek).

Presenter (Oral presentations)

4. **Ravendra Singh**, et al. (2015). Implementation of Advanced Multilayer Plant-Wide Control Architecture into a Direct Compaction Continuous Tablet Manufacturing Process.
   Session (#183d): Application of Quality By Design in Drug Product Formulation Design & Process Development.
   Monday 4:18-4:39 PM, Ballroom B (Salt Palace Convention Center).
5. **Ravendra Singh**, et al. (2015). Dynamic Modeling and Advanced Control of Tablet Press.
   Session (697f): Model Based Integrated Design of Pharmaceutical Drug Substance and Drug Product Processes II.
   Thursday 2:10-2:30 PM, Ballroom B (Salt Palace Convention Center.

Presenter (Poster presentations)

6. **Ravendra Singh** (2015). Applied Pharmaceutical Process System Engineering.
   Session (6hp): Meet the Faculty Candidate.
   Sunday 2-4 PM, Salt Palace Convention Center, Exhibit Hall 1.
7. **Ravendra Singh**, et al. (2015). Sensor modeling.
   Session (#243d): Information Management and Intelligent Systems.
   Monday 6:00 PM, Exhibit Hall 1 (Salt Palace Convention Center)

Co-author (Oral presentations)

8. M. Sebastian Escotet-Espinoza, Eric Jayjock, **Ravendra Singh**, Aditya Vanarase, Fernando J. Muzzio and Marianthi Ierapetritou (2015). Characterization and Modeling of Feeders: A Critical Component in Continuous Pharmaceutical Manufacturing. *Session (#572a): Applications of Continuous Processing in the Manufacture of Pharmaceuticals: Drug Product II.*
   Wednesday 3:20-3:40 PM, Ballroom B (Salt Palace Convention Center.
9. Zilong Wang, M. Sebastian Escotet-Espinoza, **Ravendra Singh**, Fernando J. Muzzio and Marianthi Ierapetritou (2015). Flowsheet Modeling for Oral Solid Drug Product Manufacturing. *Session (#697b): Model Based Integrated Design of Pharmaceutical Drug Substance and Drug Product Processes.*
   Thursday 12:50-13:10 PM, Ballroom B (Salt Palace Convention Center.
10. Jun Zhang, Frances Pereira, **Ravendra Singh**, Sean Bermingham, Rohit Ramachandran, Fernando J. Muzzio and Marianthi Ierapetritou (2015). A Systematic Approach of Using Material Properties Data for Pharmaceutical Process Simulation. *Session (#721b): Advances in Information Management.*
    Thursday 3:33-3:51 PM, Salon G (Salt Lake Marriott Downtown at City Creek).
11. Ashish Shah, Rohit Ramachandran, **Ravendra Singh*** (2015). Moving Horizon Based Real Time Optimization and Advanced Hybrid Model Predictive Control of Continuous Pharmaceutical Manufacturing Process.
    Session: 10B04 Economics and Process Control.
    Thursday 5:20-5:45 PM, Salon D (Salt Lake Marriott Downtown at City Creek).
12. Bob Engel, Paul Brodbeck, **Ravendra Singh** (2015). Applying Batch Data Principles to Continuous Manufacturing for the Purposes of Data Management, Batch Reporting, Analytics and Traceability. *Session (#721h): Advances in Information Management.*
    Thursday 5:21-5:39 PM, Salon G (Salt Lake Marriott Downtown at City Creek).

433711 Computer-Aided Molecular Engineering of Crystallization: From Colloidal Assembly to Geoengineering

Sunday, November 8, 2015
Exhibit Hall 1 (Salt Palace Convention Center)
Amir Haji-Akbari, Department of Chemical and Biological Engineering, Princeton University, Princeton, NJ

Crystallization is the oldest, and yet one of the most important, unit operations in chemical engineering, utilized for making a wide range of products from table salt to silicon wafers. In addition, crystallization is versatile in nature, and plays a pivotal role in determining the behavior of many atmospheric, geophysical and biological systems. The ability to understand and engineer these systems will be vastly improved by obtaining a molecular-level understanding of the underlying crystallization processes. Computer simulations are valuable tools in this quest. Most notably, they enable us to access time and length scales that are otherwise inaccessible in experiments. This latter aspect allows us to extract mechanistic information that can be further utilized in designing new building blocks, and in optimizing the existing crystallization processes, e.g. by increasing the structural and functional quality of the arising crystalline structure.

As a computational statistical thermodynamicist, I am interested in the application of the state-of-the-art molecular simulation techniques to studies of crystallization with the aim of engineering crystallization on a molecular level. During my doctoral studies in **Prof. Sharon Glotzer**'s Group at University of Michigan, I studied the self-assembly of nano- and colloidal hard particles. My work culminated in the discovery of the first hard particle quasicrystals. During my postdoc in **Prof. Pablo Debenedetti**'s Group, I have been studying ice nucleation both in the bulk and in confined geometries. In particular, I have successfully performed the first direct calculation of homogeneous nucleation rate for a molecular model of water, a problem commonly regarded as one of the most challenging open problems in computational statistical physics. I have also been working on quantifying dynamical anisotropies in confined liquids and ultrastable glasses.

As an assistant professor, I plan to use my expertise to study crystallization at different time and length scales using state-of-the-art advanced molecular simulation techniques. The overall objective of my research program is to understand the mechanism of disorder-order transitions in molecular, colloidal and biological systems, and to use this mechanistic knowledge for manipulating and optimizing such processes on a molecular level. This can, for instance, lead to the discovery of more potent ice nucleating agents, better hydrate inhibitors, or more effective antifreeze proteins. It can also help us identify therapeutic strategies for illnesses such as cataract and Alzheimer's disease that are caused due to processes such as aggregation or unwanted crystallization, by targeting the corresponding nucleation pathways. The individual research projects that I plan to pursue in the initial stage of my career are as follows:

(i) Self-Assembly of Functional Materials at the Mesoscale.

(ii) Ice Nucleation Under Atmospherically Relevant Conditions.

(iii) Kinetic Inhibition of Hydrate Formation.

(iv) Crystallization and Aggregation in Biological Systems.

HIGHLIGHTS [AS OF MAY 2015]:

- Seven first-author publications.

- One manuscript under review.

- Four manuscripts under preparation.

433735 Functional Polymers for Widespread Energy Applications

Abstract:

My current research as postdoctoral researcher at UC Santa Barbara (since May 2013) focuses on structure-property relationships of semiconducting polymers for organic field-effect transistors (OFETs) and organic thermoelectrics. One of the research directions under the advisement of Prof. Ed Kramer has been the structural characterization of aligned semiconducting polymers where the measured mobility has been greater than 20 cm^2/Vs. A combination X-ray scattering and electron microscopy techniques provides evidence for exceptional alignment of the polymer chains. This allows for an efficient intrachain charge transport between source/drain contacts, thus resulting in the measured high mobility. Another research direction under the advisement of Prof. Michael Chabinyc has focused on how various processing conditions effect the thermoelectric properties (Seebeck coefficient, electronic conductivity, and thermal conductivity) of highly doped semiconducting polymers.

My Ph.D. research at UC Berkeley under the advisement of Prof. Nitash Balsara focused on the synthesis and characterization of simultaneous electronic and ionic conducting block copolymers (P3HT-*block*-PEO) for lithium battery electrodes. Block copolymers can self-assemble and form co-continuous nanoscale domains, which are necessary for enabling redox reactions of the active material ($LiFePO_4$). In addition, the block copolymer serves as a binder to hold the redox-active material particles together. This simplifies the electrode design as one material serve both binding and charge transport functions. The application of this material in a lithium battery with $LiFePO_4$ showed specific capacity reaching the theoretical limit and with minimal capacity fade after ten cycles, thus demonstrating feasibility of P3HT-*b*-PEO as a conductive binder material. Furthermore, the ability of the conductive binder to switch between electronically conducting and insulating states in the positive electrode provides an unprecedented route for automatic overdischarge protection within the battery. This is in stark contrast to traditional lithium ion batteries where external electronics is used to provide overdischarge protection.

My Ph.D. and postdoctoral research provides a unique background in electrochemistry and polymer physics. My proposed research will build upon my diverse Ph.D. and postdoctoral research background. In particular, the proposed research topics focus on polymer-based thermal energy harvesting. The fundamental science underlying the application of conducting polymers for thermal energy harvesting focuses on the transport of charge carriers in the form of ions or electrons for electrical conduction and phonons with respect to thermal conduction. A fundamental understanding between these transport pathways and polymer morphology is crucial in the development of high efficiency polymer-based thermal energy harvesting devices. Furthermore, the work in the Patel Research Lab will be inherently collaborative spanning synthetic chemistry, soft-material science, polymer physics, and electrochemistry.

432914 Accelerating the Onset of the Hydrogen Economy

Abstract:

As a PhD student at UCLA, my research has tackled some of the technological challenges of the hydrogen economy such as vehicle integration, fill-up of hydrogen fuel cell cars, and production of green hydrogen. As part of the first prong of my research, I was in charge of designing, building and testing the hydrogen feed and cooling systems for the UCLA fuel cell go-kart that participated in different races and events through 2011. I also served as safety inspector for the team and project leader. The second prong of my research yielded new methodologies and systems for the optimum fill-up of hydrogen fuel cell cars since one of the key challenges for the deployment of this type of cars is the lack of fill-up practices that are time efficient and safe. Moreover, hydrogen fueling station operators do not employ an optimal mass flowrate policy that can satisfy the aforementioned constraints. A hydrogen vehicle user will like that the replenishment of fuel to his/her vehicle take the same time as a gasoline car. On the other hand, the type IV tanks, based on carbon fiber wrapping, have a maximum temperature limitation of 85 degree Celsius, so the fill-up process must ensure that this temperature is not surpassed. The research I have conducted has led to the development novel mathematical models and software that can simulate and optimize the fill-up process of gas fuel such as hydrogen, and even compressed natural gas (CNG). The models are based on a self-consistent thermodynamic model for real gases, transport phenomena principles, and conservations laws, which predict hydrogen's temperature and pressure time evolutions inside the vehicle tank, the station storage tank, the piping connecting the tanks, and the dispenser isenthalpic valve. The model results have proven to match, within 2%, corresponding experimental data from an actual fill-up process. Furthermore, in order to minimize the fill-up time and to determine an optimum mass flowrate policy, the process was formulated as a minimum time optimal control problem, which through the use of a novel decomposition of the problem into a process simulation problem, that checks for feasibility of the fill-up independent of time, and a simple minimum time optimal control problem, the global minimum fill-up time can be obtained analytically along with an optimal control policy on the mass flowrate. The optimization studies results lead to a fill-up time strategy that can improve current fill-up practices. The third prong of my research have been the fundamental studies for the realization of a concentrated solar power driven thermochemical cycle for the co-generation of electricity, oxygen, and hydrogen based on the thermal decomposition of alkali carbonates. First, thermodynamic calculations were employed to find the optimum reaction conditions for the individual reactions to be carried out. Then, a realization of the process' flowsheet was simulated based on Gibbs minimization method, so I could be subjected to optimization studies that maximizes hydrogen output.

In addition to my core research, I have also performed as the safety officer for Prof. Manousiouthakis experimental lab, having to go to intensive safety training and yearly inspections. I am a very active member of UCLA Center for Excellence in Engineering and Diversity efforts to motivate students coming from minority populations to attend the UCLA engineering school. I have been participating in the Mathematics Engineering Science Achievement (MESA) program, laboratory tours, panels, talks, tutoring, and volunteering and organizing events since 2006. I have been a research mentor to more than 50 middle and high school, undergraduate and graduate students yield several presentations. My research interests are alternative energy sources for vehicles, production of "green" hydrogen, design of hydrogen fueling stations, hydrogen safety, and hydrogen fuel cell vehicles integration.

433049 Engineering a Nano Display Platform from Bacterial Spore Coat Proteins

Abstract:

Bacterial endospores are encased in an assembly of coat proteins that protect the genetic material within from a number of environmental insults, including extreme heat, desiccation, caustic chemicals, and UV radiation. In addition, the protein-based coat provides remarkable structural integrity. We propose reconstituting the basement layer of the coat assembly, consisting of the proteins SpoVM and SpoIVA, onto spherical membranes to create synthetic spore-like particles that may be used in applications such as drug delivery or environmental remediation. Towards this goal, we have coated spherical micron-sized silica beads with a lipid bilayer to first characterize the localization behavior of protein SpoVM, then demonstrate that the structural protein SpoIVA can bind and polymerize atop of the SpoVM-coated beads.

During sporulation in Bacillus subtilis, SpoVM is one of the first proteins to localize to the forespore and was shown to localize via a mechanism previously unobserved in prokaryotes: recognition of convex membrane curvature. To understand the underlying mechanism of membrane curvature recognition, we use lipid membranes supported on silica beads of different radii of curvature. In a competition between 2um-diameter beads and 8um beads, microscopy shows that fluorescently labeled SpoVM-FITC preferentially binds to 2um beads, as the radius of curvature is closest to that of the Bacillus forespore. In contrast, the mutant SpoVM^{P9A}-FITC, previously shown to bind promiscuously to all membranes *in vivo*, binds equally well to both 2um and 8um beads *in vitro*. To quantify the binding of SpoVM-FITC to the beads, we use flow cytometry to analyze the fluorescence from the 2um and 8um beads separately from the mixed population. When normalized per surface area, binding of SpoVM-FITC at low concentrations is up to eight fold greater in the 2um beads compared to the 8um beads. Near saturation concentrations, fluorescence per surface area is nearly equal. In contrast, SpoVM^{P9A}–FITC has equal fluorescence per surface area even at low concentrations, due to the promiscuity of the mutant protein. Interestingly, at low concentrations of SpoVM-FITC, flow cytometry reveals two distinct population of 2um beads – one population low in fluorescence, and the other population high in fluorescence. These two populations are also observed with fluorescence microscopy, and indicates a cooperative binding behavior of SpoVM. Monte Carlo simulations show that binding cooperativity, coupled with a small difference in the rate of association of SpoVM between the 2um and 8um beads, is sufficient to achieve the preferential binding to 2um beads at equilibrium. Indeed in a kinetic association assay using flow cytometry, the k_on of SpoVM was higher for 2um beads than for 8um beads. Furthermore, evidence of cooperatively is again observed in the emergence of two distinct populations at intermediate timepoints, while later timepoints show one population of high fluorescence. Taken together, these results highlight a mechanism for curvature-sensing driven by cooperative binding of SpoVM and differences in the association kinetics between surfaces with different curvature.

We then show that the silica beads can be coated with SpoIVA. Using fluorescence microscopy, 2um beads coated with SpoVM and GFP-SpoIVA show an even distribution of fluorescence the bead, while a control in which the beads are incubated with GFP-SpoIVA in the absence of SpoVM shows fluorescence as punctate on one side of the spherical bead. This punctate fluorescence of GFP-SpoIVA is also observed in-vivo in sporulating B. subtilis cells in the absence of SpoVM. SEM of SpoVM/SpoIVA-coated beads show that SpoIVA has polymerized on the surface of the beads when compared to a SpoIVA variant with a mutation preventing ATP-hydrolysis, and thus polymerization. Finally, we have demonstrated that this platform can be used to decorate the surface of these beads using click chemistry. Here, we have incorporated one partner of a click reaction to SpoIVA via a C-terminal cysteine, with the reciprocal partner of the click reaction to the fluorophore Cy3, and confirmed the ligation via fluorescence microscopy. We then demonstrate that two different fluorophores can be displayed on the surface, independently of each other, using two orthogonal sets of click chemistry reaction pairs.

Taken together, these results demonstrate that bacterial endospore coat proteins can be used to engineer a nano display platform.

Understanding Gas-Metal Interactions for Clean Energy Applications

Kyoungjin Lee, PhD Candidate
Department of Energy Resources Engineering, Stanford University, Stanford, CA94305

RESEARCH KEYWORDS:
Metallic Membranes | CO_2 Capture | Hydrogen Production | Nitrogen Separation |
Material Synthesis and Characterization | Atomic Scale Simulation (Density Functional Theory)

RESEARCH INTERESTS:

Understanding fundamentals of atomic absorption and diffusion in inorganic bulk materials may provide versatile opportunities in improving the performance of energy-related materials. Hydrogen, nitrogen, carbon, and oxygen can be absorbed into bulk materials, which leads to unique applications such as gas separation membranes, catalysts, and electronic materials as insulators or conductors, but also causes undesirable consequences including metal corrosion and embrittlement. The goal of my research is to understand the effect of those absorbed species on the mechanical and electronic properties of bulk materials, thereby guiding material choices for future energy applications. By understanding the fundamental mechanisms, material properties may be tuned for enhancing desired properties while detrimental impacts being avoided. Theoretical calculations using density functional theory (DFT) is combined with controlled material processing experiments to provide the full understanding of bulk absorption, diffusion and transformation.

PHD DISSERTATION: ""Nitrogen- and Hydrogen-Selective Metallic Membranes for Clean Coal Utilization"

Carbon dioxide (CO_2) capture and storage (CCS) is an indispensible component for the mitigation of global warming. Metallic membrane (MM) approach is environmentally friendly and can be easily applied to dispersed carbon sources at small scales, competing to conventional methods using toxic amine solvents. The MM approach actively removes non-CO_2 gases (i.e., mainly nitrogen and hydrogen for post- and pre-combustion CO_2 capture, respectively) in order to concentrate CO_2 in the downstream. My study found that this *indirect* CO_2 capture approach can be more energy efficient compared to directly capturing dilute CO_2 from a conventional flue gas based on the minimum work analysis. By applying nitrogen-selective MMs in power plants for *ca.* 80% of nitrogen capture, the cost of carbon capture could be 3 times less than the cost of conventional capture. In addition, this approach may lead to more efficient ammonia production, which has been highly energy-intensive but heavily required for fertilizer production.

Gas permeation properties of an MM can be tuned by alloying metals with different reactivity toward nitrogen or hydrogen in order to achieve an ideal membrane property. In the first-principles calculations based on density functional theory (DFT), a high affinity of nitrogen toward vanadium and their alloys was confirmed and a high activation barrier was identified for subsurface diffusion. Further investigation on alloy compositions revealed that a dilute vanadium component is essential to achieve not only the

sufficient solubility but also the diffusivity. In order to screen materials for nitrogen-selective MMs, I have developed a calculation scheme for nitrogen solubility predictions, which enables the predictions in dilute solution systems including molybdenum, tungsten, and chromium. A similar scheme was also verified to be valid in predicting hydrogen solubility in various metals.
Following the theoretical efforts, I have fabricated thin MMs of Group V and VI metals supported on porous ceramic substrates to evaluate nitrogen permeability as well as various thermal and chemical impacts. While thin membranes lead to a high flux, fabrication of defect-free membranes has been extremely challenging in the literature. In my project, modified porous ceramic substrates were used to reduce the pore size and roughness, and a thin metal layer was sputter deposited to create a dense membrane. Coupled with characterization techniques including SEM, AFM, XRD, and XPS, it was concluded that nitrogen was absorbed to the metal surface with little structural deformation under a mild thermal condition. Using the same substrates, Mo-V alloy deposition for enhanced nitrogen flux will be continued.

TEACHING EXPERIENCE:

I served as a teaching assistant twice (2013 and 2015) for "Electronic Structure Theory and Chemical Kinetics" offered by both Chemical Engineering and Energy Resources Engineering, where I contributed to hold TA-led problem sessions and hands-on DFT calculation sessions, and helped make problem sets and grade exams. I also instructed three students to walk through DFT computations on a supercomputing cluster. In addition, I have mentored many students in my group for membrane experiments and material characterizations.

FUTURE RESEARCH PLANS:

My experiences on both experiments and theories for hydrogen and nitrogen in metal composite systems lend itself to studying nitride materials for different applications such as electrocatalysts for hydrogen evolution reactions (HERs) and lithium ion batteries (LIBs) in addition to MM applications. I believe that these topics have similarities to each other so that the lessons from MM systems could help resolve the challenges in those electrode systems. For example, the principles of enhancing catalytic reactions occurring on the membrane surface in the case of membrane reactors may be similarly applicable to the HERs on the electrode surface. Also, the lithium interaction with composite electrodes of LIBs has similar challenges as the hydrogen/nitrogen interaction with composite membranes, such as volume expansion, mismatch in the interface of composites, limited ion diffusivity, etc. The ultimate goals of my research are 1) to search materials with desired thermodynamic stability and kinetic properties by understanding fundamental electronic structures of materials, and 2) to synthesize and characterize those materials in the lab to prove that the new materials can be practically used. Through my research, I pursue the sustainable future of environment and energy landscape by mitigating CO_2 emissions, producing clean hydrogen and better utilizing renewable-sourced energy.

SELECTED PUBLICATIONS:

Research Articles

Lu, C.-Y.; Lee, K.; Psarras, P.; Wilcox, J. Mo-V alloys for nitrogen-selective metallic membranes, ***in preparation***

Wilcox, J.; Liguori, S.; Lee, K.; House, K. Nitrogen separation for carbon capture: thermodynamic minimum work analysis, ***in preparation***

Lee, K.*; De Lannoy, C. F.*; Liguori, S.; Wilcox, J. Thin metal composite membranes of molybdenum for selective nitrogen separation, ***to be submitted*** (*: equal contribution)

Lee, K.; Yuan, M.; Wilcox, J. Predicting hydrogen solubility in transition metals via first-principles calculations, ***Journal of Physical Chemistry C*** (2015) *in press.*

Rochana, P.; Lee, K.; Wilcox, J. Nitrogen adsorption, dissociation, and subsurface diffusion on the vanadium (110) surface: a DFT study for the nitrogen-selective catalytic membrane application, ***Journal of Physical Chemistry C*** 118 (2014) 4238–4249.

Kyung, D.; Lee, K.; Kim, H.; Lee, W. Effect of marine environmental factors on the phase equilibrium of CO_2 hydrate, ***International Journal of Greenhouse Gas Control,*** 20 (2014) 285–292.

Lee, K.; Lee, S. H.; Lee, W.; Stochastic nature of carbon dioxide hydrate induction times in Na-montmorillonite and marine sediment suspensions, ***International Journal of Greenhouse Gas Control,*** 14 (2013) 15–24.

Reviews

Wilcox, J.; Haghapanah, R.; Rupp, E. C.; He, J.; Lee, K. Adsorption and Membranes for CO_2 Capture, ***Annual Reviews of Chemical and Biomolecular Engineering*** 5 (2014) 479–505.

Wilcox, J.; Rupp, E.; Ying, S.; Lim, D.; Negreira, A.; Kirchofer, A.; Feng, F.; Lee, K. Mercury adsorption and oxidation in coal combustion and gasification processes, ***International Journal of Coal Geology***, 90-91 (2012) 4–20.

HONORS:

2015 | Honorable Mention in Student Poster Competition, North American Membrane Society
2015 | Elias-Klein Travel Award, North American Membrane Society
2015 | First place in poster competition, Earth Sciences Annual Research Review, Stanford University
2013 | Third place in poster competition, Earth Sciences Annual Research Review, Stanford University
2012 | Delegates of IEAGAG Carbon Capture and Storage Summer School (full travel grant)
2011–2013 | Korean Government Overseas Scholarship (full tuition for 2 years during PhD study)
2010 | Best Oral Presentation, Korean Society of Environmental Engineers

SERVICES:

2015–present | Peer Reviewer, Energy & Fuels
2014–present | Peer Reviewer, Journal of Physical Chemistry
2013 | Organizers of a workshop "Negative Emissions with the Virgin Earth Challenge" at Stanford University (assisted in planning an agenda, arranging student panels, inviting speakers and advertising to the public)

434007 Rational Design of High-Performance Catalysts for Sustainable Energy Conversion and Storage

Abstract:

It is widely known that in the not-too-distant future the world's energy dependence on fossil fuels resources will have to be replaced by a more sustainable energy source. However, the energy supply from these sources is intermittent. Moreover, the fact that the usage of electricity fluctuates independently leads to an unavoidable continuous mismatch between supply and demand. One solution would be the use of sunlight to make solar fuels by artificial photosynthesis, such as the water splitting into gaseous hydrogen and oxygen, or the water reduction of $CO_2$to methanol or hydrocarbons (Eqs. 1-3).

$2H_2O - 4e- \leftrightarrow O_2 + 4H^+$ (Eq. 1)

$CO_2 + 6\,H^+ + 6\,e- \leftrightarrow CH_3OH + H_2O$ (Eq. 2)

$CO_2 + 8\,H^+ + 8\,e- \leftrightarrow CH_4 + 2H_2O$ (Eq. 3)

A key half reaction in all the above processes is the oxidation of water to molecular oxygen, also known as oxygen evolution reaction (OER). This reaction is particularly difficult as it involves the removal of four electrons and protons from two water molecules and the formation of an O−O bond. Hence, there is an urgent need for finding a material that can catalyze this reaction in a very efficient way. This is precisely the main focus of my research, that is, the development of economic and highly efficient catalysts for the OER. To tackle this challenge, I use state-of-the-art theoretical calculations in order to get a fundamental understanding of the overall process, e.g. reaction mechanism, electronic structure of different catalysts, etc. This allows me to identify what are the main requirements that a potential high-performance OER catalyst needs to fulfill. The idea of this rational catalyst design is to be able to propose some candidates that can be tested by my experimental collaborators and that eventually will lead to a promising OER catalyst that can be scaled up for its industrial application.

434009 Investigation of Materials, Interfaces, and Processes Promoting Efficiency in Solar Energy Conversion Technologies

Abstract:

The discovery of new materials, and their organization into efficient nanoscale architectures, has the potential to transform and modernize energy conversion technologies. This is especially true for new devices for solar fuel and electricity generation based on photoelectrochemistry, which at present lack the required efficiency and product selectivity for commercialization.

I study the design of materials for solar energy conversion devices using two general methodologies. The first involves the discovery and characterization of new materials and nanoscale architectures that enable the water oxidation reaction, a key half reaction for solar fuel production, as well as the electrochemical reduction of CO_2 to fuels. The second approach presented emphasizes the use of simplified model systems to assess the physical origins of efficiency. Here primarily spectroscopic techniques are used to (1) study the electronic structures of active interfaces, and (2) characterize the interaction of molecular catalysts and reactants with relevant crystal surfaces.

When a new promising material is developed and applied in an optoelectronic device, open questions inevitably present themselves regarding the atomic- or nanoscale nature of the associated new interfaces. Much of my work in model systems is aimed at elucidating the role of active interfaces in promoting efficiency in photoelectrochemical and photovoltaic applications. Understanding the electronic structure and chemical nature of interfaces is critical because these are the source of electrochemical potential gradients for charge separation (which influences efficiency) and of active sites for electrocatalytic reactions (which influences reaction selectivity and efficiency).

Oxides are emphasized for the design of water oxidation photoanodes because they are among the only materials that remain stable during light irradiation and anodic polarization in aqueous electrolytes. In designing photoanodes for this application, I leverage the diversity of metal oxides to combine materials into unique heterostructures to create materials specialized to perform the optical, electronic, and chemical functions required of photoelectrochemical cells. I will present research on photoelectrochemical water oxidation using anodes comprised of alpha-Fe_2O_3, ZnO, and WO_3.

For the study of electrochemical CO_2 reduction, my experiments are designed to explain the role of the semiconductor electrode surface in driving the high selectivity observed in pyridine and other N-heterocycle-catalyzed CO_2 reduction systems. These systems have been shown to selectively reduce CO_2 to methanol using only solar energy and water. I use a surface science approach to examine the preferential adsorption sites, bonding interactions, and reactivities of these molecular catalysts and other relevant adsorbates on the surfaces of photoactive III-V semiconductors. Surface-sensitive core-level and vibrational spectroscopy measurements on single crystals are used to supplement electrochemistry investigations, and generate unique information describing the role of the electrode surface in CO_2 reduction. Conclusions from experimental results on these well-defined systems are supported by calculations using density functional theory.

434187 Developing Noble Metal/TiO2 and Swnt/TiO2 Composites to Improve Light Harvesting and Carrier Collection of Solar Cells

Abstract:

The performance of photovoltaic devices could be improved by using rationally-designed nanocomposites with high electron mobility to efficiently collect photo-generated electrons. Single-walled carbon nanotubes (SWNT) exhibit very high electron mobility, but the incorporation of such nanotubes into nanocomposites to create efficient photovoltaic devices is challenging. Here we report the synthesis of SWNT-TiO_2 nanocrystal core-shell nanocomposites using a genetically engineered M13 virus as a template. Using this method, SWNTs are stabilized without surfactants and surface modifications, and their electronic properties are preserved. In addition, close contact is achieved between SWNTs and TiO_2 nanocrystals. With the developed biological template approach, we demonstrate that well-dispersed semiconducting SWNTs can improve the power conversion efficiency of DSSCs. Even small fractions of nanotubes (typically 0.1 wt%) improve the power conversion efficiency by increasing the electron collection efficiency. We also show that both the electronic type and degree of bundling of the nanotubes in the SWNT/TiO_2 complex are critical factors in device performance. Semiconducting SWNTs improve the electron collection and metallic SWNTs decrease the electron collection in DSSCs. Moreover, debundled SWNTs affect the device performance more than bundled SWNTs. By optimizing the electronic type and aggregation state of SWNTs, we achieve a power conversion efficiency in the dye-sensitized solar cells of 10.6%.

Because SWNTs have good thermal conductivity in addition to high electron mobility, this approach might improve the stability of large DSSC modules. Moreover, biological engineering of multiple genes of the virus can extend this approach to creation of more complex structures. Though the route to DSSC improvement lies in the development of dyes with absorption extending into the infrared and better redox couples for higher voltages, we believe that our approach will enable the utilization of SWNTs in many practical photovoltaic devices that require efficient electron diffusion and reduced electron recombination, for instance, quantum dot solar cells, organic solar cells, and photoelectrochemical cells.

In photovoltaic devices, light harvesting (LH) and carrier collection have opposite relations with the thickness of the photoactive layer, which imposes a fundamental compromise for the power conversion efficiency (PCE). In addition, unbalanced LH at different wavelengths further reduces the achievable PCE. Localized surface plasmons (LSPs) are the elementary excitation states in noble metal nanoparticles (NPs), and can improve LH of photo-absorbers. Here, we report a novel approach to broadband balanced LH and panchromatic solar energy conversion using multiple-core-shell structured oxide-metal-oxide plasmonic nanoparticles (core-shell-shell titania-gold-titania nanoparticles). These nanoparticles feature tunable localized surface plasmon resonance frequencies (visible to infrared) and the required thermal stability during device fabrication. By simply blending the plasmonic nanoparticles with available photoactive materials, the broadband LH of practical photovoltaic devices can be significantly enhanced. We demonstrate a panchromatic dye-sensitized solar cell with an increased PCE from 8.3% to 10.8%, mainly through plasmon-enhanced photo-absorption in the otherwise less harvested region of solar spectrum. This general and simple strategy also highlights easy fabrication, and may benefit solar cells using other photo-absorbers or other types of solar-harvesting devices.

The general and simple strategy for broadband LH enhancement and panchromatic DSSCs is realized by matching LSP resonance (LSPR) wavelength (λ_{LSPR}) of plasmonic NPs with λ_{Lo} (wavelengths with low extinction coefficient) of existing photo-absorbers. Demonstrated by both finite-difference time-domain simulations and experiments, matching λ_{LSPR} with λ_{Hi} (wavelengths with high extinction coefficient) or λ_{Lo} affects LH differently. Although matching λ_{LSPR} with λ_{Hi} readily improves LH and PCE for optically-thin photo-absorbing layers, matching λ_{LSPR} with λ_{Lo} maximally increases LH and PCE for practical photovoltaic devices with optically-thick layers, through enhancing photo-absorption in the weakly-absorbing region. The multiple-core-shell oxide-metal-oxide plasmonic NPs developed and utilized here are advantageous over other geometries, by featuring: adjustable λ_{LSPR} of 600-1,000 nm, substantially enhanced near-field electromagnetic (EM) intensity, and preservable plasmonic properties during device fabrication. By matching λ_{LSPR} with λ_{Lo} of a common dye (N719), a panchromatic DSSC with broadband balanced LH and a PCE of 10.8% is achieved (30% increase comparing to DSSCs without plasmonic NPs).

434190 Exploring Biomolecular Engineering Strategies for Addressing Challenges in Therapeutic Design, Delivery and Purification

Abstract:

The overarching theme of my current research is peptide engineering, rational design of peptides and understanding protein-peptide interactions for applications in bio-separations, drug delivery, and biosensing. Peptides are an attractive class of biomolecules that can be employed as affinity reagents, drugs, drug carriers, biomaterials and biosensors. During my PhD in Prof. Pankaj Karande's lab at Rensselaer Polytechnic Institute (RPI), I explored several rational and combinatorial strategies for the design of affinity peptides against biomolecular and bio-inorganic targets. Specifically, I worked on designing peptides for the following applications – (1) as carriers for delivering neurotherapeutics across the blood-brain barrier (BBB), (2) as ligands for biosensing of cardiac troponin I, a biomarker of acute myocardial infarction, and (3) as modulators of calcium oxalate monohydrate crystals.

As a postdoctoral researcher in the labs of Prof. Steven Cramer and Prof. Pankaj Karande at RPI, I am expanding my expertise in the area of biomolecular design and engineering by specifically focusing on the problem of designing affinity ligands for the purification of non-mAb biologics. The long-term goal of this project is on-demand manufacture of biologically-derived medicines. My role in the project is to apply rational, biomolecular design strategies towards creating robust, affinity peptides for affinity-based purification of therapeutic proteins. The problem poses an exciting biomolecular engineering challenge since affinity ligands for bioprocessing applications need to possess several key features such as high affinity, selectivity, and elutability for their target along with optimum dynamic binding capacity, efficient regenerability and resistance to base and proteolytic cleavage. Further, I have also investigated into the design of peptides with specific molecular recognition properties such that they can distinguish between native and unfolded states of a protein. The multidisciplinary nature of the project has given me the unique opportunity to collaborate with several top researchers from academia and industry and understand first-hand the challenges faced by the pharmaceuticals industry in bio-manufacturing of therapeutic proteins.

My future research aims to address critical challenges in therapeutic design, delivery and purification by exploring novel biomolecular design and engineering strategies and employing experimental as well as computational approaches to this end. I aspire to build a world-class research program that will focus on finding creative and affordable solutions to the abovementioned healthcare challenges. I envision my research to address not only some fundamental questions related to these problems but also foster extensive collaborations with the industry and deliver practical solutions. As an independent researcher in academia, I also see myself in an active role as a teacher and contributor to the profession of chemical engineering as well as a mentor to the future generations of high school, undergraduate and graduate students and postdocs.

434397 Synthetic Modification of Proteins to Create New Biomaterials

Abstract:

The synthetic modification of proteins plays an essential role in the fields of biomaterials science and chemical biology. Conjugation of synthetic molecules to proteins enables the incredible diversity of protein structure and function to be harnessed in new protein-based materials. Protein bioconjugates have been used for a wide array of applications from the investigation of biological function to use in new biomaterials such as targeted therapeutics and biocatalysts. The synthesis of these materials requires well-defined protein bioconjugation reactions. However, a significant challenge is that these reactions must take place in water, at ambient temperature, near neutral pH and in the presence of a wide array of functional groups.

While many methods for the modification of native and artificial amino acids exist, they often require long reaction times, lengthy syntheses of reactive substrates, or lack specificity and selectivity. To address these issues, a suite of bioconjugation reactions that utilize *ortho*-aminophenols was developed. The oxidative coupling of non-canonical aniline residues with *o*-aminophenol substrates was developed to enable use with a wide range of protein substrates, including those containing free cysteines and glycosylation. In the course of these studies, it was discovered that under certain conditions aminophenols were reactive with native residues on protein substrates in addition to artificial aniline moieties. The reactive native residues were identified and oxidative coupling of *o*-aminophenols with the N-terminus was optimized to achieve high levels of modification on peptide and protein substrates. This new reactivity dramatically expands the bioconjugation reaction toolkit, enabling fast, site-selective modification in a single step using low concentrations of reagent.

In addition to the development of methods for protein modification, the use of modified proteins for targeted delivery and biocatalysis was explored. Two methods for the self-assembly of protein biocatalysts were explored, both relying on the ability of block copolymers to direct self-assembly. First, site-specific functionalization of cytochrome P450 BM3, a highly pliable enzyme, with poly(*N*-isopropylacrylamide) to create a diblock copolymer generates a material that can be easily processed to create a self-assembled heterogeneous biocatalyst. The self-assembly of the bioconjugate in concentrated solutions was characterized by small-angle x-ray scattering (SAXS) and depolarized light scattering (DPLS). The thin film nanostructure was also investigated by grazing-incidence small-angle x-ray scattering (GISAXS). Catalytic activity of the thin films for C-H oxidation was measured using a variety of substrates. An additional method for controlling the self-assembly of proteins was probed. Non-specific modification of proteins to vary their surface charge was also examined as a means for directing protein self-assembly. Using model proteins and lysine succinylation, a panel of proteins with varying charge density was synthesized. The ability of these proteins to form complex coacervates with oppositely charged polyelectrolytes was studied in order to identify parameters that governed complex coacervation of proteins. The fundamental parameters uncovered allow any proteins of interest to be successfully modified to enable complex coacervation. These findings were applied to the synthesis of complex coacervate core micelles with protein cores. Encapsulation of enzymes in these ionic micelles enables their use as biocatalysts and future work will investigate the use of these micelles for protein drug delivery.

434011 Functional Polymeric Materials for Sustainable Energy and Biomedical Applications

Sunday, November 8, 2015
Exhibit Hall 1 (Salt Palace Convention Center)

Shudipto Konika Dishari, Chemical Engineering, The Pennsylvania State University, University Park, PA

With a view to facing the not-so-far-away fossil fuel crisis, research focuses have been diverted towards sustainable energy based high efficiency energy conversion and storage devices. To achieve nation's vision towards hydrogen economy, 2010-2040 is the timeframe during when DOE's Energy Efficiency and Renewable Energy (EERE) program (FY16 budget $2.72B) plans the critical development of fuel cell and mass production of fuel cell based vehicles. The nanoscale understanding of polymer-catalyst interfaces on electrodes is crucial to lower the cost of fuel cell (target $30/KW) and improve the performance of sustainable energy propelled applications. Ionomer thin films behave very differently from bulk membranes at hydrated state due to severe confinement effect at interfaces. Although a great deal of work has been done so far to explore the material properties of ionomers in bulk membranes, the mobility, hydration and transport properties of functional ionomers in several nm-thick films are relatively unexplored. The challenges in characterizing thin films start when the conventional bulk characterization techniques fail to predict the nanoscale changes in materials properties. Fluorescence, when used in conjunction with other surface characterization techniques, overcome many of these issues and provides a powerful platform to characterize bulk membranes as well as thin films. Fluorescent probes sensitive to local viscosity, water-polymer mobility and proton concentration has been incorporated into ionomer thin films and membranes and the changes in fluorescence properties of the dyes as a function of film thickness and hydration has been studied. The changes in steady state and time-resolved fluorescence has offered valuable insight into stiffness and proton mobility of hydrated confined systems consistent with hydration behavior explored by QCM and Ellipsometry. In addition, the water-polymer and density distributions along the film thickness have been thoroughly investigated in hydrated thin polymer films using neutron reflectometry (Post-doc at MatSE, Penn State, PI: Michael Hickner).

Virus filtration membranes function predominantly through robust size exclusion mechanism and have potential applications in bioseparation and water purification. Current target is to achieve >4-log removal of virus particles from product streams with minimal membrane fouling and good capacity (>200 L/m^2). The biggest challenges in removing viral contaminants from therapeutic drugs are lack of sufficient understanding of factors controlling virus retention within these membranes and unexpected decay in log reduction value (LRV) in course of filtration. Apart from the conventional quantitative studies of virus retention, the application of fluorescence confocal microscope enables the visualization of virus retention zones along the thickness of the membranes. This exciting and new concept has been taken one step forward by adopting a multiple dye approach to track model bacteriophage retained inside different virus filtration membranes (with varied pore structure and morphology) at different stages of membrane operations with intermittent pressure releases. Also the performance of these membranes has been improved by enabling dual mode virus capture (electrostatic adsorption and physical size based entrapping) (Post-doc at ChE, Penn State, PI: Andrew Zydney).

In addition, fluorescent p-conjugated oligoelectrolytes have been synthesized and characterized in solution and solid state utilizing their light-harvesting properties. Layer-by-layer self-assembled thin films of fluorescent oligoelectrolytes have been investigated to develop fluorescence based bioassay platforms (PhD at ChBE, National University of Singapore).

As a future faculty, I would like to address the critical concerns of unexplored confined systems and fill in the knowledge gaps in nanoscale polymeric and biomolecular systems utilizing my multi-disciplinary research experience. New and simpler techniques will be innovated to probe zone-specific material properties within thin film systems. A great emphasis will be given on synthesis of new molecules with varied geometry to improve confined state properties and develop next-generation materials for energy and biomedical applications.

434272 Heterogeneous Catalysis: Synthesis and Spectroscopy of Supported Metal Oxide Catalysts for Natural Gas Upgrading

Abstract:

My overarching goal is to understand the factors controlling reactivity and selectivity of supported transition metal oxide catalysts in the activation of light *n*-alkanes. I have extensively studied the oxidative dehydrogenation of propane on well-defined supported metal oxide catalysts (ACS Catalysis 2014). The thorough use of in situ Raman and UV-vis (Catal. Sci. Technol. 2012), as well as electron paramagnetic resonance spectroscopy (ChemCatChem 2012 – J. Phys. Chem. C 2013) allowed me to complement the conclusions I gained from kinetic studies (Dalton Trans. 2013). Currently, I am synthesizing and characterizing a pool of supported metal oxide catalysts, aimed to find synergetic effects for industrially attractive reactions, followed by a fundamental understating of the kinetics, catalyst structure and reaction mechanism. Linking high quality research and mentoring is one of my strengths. For instance, recently I patented (several publications under preparation) a new method to disperse transition metal oxides on SiO_2. I also was awarded with the 2015 Mentor Award of the Chemistry department at UW Madison. In parallel, I am developing a new approach to run operando Raman spectroscopy for liquid phase reactions. As a faculty at one of the leading universities in the US, my research would focus on developing novel designs and synthesis of catalysts for more sustainable routes to important bulk and fine chemicals. For this, in situ spectroscopy will be one of my main pillars. My education and research experience in organic and inorganic chemistry, as well as chemical engineering give me the scientific and academic background to immediately teach courses in general, organic and inorganic chemistry, physicochemistry, spectroscopy, reaction engineering and catalysis.

434541 Morphology and Ion Transport in Polymer Electrolyte Membranes

Sunday, November 8, 2015
Exhibit Hall 1 (Salt Palace Convention Center)

Xi Chelsea Chen, Materials Sciences Division, Lawrence Berkeley National Laboratory, Berkeley, CA

Polymer electrolyte membranes have broad applications in energy storage and conversion, clean-water-related technologies and medicine. Common examples include fuel cells, redox flow cells, electrodialysis, reverse osmosis, hemofiltration and hemodialysis. My current research as a postdoctoral fellow at Lawrence Berkeley National Laboratory is focused on the fundamental investigation of morphology-conductivity relationships in proton-conducting block copolymer electrolyte membranes, comprising ionic and non-ionic blocks.

Membranes currently used for proton transport in fuel cells are nonporous (e.g. Nafion). In this poster, I show that the introduction of pores into block copolymer electrolyte membranes provides fine control over water uptake and conductivity. We start with a membrane comprising a mixture of homopolymer polystyrene (hPS) and a polystyrene-b-polyethylene-b-polystyrene (SES) copolymer. Porous membranes are fabricated by rinsing the hPS/SES mixture membranes in tetrahydrofuran and methanol. The polystyrene domains in the porous SES membranes are then sulfonated to give a porous membrane with hydrophilic and hydrophobic domains. The porosity is controlled by controlling $\square_v$, the volume fraction of hPS in the blended membrane. The morphology of the membranes was studied by scanning transmission electron microscopy (STEM), electron tomography and resonance soft X-ray scattering (RSoXS). The porous structures before and after sulfonation are qualitatively different. Water uptake of the membranes increases with increasing $\square_v$. Proton conductivity is a non-monotonic function of the water content of the membranes.

Using this novel method, we were able to control the water uptake of block copolymer electrolyte membranes without changing the chemistry or chain architectures. We believe that the novel processing method described above has created microscale pores in the block copolymer electrolyte membranes which allows us to tune the water uptake of these membranes to optimize proton conductivity.

As a faculty candidate, my proposed research will be initially focused on ion-containing polymers. I propose to build a program aimed at establishing the design principles of functional ion-containing polymers with a scientific focus on obtaining a fundamental understanding of the structure – property relationships in specific systems.

My proposed research is broadly organized into three interrelated projects:

1. Hybrid ion-containing polymer membranes for clean energy and water;
2. Morphology and transport properties of thin film ion-containing polymers;
3. Ion-containing polymers for endovascular drug filtration.

434542 Rational Design of Catalytic Sites for Energy Applications

Abstract:

The rapidly increasing global demand for energy coupled with large population growth necessitates the development of increasingly efficient systems for utilizing our finite energy and chemical resources. While no single approach will address these complex issues, I strongly believe that catalysts have an important role to play in addressing these challenges. Our society is already heavily reliant on these materials to produce the chemicals that we need to sustain our lifestyles. Catalysts are incredible because of their ability to facilitate selective chemical transformations while converting between chemical, thermal, radiant, and electrical forms of energy. Coupling catalytic systems with renewable energy sources presents a unique opportunity to regenerate useful high value chemicals from low value feed stocks like CO_2 and H_2O. In order to develop these technologies, it is vital to understand the pathway through which these chemical transformations proceed and how catalytic systems respond to different energetic stimuli. It is also vital to identify key characteristics of active and stable catalysts to enable the design of materials with optimal sites for a particular reaction. This strategy was implemented during my PhD studies for the development of hydrogen fuel cell electrocatalysts.

Over the course of my graduate studies under Prof. Suljo Linic, we have investigated how geometric and chemical manipulations of catalytic sites influence overall catalytic performance. Specifically, we focused on understanding how changing the size, shape, and composition of Ag and Pt alloy electrocatalysts affects their activity towards the oxygen reduction reaction ($O_2+4H^+/e^-\rightarrow 2H_2O$, ORR) in alkaline and acidic electrolytes. First, kinetics studies of the oxygen reduction mechanism on these surfaces allowed us to identify key chemical descriptors (like OOH or OH adsorption energies) that are correlated with catalytic activity. We then utilized quantum chemical calculations to determine how these descriptors could be selectively influenced in platinum based catalysts through perturbations such as alloying or inducing geometric strain. After identifying promising catalysts, we developed sophisticated synthesis strategies to prepare these materials. We then tested their activity, with rigorous experimental protocol allowing us to obtain valid kinetic rates from the observed rates, and found that improved activity was achieved. Finally, we performed extensive in situ and ex situ characterization to confirm both that the atomic structure of the electrocatalyst matches that of our model and that the observed activity enhancement is a direct result of the improved key chemical descriptors of this particular nanostructure.

My future research interests focus on the design of cheap, active, and stable catalysts for hydrogen evolution, CO2 reduction, and nitrogen fixation. My knowledge and experience with rational design, chemical synthesis, activity testing, and material characterization equips me with the skills to tackle these challenges.

434408 First-Principles Computational Chemistry Research in Sustainable Energy and Catalysis

Abstract:

The focus of my research is applying first-principles methods for sustainable energy research and catalysis. Specifically, in my graduate research at Princeton University, we applied first-principles methods to advance the performance of oxide-based solar cell. We studied different aspects of hematite (Fe_2O_3) as a photocatalyst: (i) light adsorption; (ii) electron/hole transport; (iii) band edge alignment relative to reaction potentials; (iv) surface reaction mechanism for water oxidation. Based on the computational results, we proposed design strategies for improving the efficiency of oxide-based photocatalysts.

In my current postdoctoral research, I expand my expertise to study catalysis with metals/metal oxides and metal-organic frameworks (MOFs): (i) CO_2 reduction on modified TiO_2 surface; (ii) hydrolysis of chemical warfare agents on MOFs; (iii) hydrogenation of ethylene and C-H activation of ethane on metal/metal oxide clusters.

Throughout my research, I have contributed to method development: (i) evaluation of U-J values for the DFT+U method; (ii) embedding model that treats subsystems at different levels of theory; (iii) fast implicit solvation model for large systems.

In the future, I will continue to focus on applying first-principles methods for research in sustainable energy and catalysis. The topics of interest will include photocatalysts and MOF materials, and extend to fuel cells and batteries, to make sustainable energy more affordable and more efficient.

434574 Nano-Structured Catalysts for Clean Fuels and Chemicals: Directing Activity and Selectivity By Design

Abstract:

One of the ultimate goals of catalyst-related research efforts is the rational design of active, selective and stable catalytic materials. To this end, the rapid advancement of state-of-the-art characterization techniques (e.g. *in-situ* and *operando* electron microscopy and X-ray spectroscopies) provides new opportunities for identifying the critical properties of nanomaterials that drive various catalytic processes. My previous work on metal-based catalytic materials for clean fuel and green chemistry applications lies at the intersection of catalyst design, synthesis, testing, and characterization. This puts me in a unique position to apply powerful techniques to emerging challenges in catalyst development for today's changing energy landscape. I will present my work on these topics and plans for future research.

My Ph.D. thesis focused on developing our understanding of alkali-promoted noble metal catalysts for alternative fuel production. The ability of alkali metals to promote the water-gas shift activity of platinum-based catalysts has been reported on a number of supports, although the nature of the active site remains controversial. Water activation is thought to be the rate-limiting step, making the surface hydroxyl concentration fundamentally linked to the reaction rate. The use of oxide-free (carbon) supports and state-of-the-art microscopy and spectroscopy techniques allowed us to confirm the stabilization of atomically dispersed $Pt\text{-}OH_x$ species on Na-modified Pt catalysts. The approach for stabilizing such species has resulted in further development of promoted noble metal catalysts for the water-gas shift reaction.

My postdoctoral research aims to elucidate the nature of the interplay between Ag and Au in nanoporous gold architectures, which contain only dilute quantities of Ag. Ag is believed to act as a promoter that activates dioxygen, allowing for selective oxidation (e.g. esterification) reactions to take place with high (Au-like) selectivity. An activation procedure resulting stable catalyst activity was developed. With the use of *in-situ* spectroscopy and microscopy techniques, atomic-scale structural changes in the catalyst were related to activity and selectivity in alcohol coupling reactions. These principles can be used to design new catalyst formulations for other important reaction systems.

By utilizing my experience in catalyst characterization and design, my research plan aims to develop a framework for understanding the dynamics of multi-component catalysts in order to design novel materials for today's energy and green chemistry challenges.

Ph.D. Advisor: Professor Maria Flytzani-Stephanopoulos

Postdoc Advisors: Professor Cynthia M. Friend and Professor Robert J. Madix

434614 Emerging Patterns in Soft Materials from Geometric Confinement

Abstract:

Geometric confinement of soft materials has a significant impact on how they behave and organize. Different from a sphere, a toroidal confinement is characterized by having a handle and a varying curvature. I will describe a recently developed templating method that enables rapid generation of stable 3D/toroidal confinements. I will discuss the effect of geometry and topology on the structure of several out-of-equilibrium systems. Biofilms of B. subtilis bacteria, when confined in toroidal droplets, grow on both the inner saddle and the outer regions. Compared to biofilms growing on a flat substrate, the relative coverage is an order of magnitude higher and the biofilm roughness almost doubled. Biofilm architecture is also quantitatively distinctive. The final biofilm structure demonstrate a heavy dependency on geometry-dependent oxygen accessibility. The templating method also allow us to generate soft solid structures like toroidal hydrogels. Thermo-responsive toroidal hydrogel swell and deswell isotropically via uniform shrinkage in all dimensions in the quasistatic case. In contrast, non-quasistatic shrinkage leads to mechanical instabilities and shape changes. Finally, I will discuss some experimental results on the organization and proliferation pattern of tissue cells on curved toroidal surfaces.

(6if) Multi-Scale Modeling to Study Soft Matter

Rajput, N. N., Lawrence Berkeley National Laboratory

Molecular simulations are indispensible tools to capture molecular level properties of bulk solutions as well as interfaces. My goal is to couple Quantum mechanical (QM) calculations with molecular dynamics (MD) to understand the electronic structure, energies as well as bulk structural and dynamical properties of soft matter. QM calculations are well suited to study accurate energies, forces and reaction chemistry. On the other hand, MD simulations are capable to study many body properties of thousands of molecules at longer time scale, which is essential to capture realistic properties of bulk solutions as well as interfaces. ILs have gained immense attention recently due to great thermal, chemical, and electrochemical stability and their non-volatile and non-explosive nature. Because of their numerous applications and unique properties, there is a rapidly growing scientific and commercial interest in ILs. In particular, some experimental studies have shown their use as potential electrolytes for batteries. I plan to study electrolytes doped ILs for multivalent batteries using QM and MD simulations. The current state-of-art Lithium ion batteries are suffering with their limited performance, short lifetime and safety concerns. An approach to achieve energy density higher than the existing Li-ion batteries is to replace monovalent Li-ion by multivalent ions such as Mg^{2+}, Ca^{2+}, Zn^{2+} and Al^{3+}.[1,2] Using QM calculations we can study the electrochemical stability window, bond dissociation energy and dissociation constant of ion pairs. The optimized structure and electrostatic potential from QM calculations can be used to generate force filed parameters for MD simulations. MD simulations can then be used to predict solvation structure, diffusion coefficient and viscosity of electrolytes. It is important to understand the cation-anion distribution, bonding structure and ion exchange mechanism in these solutions. The effect of temperature and concentration will be considered on these properties. Interfacial properties play important role in determining performance of electrochemical devices and it is hard to study interfaces experiments. In the next step these properties will be studied at the electrode/electrolyte interface. Significant information about the interfaces can be obtained by studying solvation structure, charge transfer and dynamical properties of electrolytes at the electrode/electrolyte interface. Molecular simulations will be the key for studying the molecular level properties of IL doped electrolytes at the interface leading to advancement in designing tomorrow's high performance electrochemical devices

(1) Rajput, N. N.; Qu, X.; Sa, N.; Burrell, A. K.; Persson, K. A. *Journal of the American Chemical Society* **2015**.

(2) Lin, M.-C.; Gong, M.; Lu, B.; Wu, Y.; Wang, D.-Y.; Guan, M.; Angell, M.; Chen, C.; Yang, J.; Hwang, B.-J.; Dai, H. *Nature* **2015**, *520*, 324.

434716 Solids and Particulate Processing Applications

Sunday, November 8, 2015
Exhibit Hall 1 (Salt Palace Convention Center)
Juan G. Osorio, Department of Chemical Engineering, Massachusetts Institute of Technology, Cambridge, MA

Solids and particulate processing and handling are important aspects in industrial applications. Industries, such as pharmaceuticals, chemicals, food, catalyst and construction, process their raw materials and final products as solids, with >75% and >50%, respectively. Yet particulates and their processing are not well understood, lacking engineering and design principles, which commonly cause failures in intermediate or final products. This leads to a product yield averaging 40% for solids, while the production rate is about 90% for fluids. Hence, there is a strong need to study solid particulate systems and to develop undergraduate and graduate chemical engineering programs in this field of particle technology.

My doctoral work at Rutgers University focused on understanding material and bulk properties of solids for pharmaceutical applications. I also developed and studied batch and continuous manufacturing processes including feeding, mixing, compression and capsule filling while using process analytical technologies to monitor and control such processes. During my first postdoctoral appointment at Purdue University, I studied granule breakage mechanisms and developed scaling principles in wet particulate systems, which included continuous wet granulation. At my current postdoctoral appointment, I am developing small-scale methods for manufacture of pharmaceutical solid dosage forms.

The common theme in my research work is the understanding and development of solids processes and products through investigation of material and bulk properties, their interactions with equipment and their effect on final products. As a professor, I plan to develop a world-class leading research program in solid particulate systems, not limited to pharmaceutical applications, but an extension into several other industries and areas. I am particularly keen to grow a large portion of my research program to study biomass processing and handling. The latter is a "New Frontier," as described by the AIChE this year, in solids flow, handling and processing and is needed in the alternative energy realm to increase biofuel yields.

434828 Integrated Simulation Methods for Protein-Nanoparticle (NP) Interactions with Complex Surface/Solvent Environments

Abstract:

Understanding the fundamental biophysics behind protein-NP interactions would be the essential of designing and engineering protein-NP systems for better performance in applications such as biosensors and NP drug carriers. To address this point, we developed a coarse grain model, which evolves from a well parameterized and validated surface potential for protein-(flat) surface interactions I developed as part of my Ph.D research. The main advantage of this method is that using a single-bead-per-residue model it quantitatively accounts for the hydrophobic interactions of residues and the NP surface and the dependency of protein stability on the curvature of NPs. As validated by the reference experimental study, the coarse grain model successfully reproduces the protein GB1 structure , orientation, and binding affinity on the NP that are consistent with experimental observations such as circular dichroism (CD) and fluorescence spectroscopy. Further, this method has been successfully applied to study the effect of NP on the nucleation of protein fibrillation, which is one of the main concerns of the potential toxicity of NP in vivo. The accelerating or inhibiting effects on the protein fibrillation is depending on the surface hydrophobicity and curvature of the NP and the inherent protein stability, which can be captured by this coarse grained simulation method.

Furthermore, denaturants such as urea and guanidinium chloride (GdmCl) are widely used to study protein folding/denaturing processes in solution and at interfaces such as protein-NP complexes. Denaturants especially have their advantages in studying protein folding over heating due to the convenience in experiments. To extend the ability of the coarse grained simulation system for the direct comparison to experiments, we implemented a model that simulates protein folding in urea and GdmCl. In this work, by using the similar idea and rescaling the parameters of O'Brien's molecular transfer model, we implement a Tanford's transfer free energy method for the KB Go-like model. More importantly, instead of reconstructing the all-atom structure as the MTM, we employ an optimized fast solvent accessible surface area estimator to directly calculate the residue-based SASA based on the Cα information only.

Therefore, the free energy contribution of the solvent is able to be measured and included in the simulation sampling on the fly. The results show great consistency in the predicted change of protein melting temperature by this simulation method and the experimentally measured values. Different thermally stable intermediates of protein folding in denaturants are captured by this simulation method, which may be out of limit of experimental measurements.

Optimization and expansion of this simulation system is on-going, which aims to form an integrated modeling system that accurately and efficiently predicts proteins' stability (both thermal and in denaturants), binding affinity, aggregation rate, structure, and orientation at various NP surfaces and solvent conditions. Therefore, together with collaborations with experimental methods, this simulation system has a strong potential in leading to the better design of NP-protein complexes for biosensors and drug delivery systems.

434994 Design and Optimization of Nano- and Macro-Scale Biomaterials for Vaccines and Immunomodulation

Abstract:

Since the development of modern vaccines, many diseases have been globally eradicated. Yet vaccines for several widespread diseases, including HIV, malaria, and tuberculosis, remain elusive. While traditional vaccine approaches have failed against these infectious diseases, researchers have turned towards the development of engineered subunit vaccines. Specifically, my postdoctoral work in the Irvine lab at MIT has focused on the design and optimization of general B-cell targeted and disease-inspired nanoparticulate vaccines.

Lipid-based nanoparticles have been widely characterized for the delivery of small molecules, peptides and proteins. Liposomes containing surface-bound antigens are similarly being developed as vaccines and have been shown to induce strong immune responses when compared to soluble antigens by facilitating cross-presentation and enhancement of humoral responses. Yet so far no studies have focused on how specific liposome design considerations modulate immune responses at the single cell level to enhance CD4+ T cell help and antibody production through interaction and activation of specific B cells. In the first half of my poster, I will describe how we evaluated the role of liposome composition on the direct binding and internalization of liposomes with surface-displayed peptide antigens to antigen-specific B cells. Focusing on the roles of surface PEGylation and antigen tethering, we optimized a particle formulation which bound to antigen-specific B cells with the highest intensity. B cell activation in vitro, assessed at the single-cell level using flow cytometry, was shown to be a direct function of antigen-specific binding. The incorporation of TLR4 agonist, monophosphoryl lipid A (MPLA), into the liposome bilayer also significantly enhanced early activation (day 1) and helped sustain activation and B cell viability at later times (up to 4 days) in an antigen-specific manner. Interestingly, antigen-specific antibody titers in mice immunized with antigen-bearing liposomes containing MPLA were found to correlate directly to in vitro B cell binding profiles for a variety of liposome formulations. The results obtained in these studies will further the development and understanding of effective particle-based peptide vaccines.

Next I will describe a lipid-based nanoparticle vaccine designed specifically for HIV prophylaxis. The HIV envelope spike, comprised of gp41 and gp120 proteins assembled as a homotrimeric complex, is the only exposed target on the viral membrane for neutralizing antibodies. Over the past 15 years, a number of HIV envelope trimer mimics have been engineered as potential immunogens for the induction of broadly neutralizing antibodies (bNAbs) against HIV infection. Recently, a stable and homogeneous HIV env trimer, BG505 SOSIP.664 (SOSIP), has been developed and shown to express epitopes for bNAbs and not for non-NAbs. It is generally thought that presentation of env trimers in a highly multivalent form from the surface of a particle may contribute to the development of strong and durable bNAb responses in immunization. To this end, we coupled SOSIP to the surface of liposomal nanoparticles in order to (1) multivalently present SOSIP in a fashion mimicking its display on the viral surface, (2) control SOSIP orientation and prevent antibodies from developing to the SOSIP base, (3) improve draining to lymph nodes and (4) incorporate various adjuvants and helper epitopes to enhance humoral responses. SOSIP trimer was efficiently (~50-65% loading) and reproducibly coupled to the surface of 5% Ni-NTA containing liposomes of various sizes via C-terminal 6xHis tags. CryoEM imaging revealed homogeneously distributed and oriented SOSIP on the surface of liposomes irrespective of particle size and lipid composition. One key aspect in particulate vaccine design is to ensure the desired antigenic profile of the antigen is not compromised by the conjugation or encapsulation methods used. SOSIP trimer antigenicity, determined by ELISA on intact liposomes, was maintained upon liposome coupling, displaying epitopes targeted by known bNAbs (i.e. VRCO1, PGT145, PGT151) and not epitopes targeted by non-NAbs (i.e. B6, 447D). Upon immunization in mice, 100 nm liposomes resulted in the highest serum SOSIP-specific IgG titers compared to 50 and 200 nm liposomes or soluble SOSIP. Ongoing studies will investigate the role of lipid composition and bilayer fluidity, as well as antigen density to determine optimal particulate vaccine conditions in mice. These results show promise for the use of SOSIP-coupled liposomes in larger animal models and as a potential component of an eventual prophylactic HIV vaccine.

435076 Linear and Non-Linear Programming Techniques for Process Intensification

Abstract:

My research focuses on linear programming (LP) and non-linear programming (NLP) techniques for process intensification applications. To this end, I have explored interval analysis-based branch-and-bound techniques for parameter identification, assessment/synthesis of sustainable systems, graphical/analytical techniques for the global solution of the total annualized cost problem for a series of compressors, and the introduction of the novel concept of the attainable region for process networks. The latter concept has been applied to identifcation of the attainable region for VLE separator networks in which the process network's outlet flowrates are allowed to vary. Throughout the course of my work, I have experienced first-hand the "curse of dimensionality" prevalent in non-linear optimization algorithms. Thus, I have applied analytical techniques, dimensionality reduction, convexification strategies, and functional analysis to identify global optima and to enhance the efficiency of their computation.

My global optimization expertise has prepared me well to pursue new research avenues in several areas, particularly optimal system identification, initial project feasibility assessment via attainable region concepts, and sustainability assessment and decision making. My strong mathematical and computational background will also allow me to contribute to collaborations that can benefit from this kind of expertise. Furthermore, I will disseminate all developed concepts and lessons learned into course-friendly materials to provide students opportunities to apply basic chemical engineering and optimization concepts to real-world green engineering problems.

435158 Membranes for Energy-Efficient Separations

Abstract:

Gas and vapor separations in the petrochemical industry are often accomplished with energy-intensive and environmentally harmful processes such as distillation and absorption, so a need exists to identify new, environmentally benign, and more energy-efficient separation technologies. Of particular promise are membrane-based separations, which, unlike distillation, do not require a thermal driving force, and, unlike absorption, operate under steady-state conditions without the need for regeneration. However, in their current form, membranes often do not possess the requisite productivity, efficiency, and chemical stability needed to compete with traditional separation technologies. To overcome these challenges, new polymer membranes and composite membranes, containing highly-structured, inorganic materials that far surpass the separation performance of pure polymers, are needed.

Separations with dense polymer membranes are accomplished by selectively permeating molecules, and permeation, in turn, is governed by the product of molecular diffusion rates and gas sorption in membranes. Both of these factors can be tuned to control separation performance. My graduate research focused on two distinct families of polymers, both of which showed outstanding separation properties for certain gas pairs. One family of polymers achieved their separation performance based on selective diffusion (i.e., thermally rearranged polymers) and the other on selective sorption (i.e., perfluoropolymers). My postdoctoral research focuses on extending the limit of membrane performance by synthesizing and characterizing mixed-matrix membranes, whereby metal-organic framework (MOF) nanoparticles with unsaturated metal sites are suspended into polymer matrixes. MOFs are porous, crystalline materials that have fundamentally different separation property sets compared to those of pure polymers. Additionally, the metal cations and organic linkers used to build MOFs can be modified to precisely tune porosity, crystalline space groups, and MOF-molecule adsorption and diffusion characteristics, thereby providing a novel platform of porous materials for achieving more efficient separations.

My research interests are in understanding the fundamental structure-property relationships of gas sorption and transport in both polymers and organic/inorganic composite materials. These interests interface between chemical engineering, chemistry, and materials science, where fundamental research expertise is needed to synthesize and characterize novel polymers and nanomaterials, and applied research expertise is needed to characterize permeation, sorption, and diffusion rates of molecules through these diverse systems. Creating new polymeric and composite materials needed to revolutionize separation practices currently used in industry will remain a primary focus in my future research. Synthetic work will focus on designing new polymers and new MOF nanocrystals, and characterization work will focus broadly on gas separations and exploring membrane performance under aggressive, high-temperature, gas-phase vapor-vapor separations.

435270 Data Science and Omics Approaches for Network Biology

Abstract:

Biological cells are complex dynamic systems that exhibit both extreme robustness and fragility. Robustness can be characterized as a cell's ability to isolate perturbations locally, which is a useful feature in the context of drug discovery, as treatments need to be efficacious with minimal adverse side effects. On the other hand, cells also possess fragile points that can stimulate apoptoic (cell death) pathways, providing intuitive targets for anti-cancer therapy. My overall goal is to develop a strong engineering research platform to study these two system properties at a fundamental level and ultimately facilitate the discovery of novel drug targets in diseased mammalian cells. My research involves the integration of large-scale omics data with metabolic and signaling networks to quantify cellular dynamics in response to various stimuli. In particular, I plan to develop state-of-the-art experimental workflows for efficient metabolomics and proteomics data collection using LC/MS-MS (liquid chromatography-mass spectrometry). I also have a strong interest in applying 'Big Data Analytics' and parallel cloud computing for systems biology applications to ensure that the algorithms we develop are scalable with the increasing complexity of biological networks. As a primary application of my work, I plan to explore novel treatments for liver disease, specifically nonalcoholic fatty liver disease (NAFLD) and hepatocellular carcinoma (liver cancer).

435272 Membrane Materials and Transport Studies for Sustainable Water, Energy and Life Sciences

Abstract:

Membrane Materials and Transport Studies for Sustainable Water, Energy and Life Sciences

Ngoc Bui, *Lawrence Livermore National Laboratory, Livermore, CA*

2nd Year Postdoctoral Fellow

Research Interests and Future Direction:

Overcoming the emerging global scarcity of water, energy, and other natural resources (e.g., nutrients, metals, minerals, etc.) are grand challenges faced by humanity today. These critical resources are inextricably linked, and therefore, must be considered together as new technologies are developed. Membrane-based processes promise to provide more efficient and economical separation solutions which are then applicable at large scale for sustainable water and energy production. Recent advances in controlled fabrication and characterization of materials at nanoscale allow for creative designs of functional membranes with effective nanostructure and chemistry. Furthermore, they provide effective tools to unravel transport mechanisms of various molecules across different membrane nanostructures which are crucially important for the development of next-generation membranes with enhanced permselective performance at lower energetic and environmental costs.

In a big picture, my research interests lie in tackling the energy - water nexus by seeking a robust and cost-effective solution to produce water sustainably using renewable energy. Forward osmosis (FO) is a unique and emerging platform technology that may ultimately help address water, resource, and energy scarcity by enabling the harvesting of salinity gradients for electricity generation (pressure retarded osmosis), concentrating high-value dissolved solids for recovery and beneficial reuse (direct osmotic concentration), and converting saline waters to fresh water (FO). Such a versatile technology platform has inspired an increasing number of imaginative uses for both broad and niche applications. Challenges remain, however, as widespread adoption of this platform technology is hindered by the lack of an appropriately designed membrane. Therefore, I would like to spend efforts on innovating new membrane structures by interdisciplinarily bridging nanotechnology and membrane science for the next generation of membrane with high permselectivity, mechanical integrity, anti-fouling and scalability. Also, I am interested in leveraging the advanced characterization tools to study the correlations between membrane nanostructure and chemistry with the transport properties of chemical species across membranes which, in turn, offer opportunities for better predictive models and system designs with precise control to be proposed. I would like to unite the gathered knowledge from my master degree (*synthesizing activated/porous carbon nanofibers as efficient electrodes for energy storage devices*), my Ph.D. (*synthesizing thin-film composite membranes supported on nanofiber-related materials for sustainable water and energy production via forward osmosis; and developing osmotic transport model across the semipermeable membrane*) and my postdoc (*developing robust vertically-aligned carbon nanotube membranes for breathable and protective fabrics; and unraveling the transport mechanisms through nanoconfinement*) as a toolkit in further exploring my direction. Furthermore, I am also interested in learning and exploring more niche applications (in drug delivery, protective fabrics, wound dressings, biomolecule separation devices, adsorption, bio-inspired materials and so on) for functionalized nanomaterials (nanofibers, nanotubes) and their composites. Based on the proposals that I have assembled, I foresee myself collaborating with other experimentalists, the simulation teams and industry.

Proposals: LDRD-Labwide

Postdoctoral Project: "*Breathable and Protective Fabrics based on Vertically-aligned Carbon Nanotube Membrane*". Under supervision of Dr. Kuang Jen Wu and Dr. Francesco Fornasiero, Physical and Life Sciences, Lawrence Livermore National Laboratory, Livermore, CA.

-

PhD Dissertation: "*Nanofiber-supported Thin-film Composite Membrane Synthesis and Transport Study for Forward Osmosis*". Under supervision of Professor Jeffrey R. McCutcheon, Chemical and Biomolecular Engineering Department, University of Connecticut, Storrs, CT.

-

Research Experience:

· Pioneered in the development of a new and effective generation of nanofiber-supported thin-film composite membrane platform for sustainable water and energy production by harnessing salinity-gradient energy *via* engineered osmosis (forward osmosis and pressure-retarded osmosis).

- Synthesized submicron-thick polyamide film onto highly porous nanofibrous web.

- Studied the correlation between membrane structures and properties (e.g. functionalized nanofiber structure, interconnected porous structure, pore size distribution) with their performances (e.g. water permeability, osmotic water flux, water/salt selectivity, power density harnessed from salinity-gradient energy).

- Characterization/testing techniques: FE-SEM, TEM, thermal analyses, tensile strength, contact angle, FT-IR, BET specific surface area, reverse osmosis, microfiltration, forward osmosis, pressure-retarded osmosis.

· Expertise in electrospun nanofiber

- Developed various nanofibrous support platforms for thin-film composite membrane from different polymers, solvents and nanocomposite.

- Controlled nanofibers structures and properties (e.g. mechanical properties, fiber surface porosity/roughness *via* electrospinning process control).

· Integrated expertise in (activated) carbon nanofiber synthesis with their potential in water treatment and sustainable energy production.

- Introduced concept and skillset learnt from South Korea to the new laboratory at UCONN, indirectly leading to several funding awards from National Science Foundation and others in which carbon nanofiber is the core material.

- Developed synthesis procedure and preliminary results as well as trained lab-members on fabricating activated carbon nanofiber (received *student proposal award* for *Multidisciplinary Environmental Research Proposal* by UCONN Center for Environmental Sciences and Engineering (CESE), PI: Dr. Jeffrey McCutcheon, Co-PI: Dr. Baikun Li).

· Developed a new mathematical osmotic transport model.

· Systems design

- Designed and custom-built electrospinning systems and trained lab-members on nanofibers fabrication.

- Custom-built cross-flow forward osmosis system (working with another lab member).

Master Dissertation: *"Activated Carbon Nanofibers as Electrodes for Supercapacitors"*

Chonnam National University, *Gwangju, S. Korea.*

Research Experience:

· Developed high specific surface area activated carbon nanofibrous electrode for supercapacitor (aim is to increase energy density and specific capacitance of supercapacitors by controlling the porous structure and specific surface area of carbon nanofiber electrode derived from decorated polymeric nanofibers).

· Studied electrochemical performance of porous carbon fiber-based electrodes.

· Characterization/Testing techniques: FE-SEM, TEM, BET specific surface area, Nitrogen adsorption, thermal analyses, impedance spectroscopy, cyclic voltammogram, charge - discharge behaviors.

Teaching Experience:

Guest Lecturer – University of Connecticut *Storrs, CT*

CHEG 5395 – Investigation of Special Topics – Membrane Separations Winter 2011, Winter 2012, Fall 2013

Teaching Assistant – University of Connecticut *Storrs, CT* ENGR 1166, Foundations of Engineering Spring 2009

Teaching Assistant – University of Connecticut *Storrs, CT* CHEG 4147, Introduction to Process Dynamics and Control Fall 2008

Teaching Assistant – University of Connecticut *Storrs, CT* CHEG 3123, Transfer Operations I Fall 2008

Selected Publications:

PATENTS

1. McCutcheon, J.R., **Bui, N.N**, Hoek, E.M.V., Lind, M.L, Nano-Structured Membranes for Engineered Osmosis Applications, WO 2011/060202 A1, Application Number PCT/US2010/056426.

2. Kim C., Balathanigaimani M.S., **Bui, N.N.**, Moon H., Yang K.S., Lee J.W., Activated carbon with ultra-high specific surface area from corn and Production method thereof, KO 10-2008-0085605, application number 10-2007-0027360.

BOOK CHAPTER(S)

McCutcheon, J.R., **Bui, N.N**: *Forward Osmosis*, in "*Desalination: Water from Water*", Ed. Jane Kucera.

JOURNAL PUBLICATIONS (H-Index: 10, Total citations: 548)

1. **Bui N.N.**, McCutcheon J.R., Nanoparticle-Embedded Nanofibers in High Permselectivity Thin-film Composite Membranes for Forward Osmosis, *ACS Applied Materials & Interfaces*, under revision.

2. **Bui N.N.**, Arena J., McCutcheon J.R., Proper Accounting of Mass Transfer Resistances in Forward and Pressure Retarded Osmosis: Improving the Accuracy of Model Predictions of Structural Parameter, *Journal of Membrane Science*, in press.

3. **Bui N.N.**, McCutcheon J.R., Impacts of Nanofiber Size on Osmotic Flux Performance of Thin-film Composite Membrane for Engineered Osmosis (in preparation).

4. **Bui N.N.,** McCutcheon J.R., Nanofiber Supported Thin-film Composite Membrane for Pressure Retarded Osmosis, *Environmental Science & Technology*, 48, 2014, 4129-36.

5. **Bui N.N.,** McCutcheon J.R., Hydrophilic Nanofibers as New Supports for Thin Film Composite Membranes for Engineered Osmosis, *Environmental Science & Technology*, 47, 2013, 1761-9.

6. **Bui N.N.,** Lind M.L., Hoek E.M.V., McCutcheon J.R., Electrospun Nanofiber Supported Thin Film Composite Membranes for Engineered Osmosis, *Journal of Membrane Science*, 385-386, 2011, 10-19.

7. Huang L., **Bui N.N.**, Meyering M.T., Hamlin T.J., McCutcheon J.R., Novel Hydrophilic Nylon 6,6 Microfiltration Membrane Supported Thin Film Composite Membranes for Engineered Osmosis, *Journal of Membrane Science*, 437, 2013, 141-9.

8. Huang L., **Bui N.N.**, Manickam S.S., McCutcheon J.R., Controlling Electrospun Nanofiber Morphology and Mechanical Properties using Humidity, *Journal of Polymer Science Part B: Polymer Physics*, 49, 2011, 1734-44.

9. Manickam S.S., Karra U., Huang L., **Bui N.N.**, Li B., McCutcheon J.R., Activated Carbon Nanofiber Anodes for Microbial Fuel Cells, *Carbon*, 53, 2013, 19-28.

10. **Bui N.N.**, Kim B.H., Yang K.S., Dela Cruz M.E., Ferraris J.P., Activated Carbon Fibers from Electrospinning of Polyacrylonitrile/Pitch Blends, *Carbon*, 47, 2009, 2538-9.

11. Kim B.H., **Bui N.N.**, Yang K.S., Dela Cruz M.E., Ferraris J.P., Electrochemical Properties of Activated Polyacrylonitrile/Pitch Carbon Fibers Produced Using Electrospinning, *Bull. Korean Chem. Soc.*, 30, 2009, 1967-72.

12. Kim C., **Bui N.N.**, Yang K.S., Kojima M., Kim Y.A., Kim Y.J., Endo M., Yang S.C., Self-Sustained Thin Webs Consisting of Porous Carbon Nanofibers for Supercapacitors via the Electrospinning of Polyacrylonitrile Solutions Containing Zinc Chloride, *Advanced Materials*, 19, 2007, 2341-6.

13. Kim C., Jeong Y.I., **Bui N.N.**, Yang K.S., Kojima M., Kim Y.A., Endo M., Lee J.W., Synthesis and Characterization of Porous Carbon Nanofibers with Hollow Cores through the Thermal Treatment of Electrospun Copolymeric Nanofiber Webs, *Small*, 3, 2007, 91-95.

14. Kim C., Cho Y.J., Yun W.Y., **Bui N.N.** , Yang K.S. , Chang D.R., Lee J.W., Kojima M., Kim Y.A., Endo M., Fabrications and structural characterization of ultra-fine carbon fibres by electrospinning of polymer blends, *Solid State Communications*, 142, 2007, 20-23.

15. Cho Y.J., Kim C., **Bui N.N.**, Yun W.Y., Yang K.S., Fabrication of Electrospun Carbon Nanocomposite Fibers from PAN and PAA Blended Solutions, *Carbon Science*, 8, 2007, 49-51.

435393 Programmable Dynamic Materials As Information Carriers

Abstract:

Interfacial properties dictate interactions and communications of every existing specie with its surrounding environment. Controlling interfacial chemistry of materials and its response to external stimuli such as light, stress, humidity, temperature, electric or magnetic field, pH, or ionic strength can be a pathway to attain unique functionalities. Nature provides abundant examples of processes that are triggered by interfacial stimulants. Interfacial dynamic response to the external stimuli could lead energetic and conformational transformation or might yield highly controlled material fabrication. Examples are plentiful; some instances include: plants capable of dynamically reform their conformation by light or subtle physical forces, animals that harmonize their skin color with their environment, human eyes pupils' expansion and dilation with light, hierarchical microorganisms formed under perfect control of layers' interfacial chemistry, and many more.

My research theme is to design systems with controllable dynamics that are responsive to external stimuli in order to deliver work, material, or information of any kind. Once stimulus is applied to a material including segments with sensory receptor, a reflex occurs via energetic transformation from one state to another. This transducing between certain eigenstates can then be transformed into a cascade of events to elicit desired functionalities.

The key to successful design of smart systems is the suitable choice of a structure core that can undergo conformational variation upon applying external trigger. Grafting head groups with contrasting properties can then lead to achieve dynamically interchangeable characteristics such as electric charge, wettability, specific affinity in order to transfer matter (at different scales), capture and release ions, for self-cleaning purposes, and many more. The energetic stability of transferrable eigenstates and the hysteresis will then determine the efficiency of these systems.

There are many possibilities to design smart systems with increasingly fascinating capabilities. Few from many include: smart membranes made of nano-valves that allow unidirectional transfer of molecules or ions "on-demand", charge inversion of colloids, smart polymeric material that are sensitive to external chemical composition usable for simple medical diagnosis, and touch-sensitive polymers applicable as artificial skin.

For all above, I will apply theoretical modeling and simulations, thermodynamics, and statistical mechanics approaches to understand surface-dynamic phenomena at the level of fundamental thermodynamics. Nonequilibrium actuation, characteristic response times, and energetic efficiencies will be studied by theoretical and simulation techniques.

430074 Accelerating Ring-Polymer Molecular Dynamics Simulation: A Parallel-Replica Dynamics Approach

Tuesday, November 10, 2015: 9:40 AM
255A (Salt Palace Convention Center)

Chun-Yaung Lu, Energy Resources Engineering, Stanford University, Stanford, CA, Arthur F. Voter, Theoretical Division, Los Alamos National Laboratory, Los Alamos, NM and Danny Perez, Los Alamos National Laboratory, Los Alamos, NM

Classical molecular dynamics (MD) method is a powerful tool for simulating materials on the atomic scale, which can be used to compute a wide range of equilibrium and dynamical properties, such as elastic moduli, diffusivity, free energy barrier and reaction rate. However, classical MD neglects quantum mechanical nuclear zero-point energy and the nuclear tunneling effects. For systems containing light elements at low temperatures, quantum corrections are necessary in order to obtain a quantitative correct description of the system.

While solving the full time-dependent Shrodinger equation is usually limited to small systems, there are several approximate methods that have been developed for dealing with those complex, many-body systems. Among them, the ring-polymer molecular dynamics (RPMD) method is a popular semi-classical approach to include quantum nuclear effect in a classical MD simulation. RPMD is based on the path-integral formalism of statistical mechanics, which yields real-time MD trajectories that preserve the exact quantum Boltzmann distribution. It has been demonstrated in several systems that RPMD method does very well when used to calculated thermal reaction rates.

A long-standing problem is that conventional MD is limited to relatively short simulation timescales, which leaves many important phenomena out of reach. These so-called activated processes are typically characterized by long periods of uneventful vibrational dynamics punctuated by rapid transitions. While very powerful, RPMD suffers from an even more severe timescale problem as trajectories are generated not on a single system, but on a ring polymer where each bead is a complete replica of the system.

In classical MD, accelerated molecular dynamics (AMD) methods address this challenge by concentrating on the sequence of transitions and shortening the inter-transition intervals. AMD methods require no apriori knowledge of the possible transitions, but to modify the dynamics so that transitions occur faster. In the present work, RPMD was coupled with Parallel Replica Dynamics (ParRep) method to accelerate the simulation of systems containing light atoms at low-temperatures. An example system of helium atom diffusing in the bulk Fe, W, and Fe-Cr alloy at various temperatures will be discussed.

(6ip) Colloids with Valence: Design, Fabrication, and Directed Self-Assembly

Wang, Y., Massachusetts Institute of Technology

Self-assembly of colloidal particles has emerged as a promising strategy to construct micrometer-scale three-dimensional ordered composite materials such as photonic crystals. Meanwhile, colloidal particles can serve as model building blocks to study the self-assembling processes of molecular systems. However, the lack of directional interactions between isotropic particles has imposed much limitation on the study of colloidal self-assembly. We design new colloidal particles, *colloids with valence*, which possess reversible, specific, and directional interactions with well-defined symmetries. The primary hypothesis is that particles exhibiting distinct surface patches can be fabricated (patchy particles). The patches can be functionalized with biological or synthetic macromolecules containing recognition units, and provide directional bonding interactions in particle self-assembly.

Based on this hypothesis, we show that patchy particles are synthesized by partially encapsulating small clusters of colloidal microspheres, leading to tailored patch size, geometries, and surface functionalities. Particles with amidinated patches are further functionalized with biotin molecules, to which DNA with single-stranded "sticky ends" are attached. Directed assembly of DNA patchy particles is demonstrated, by which a wide variety of colloidal structures mimicking molecular symmetries are formed. In addition, self-assembly of patchy colloids is viewed as chemical reactions and its kinetics has been studied. The assembling process follows a stepwise-sequential mechanism, as can be described by a model based on diffusion controlled reactions. Apart from using biological molecules and DNA hybridization, a synthetic tri-block copolymer bearing supramolecular recognition moieties are synthesized and functionalized onto carboxyl patchy particles. These surface-modified patchy colloids spontaneously form larger structures such as daisy chains induced by palladated-pincer-pyridine metal coordination. Due to the valence characteristics, patchy colloids introduced will open great opportunities to build a spectacular spectrum of functional materials as well as to study self-assembling processes in more complicated molecular and biological systems.

436317 Nano-Engineered Functional Materials for Energy Storage and Biomimetic Applications

Abstract:

Nano-Engineered Functional Materials for Energy Storage and Biomimetic Applications

Samanvaya Srivastava

Institute for Molecular Engineering, The University of Chicago, Chicago, IL 60637, USA

samsri@uchicago.edu

A multitude of novel building blocks, including macromolecules and nanoparticles of myriad shapes and sizes, are now available to researchers owing to recent progress in material synthesis and fabrication techniques. However, as technology moves towards smarter materials and complex multifunctional devices, the need for understanding and controlling the interactions and assembly of these mesoscale units also becomes increasingly important. Driven by this motivation, my proposed research program will aspire to employ fundamental approaches from colloid, macromolecular and soft matter sciences to understand and tailor the physical processes that govern the structure-property relationships in these material building blocks and their assemblies. These studies will provide the groundwork for the overarching theme of my research, which is to design bottom-up approaches for assembly of mesoscopic building blocks into nano-engineered materials with *on-demand* properties.

My graduate research with Prof. Lynden Archer at Cornell University primarily focused on developing strategies for synthesis of inorganic-organic hybrid materials with tunable properties. Specifically, I developed strategies for controlling nanoparticle dispersion in polymer hosts, and consequentially synthesizing nanoparticle-polymer composites with desired mechanical and electrochemical properties. Further, in a collaborative effort with Lipson group at Cornell University, I worked on the development of functional inks for unique direct-write applications, including 3-D printed electronic and energy storage devices.

My current research as a postdoctoral researcher with Prof. Matthew Tirrell at The University of Chicago and Argonne National Laboratory involves designing and synthesizing block copolyelectrolytes that self-assemble via electrostatic interactions into polyelectrolyte complex (PEC) based micelles and hydrogels. Our investigations reveal that PEC based hydrogel morphologies span a wide range of features and are very sensitive to external stimuli, thus allowing for facile tuning of the hydrogel properties. I am also investigating the enhancements in the protein activity upon encapsulation in polyelectrolyte complexes. These studies, in collaboration with the Gardel group at The University of Chicago, are particularly important for intelligent design of PEC based micelles and hydrogels as drug carriers and tissue growth scaffolds.

My goal is to establish an independent research program that aims at designing novel bottom-up approaches for self-assembled material. Building on my previous work and harnessing my expertise in soft matter physics, I plan to pursue three area of research that are set in a similar fundamental landscape, but will serve entirely different realms of applied research. The research themes that I intend to pursue are: 1) Developing new material platforms for redox flow batteries, which employ electroactive complex fluids that are flown in a parallel flow setup during the charging/discharging process and stored in reservoir tanks otherwise, 2) Development of smart inks for additive fabrication (also known as 3-D printing) of responsive materials and integrated devices, and 3) Synthesizing surface-modified nanoparticle based hydrogels that will have superior tunability, stimuli responsiveness and tissue supporting functionalities.

Selected Publications (12 total, 2 in review):

1. R. Mangal, **S. Srivastava**, and L. A. Archer, Phase stability and dynamics of entangled polymer-nanoparticle composites, *Nature Communications* (2015), in print.

2. **S. Srivastava**, J. L. Schaefer, Z. Yang, Z. Tu and L. A. Archer, 25th Anniversary Article: Polymer–Particle Composites: Phase Stability and Applications in Electrochemical Energy Storage, *Advanced Materials* **26**, 201 (2014).

3. **S. Srivastava**, S. Narayanan and L. A. Archer, Structure and Transport Anomalies in Soft Colloids, *Physical Review Letters* **110**, 148302 (2013).

4. **S. Srivastava**, P. Agarwal and L. A. Archer, Tethered Nanoparticle–Polymer Composites: Phase Stability and Curvature, *Langmuir* **28**, 6276 (2012).

5. **S. Srivastava**, J. H. Shin and L. A. Archer, Structure and Rheology of Nanoparticle–Polymer Suspensions, *Soft Matter* **8**, 4097 (2012).

6. P. Agarwal, **S. Srivastava** and L. A. Archer, Thermal Jamming of a Colloidal Glass, *Physical Review Letters* **107**, 268302 (2011)

For more information, please visit: http://tirrell.ime.uchicago.edu/samsri

436467 Novel Routes to the Synthesis of Fuels/Lubricants and Chemicals from Biomass Derived Synthons

Abstract:

As the world's accessible oil reservoirs are gradually depleted, it is important to develop suitable long-term strategies based on the utilization of renewable fuel that will gradually substitute the declining fossil fuel production. In this context, lignocellulosic biomass would ideally supply fuel/chemical markets and be fully compatible with existing petroleum infrastructure. Biological transformations, though efficient, typically produce bio-fuels like ethanol or carboxylic acid esters that can only be used at limited blend levels and integrate poorly with current fuel handling. Thus the development of new catalysts and methods is essential to move the energy related research forward. We recently reported different chemical routes to convert lignocellulosic products from a variety of renewable carbohydrate sources into hydrocarbons that can be used for gasoline, jet fuel, and diesel. As most natural fermentation products have lower carbon numbers coupling chemistry is needed to design larger molecules, ideally achieved by exploiting the functionalities present in the starting materials. We have shown that dehydrogenation reactions of fermented primary and secondary alcohols or ABE mixture can be readily converted to carbonyl compounds which on subsequent aldol-type condensation provides an excellent procedure for increasing the carbon number and decreasing the O/C ratio of bio-derived molecules through removal of oxygen as water. For instance, we show that the secondary amine-functionalized silica can be used in the very selective dimerization of aldehydes and methyl ketones to produce molecules that are components of jet-diesel fuel and lubricants. The catalytic activity of secondary amine supported silica sample can be altered significantly by controlling the Bronsted acidity of M-OH (M = Al, Ti, Zr, Sn) species present on the support. These least oxygenated products were subjected to mild hydrodeoxygenation conditions in the subsequent step, in order to remove unsaturation as well as the oxygen atom make them as usable fuels.

436488 Organ-on-a-Chip Platforms to Mimic Physiology for Drug Screening

Abstract:

The liver is a central metabolizing organ and is susceptible to damage by chemicals and/or their metabolites that enter the body. *In vitro* liver platforms are essential tools that can provide insights for understanding drug metabolism, metabolite generation and resulting toxicity. While relevant, most of the successful liver platforms incorporate hepatocytes alone, thus lacking organ specific metabolism – such as contribution of other, non-parenchymal cells and ultimately can lead to their failure to predict drug induced liver injury. I am interested in developing liver-on-chip platforms that not only incorporates non-parenchymal cells, but also recapitulates the function of the liver effectively.

The proposed research builds on the recent developments that enable multi-cell culture of primary hepatocytes with isolated non-parenchymal cells to enable development of *in vitro* liver models that focus on multiple cell types. I have developed a novel method to isolate and culture primary hepatocytes with primary liver endothelial cells [1], showing the favorable outcomes of cell-cell contacts for long-term stability. I have developed a novel approach that dramatically simplifies evaluation of liver communication with other organs in eliciting drug efficacy and toxicity [2]. This approach relies on capturing organ-organ interactions in a simple, static system that overcomes dilution effects that are present in the current interrogation systems. These findings provide a handle on an addressing an important question: what are the parameters essential for developing a functional *in vitro* liver platform that are essential to evaluate organ and organ-organ interactions? My research program will focus on the development of microscale platforms to integrate with other downstream organ units in a functional way. For example, is it possible to mimic the metabolic and inflammatory response to cytokine stimuli and evaluate its effect on an organ downstream (such as muscle). The ultimate objectives of this research is to design functional and physiologically relevant organ-on-chip models, to understand drug metabolism, toxicity and enable in development of effective therapeutic screening methods.

References:

1. Bale, S.S., et al., *Long Term Co-culture Strategies for Primary Hepatocytes and Liver Sinusoidal Endothelial Cells.* Tissue engineering. Part C, Methods, 2014.

2. Bale, S.S., et al., *A novel low-volume two-chamber microfabricated platform for evaluating drug metabolism and toxicity.* TECHNOLOGY. **0**(0): p. 1-8.

436108 Efficient Accumulation of Carbonhydrate in Microalgae and It's Utilization

Abstract:

Efficient Accumulation of Carbonhydrate in Microalgae and It's Utilization

Xu Jingliang, Guo Ying Zhou Weizheng Li Xiekun Yuan Zhenhong

Key Laboratory of Renewable Energy, Guangzhou Institute of Energy Conversion, Chinese Academy of Sciences, Guangzhou 510640, P.R. China

xjl@ms.giec.ac.cn http://people.ucas.ac.cn/~xjl?language=en

Abstract: As efficient photosynthetic microorganism, microalgae can convert solar energy and CO_2 into biomass, such as starch and cellulose, which is the desired feedstock for bioethanol production. As bioethanol feedstock, microalgae bioethanol is known as the third generation bioethanol because of many advantages, such as high photosynthetic efficiency, no farmland occupation, can be cultivated during all year, can absorb CO_2, high carbohydrate content and easy to be degraded. In present work, a high carbohydrate content microalgae strain was identified. Nitrogen stress was performed on microalgae cultivation. Simulated flue gas was used as carbon source to study the effect of flue gas on microalgae cell components. Liquid hot water was applied in the microalgae pretreatment process to increase the enzyme hydrolysis efficiency.

Based on the morphological characteristics and phylogenetic analysis, microalgae WZKMT was identified as *Scenedesmus raciborskii* WZKMT. Five BG11 cultures with different $NaNO_3$ concentrations (0, 0.5, 1.0, 1.5, 2.0 g L^-

[1]) were designed for microalgae culture, biomass yield, total sugar and starch content was determined every 24h. The results showed that the growth of microalgae was inhibited severely under nitrogen-free condition, while there was no difference on biomass yield under other four nitrogen concentrations, the maximum of which can attain to 4.0 g L^{-1}. The content of starch under nitrogen deficiency condition (0 g L^{-1}, 0.5 g L^{-1}) was much higher than that under the other three nitrogen conditions in the first 6 days. After day 8, the starch content showed no obvious distinction under all nitrogen conditions, which indicated that starch accumulated as short-term stress response under nitrogen deficiency condition. For long-term cultivation, the nitrogen content in culture can be economized to reduce cost. The total sugar and starch content of WZKMT can attain to 68.95% and 56.27%, respectively and is the highest starch content report in current reported articles, which indicates that *S. raciborskii* WZKMT is a desired feedstock candidate for bioethanol production.

Still the growth parameters, biomass yield, total sugar, starch, lipid composition, protein, chlorophyll of *Scenedesmus raciborskii* WZKMT cultivated with simulated flue gas (15% of CO_2, 200 ppm SO_2, 100 ppm NO with N_2 balance) and 7% CO_2 (as control) were investigated to study the effects of flue gas on the cellular components. Dissolution of SO_2 and NO in simulated flue gas led to pH decrease and toxicity to microalgae cells, resulting in the decrease of buffer capacity of culture. Microalgae growth and starch accumulation was inhibited. with simulated flue gas, the maximum biomass yield and starch content were 2.14 g/L and 36.23% while with 7% CO_2, those were 3.25 g/L and 53.16%, respectively. Microalgae cells produced more chlorophyll, protein and lipid when simulated flue gas was used as the carbon source. Fatty acids composition analysis indicated that there was no significant distinction on fatty acids relative content (fatty acid/TFA) between cells aerated using simulated flue gas and 7% CO_2.

Liquid hot water (LHW) pretreatment on the carbonhydrate-rich microalgae *S. raciborskii* WZKMT was performed to enhance glucose recovery and the glucose concentration in enzymatic hydrolysate was determined to study the effect of LHW pretreatment on enzyme hydrolysis. Temperature, time, ratio of solid to liquid in pretreatment process were analyzed. The optimal temperature (140), time (40 min), the ratio of solid to liquid (1:15) were determined by single-factor experiments. The optimization pretreatment conditions analyzed by response surface methodology were as follows: ratio of liquid to solid 13:1, temperature 147 , heating time 41 min. Under the conditions, the glucose concentration was 14.223 g/L, the glucose recovery was 89.32%, which was up to 5-fold higher than the one of the sample without LHW pretreatment (17.91%). SEM images of microalgae after LHW pretreatment showed that cells were broken and fibre bundles on cell surface were more exposed.

Key Words: microalgae, nitrogen stress, simulated flue gas, liquid hot water, starch

436115 Engineering Non-Model Eukaryotes for the Production of Sustainable Fuels, Chemicals, and Pharmaceuticals

Abstract:

The production of fuels, bulk chemicals, pharmaceuticals, and nutraceuticals in an environmentally sustainable manner and at low costs is one of the greatest challenges of this century. Biological production of these molecules can bring many benefits over the use of petroleum and other fossil fuels as feedstocks. My independent lab will focus on developing novel biological solutions for the production of bioproducts using non-model biological host systems. We will develop methodologies, tools, and computational approaches to engineer these biological systems to harness their potential for solutions to some of society's most pressing issues, including developing sustainable fuels, chemicals, and low-cost pharmaceutical precursors.

My research program will have three broad goals: (i) develop a high-throughput, robotics based synthetic biology platform for non-model, commercially relevant host organisms (i.e., eukaryotes such as algae or fungi), (ii) create a bioinformatic data mining pipeline based on machine learning for the discovery of novel enzymes and metabolic pathways, which can be quickly tested in our high-throughput synthetic biology pipeline developed in Aim I, and (iii) apply these tools (Aim I & II) to answer the most pressing engineering and scientific questions, such as the discovery of the enzymes responsible for the biosynthesis of the biopolymer algaenan, a precursor to petroleum, or to find novel plant enzymes to produce drug precursors that cannot be sustainably harvested.

With this presentation I will highlight some of my past research accomplishments and a selection of research directions I would like to pursue. Specifically, (i) results from my research into engineering algae for biofuel production (published in *Nature Communications*, *Eukaryotic Cell*, *Bioengineered*), (ii) examples of my high-throughput synthetic biology and bioinformatic discovery platforms, and (iii) specific research questions to address with these tools, including discovery of the enzymes that created the biopolymer precursors to petroleum, a pipeline for screening plant enzymes for secondary metabolite production, and the high-throughput creation of gene knock out libraries in non-model eukaryotes.

INCORPORATION OF METAL OXIDES TO ACTIVATED CARBON FOR THE ADSORPTION OF ACID GASES

Jose Luis Altamirano-Corona, Ma. del Carmen Chávez-Parga, Horacio González-Rodríguez, Ma. Aida Béjar-Ubaldo, Jaime Espino-Valencia. Universidad Michoacana de San Nicolás de Hidalgo, Morelia, México.

Keywords: Biofuels (Energy), Carbon Management, Materials.

Abstract

The incorporation of metal oxides to activated carbon was proposed to increase the adsorption capacity of these materials. An attractive option to remove acid gases, which may be less expensive, is the use of activated carbon modified with basic oxides (CaO, MgO and ZnO), which could increase the basic properties of the adsorbents, thus increasing their adsorption capacity towards acid gases such as CO_2, H_2S, NO_X, VOCs. In this study the incorporation of basic oxides to activated carbon through an *in situ* method is proposed. The synthesized materials were characterized by XRD, IR spectroscopy, N_2 physisorption, SEM and particle size determination. Materials with a hexagonal structure were predominant, which exhibited a good distribution of the oxides on the activated carbon, functional groups, like alkoxy and hydroxyl, capable to react with acid gases on the surface, were identified and particle sizes were homogeneous in all synthetized materials after washing.

Introduction

Gas adsorption on porous solids is commonly used in many processes for removal of polluting gases. In the particular case of acid gases like CO_2, H_2S, NOx, VOCs, a selective adsorbent with basic properties is usually necessary for improving their adsorption capacity. Therefore, the modification of adsorbents such as activated carbon [1], through the incorporation of basic oxides like CaO [2], [3], MgO [4], and ZnO [5], could increase the adsorption capacity of these materials, being relatively inexpensive their synthesis.

The presence of basic sites on the surface of the adsorbent [6], is essential for the adsorption of acid gases, so the addition of hydroxides, oxides or amphoteric materials to supports like activated seems to be an attractive option for the adsorption of acidic gases [7], [8]. In addition, the incorporation of more than one type of oxide, may give to the new material removal capacity of a large amount of acid gases and avoid problems of sintering [9], [10].

In this work the simultaneous addition of three basic oxides (CaO, MgO and ZnO) to activated carbon, through an in-situ synthesis, is proposed. Therefore, in a single thermal process the alkaline carbonaceous support and basic adsorption sites are obtained.

Calcium oxide has been used both in water and gas treatment, since it has the ability to react with a lot of acid compounds at ambient conditions. Furthermore, it has been reported that at higher temperatures, CaO also interacts with gases such as CO_2. It has been also reported that sintering after each adsorption-desorption cycle also decreases when the CaO is mixed with magnesium oxide [9]. On the other hand, zinc oxide has showed good properties for the adsorption of compounds containing nitrogen (NO_X) and other gaseous contaminants [11], [12], [13]. Magnesium oxide has been used for acid gas removal in the presence of sulfur [14]. In the literature are mentioned materials from precursors of carbon and oxides of calcium, magnesium, zinc, iron, nickel, etc. [15], [16], [17], [18], in which a single type oxide predominates and they are incorporated to support material usually by the impregnation method. Therefore, the simultaneous incorporation of oxides (calcium, magnesium and zinc) to activated carbon, may offer the opportunity to obtain materials with chemical and structural characteristics to perform the adsorption of acid gases after several cycles of adsorption-desorption

In this work the synthesis of activated carbon from coconut endocarp, modified in-situ with CaO, MgO and ZnO was carried out. The resulting materials were characterized through XRD, FT-IR, particle size analysis, scanning electron microscopy and N_2 adsorption. The most abundant compounds in the materials and their crystalline structure and morphology, was studied along with the identification of the main functional groups on the surface, and finally these results were associated with the adsorption capacity of the materials.

Experimental

Materials Synthesis

Coconut endocarp powder with an average particle size of 42 micrometers was used as the source of carbon. This material was heated at 115 °C for 12 hours to remove the moisture. Calcium acetate, magnesium acetate and zinc acetate were used as precursors during the synthesis of the adsorbents. For washing, distilled water was used and the carbonization reaction was carried out in carbon steel tubes with screw type lid.

Five coal samples were prepared following a standard procedure [19]; according to the mass ratios shown in Table 1. The mixed powders were introduced into the reactor and treated during four hours by carbonization at 800 °C. Then the samples were allowed to cool within the tubes to room temperature, the mixtures were sheltered in sealed containers to avoid humidification. A fraction of each sample was removed to carry out washing using 100 ml of distilled water per 3 grams of carbon formed on a filter paper. In the case of the sample with the three oxide precursors, washing was performed in aqueous solution with stirring for 1 hour in order to determine if the washing procedure provides significant changes. The washed samples were dried for 12 hours at 115 °C and protected in sealed containers.

Table 1. Samples and their mass ratios.

Samples	Precursors	Mass ratios for the experiment
Simple Activated Carbon	Coconut–NaOH	1:2
Carbon–$Ca(Ac)_2$	Coconut–$Ca(CH_3COO)_2$	1:0.5
Carbon–$Zn(Ac)_2$	Coconut–$Zn(CH_3COO)_2$	1:0.5
Carbon–$Ca(Ac)_2$–$Zn(Ac)_2$	Coconut–$Ca(CH_3COO)_2$–$Zn(CH_3COO)_2$	1:0.25:0.25
Carbon–$Ca(Ac)_2$–$Mg(Ac)_2$-$Zn(Ac)_2$	Coconut–$Ca(CH_3COO)_2$–$Mg(CH_3COO)_2$–$Zn(CH_3COO)_2$	1:0.1:0.1:0.1

**The simple activated carbon sample was prepared with a weight proportion of 1: 2 of coconut powder and sodium hydroxide. After the carbonization, the sodium was removed with a solution of HCl until pH=7 were reach and finally washed with distilled water and dried at 115 ° C for 12 hours, according to the methodology proposed by Béjar-Ubaldo and Rodriguez-Flores (2010), [19].

Characterization Stage

Samples were observed using a metallographic microscope (Van Guard) to appreciate some macroscopic characteristics. These materials were also characterized by scanning electron microscopy with a JEOL JSM-5910L equipment using the backscattered technique, to analyze the shape and distribution of particles formed on activated carbon and the overall morphology.

The crystalline phases present in the materials were analyzed through powder X-ray diffraction using a Bruker D8 Advance Davinci apparatus. The identification of the main functional groups present in the materials, was carried out in an infrared Bruker Tensor 27 equipment, carrying out the determination of percent transmittance and attenuated total reflectance (ATR).

The textural properties of the materials were determined through N_2 adsorption experiments and the surface area quantification was performed using the single-point method and the BET model. A Horiba 9600 equipment was used in these experiments with a gas flow consisted of 30% helium and 70% nitrogen. Particle size was determined in a Coulter LS100Q equipment.

Results and discussion

A metallographic microscope Vanguard with 4X of magnification, was used to observe the main characteristics of the washed and unwashed samples. In general the samples showed metallic clusters, small granule forms and some non-definite structures. Particularly those which were prepared with zinc acetate. Materials containing only calcium precursor were fragile to the touch, had a less shiny grayish color, while those containing the three precursors showed similar characteristics. The sample of activated carbon showed a shiny black color in a finer powder form.

Mass balance

In Table 2, the amounts of compounds used for the synthesis of the materials are presented, and the mass yield of the product is shown. The mass yield is the ratio of the mass of producing activated carbon and the desired theoretical mass, which is equal to the initial mass. Comparing with other activated carbons, slightly higher yields were obtained to those reported in the case of coal without metal precursors and within typical for precursors containing Ca, Mg and Zn, [20], [21], [22].

Table 2. Mass balance and mass yields.

Samples	Oxide Precursor	Total Initial Mass	Final Mass	Metal Mass	Mass Yield
Simple Activated Carbon	***	35 g	21.4 g	***	61.1 %
Carbon–$Ca(Ac)_2$	30 g	60 g	25.6 g	7.6 g	42.7 %
Carbon–$Zn(Ac)_2$	20 g	40 g	14.1 g	7.1 g	35.3 %
Carbon–$Ca(Ac)_2$–$Zn(Ac)_2$	5 g – 5.8 g	20.8 g	7.9 g	1.3 g – 2.1 g	38.0 %
Carbon–$Ca(Ac)_2$–$Mg(Ac)_2$-$Zn(Ac)_2$	3.9 g – 3.9 g – 3.9	51.1 g	16.5 g	0.99 g – 1.39 g – 0.67 g	32.3 %

*** 70 g of NaOH were used with 35 g of coconut powder, giving a final carbon mass of 21.4 g after neutralization with 300 mL of 36.6% HCl and washed with distilled water.

Characterization by XRD

Powder XRD analysis was performed to all the synthesized materials. It is noteworthy that there is almost no formation of CaO (the amount not exceeding 4%) which is the limit of the equipment to identify materials. This indicates that it is necessary to raise the calcination temperature to free the CO_2 of the calcium carbonate and transform it to an oxide. The material synthesized with the three precursors (Ca, Mg and Zn) present cubic crystal structures of metallic magnesium and hexagonal crystals of zinc oxide, calcium carbonate, magnesium oxide and calcite based of calcium and magnesium ($Ca_{0.936}Mg_{0.064}O_3$), [23], [24], [25].

Compounds in major proportion in the materials emit visible signals in both samples, which are illustrated in Figure 1. Zinc formations are easily discernible through X-ray diffraction, but not so by SEM, which leads to think that the formed ZnO, is part of the internal structure of coal.

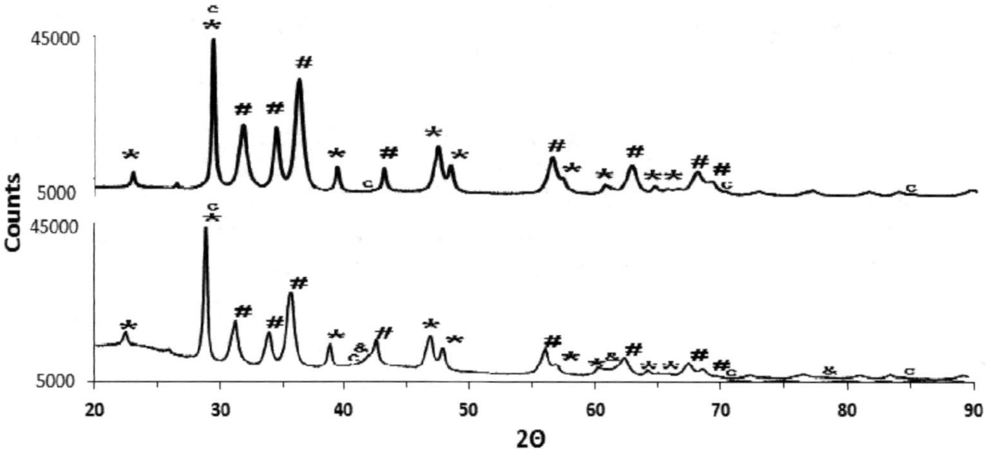

Figure 1. Comparison between Carbon-Ca-Mg-Zn and Carbon-Ca-Zn diffragtograms. c = carbon signal; * = $CaCO_3$ signal; # = ZnO signal; & = MgO signal.

Characterization by IR

Characterization of the synthesized materials through infrared spectroscopic analysis was also performed. It was possible to identify some functional groups characteristic of calcium carbonate, magnesium oxide and zinc oxide materials and some functional groups characteristic of the activated carbon.

The simple activated carbon was also analyzed by this technique and a commercial carbon was used as the reference, obtaining less intensity on the synthesized activated carbon signal in wavenumbers ~ 1422 cm^{-1}, related to groups C-H. Commercial carbon emitted a lower signal because of the vibration of the hydroxyl groups (~ 3500 cm^{-1}). Other functional groups, are identifiable in the simple activated carbon at wavenumbers of ~ 1225 cm^{-1} (CO ester; -CH2-), ~ 1584 cm^{-1} (C = C) and ~ 1691 cm^{-1} (C = O aldehydes and lactones).

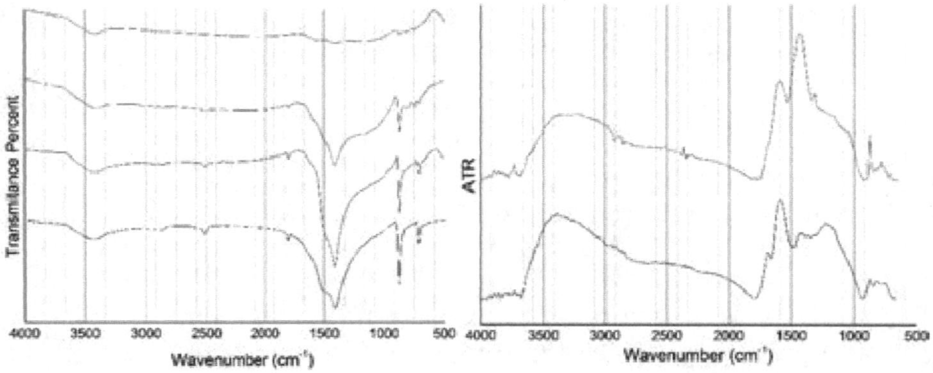

Figure 2. Infrared spectra measured for the activated carbons. Left graphic is about a comparison between the samples that contain metal precursors. Right graphic compare ATR signal of produced activated carbon without metal precursors and a commercial vegetal activated carbon (Meyer, lot K1112461).

For other samples, it can be seen that the spectra of each still much the same way, differing from them only on signal strength. The sample prepared only with zinc precursor, do not present the peak in the wave number ~ 1500 cm^{-1}, which is related to the calcium carbonate. It was possible to identify hydroxyl groups (~ 3500 cm^{-1}), carboxylic groups (~ 1790 cm^{-1}) and C-O-C alkoxy groups (~ 1100 cm^{-1}), presence of double and triple bonds carbon-carbon (~ 3000 and 2900 cm^{-1}, respectively), which are essential for the adsorption of gases. This indicates that it is possible the interaction between the surface and the acid gases, given the presence of metal-oxide formation and by the existence of basic functional groups, primarily hydroxyl and alkoxy.

Characterization by SEM

Structures formed by carbon and calcium-based components exhibit a good distribution. It can be also observed the presence of non-ordered strips of material which are not detectable in the presence of magnesium materials.

The activated carbon sample made from coconut powder and zinc acetate, has some very striking characteristics compared with the rest of the synthesized samples. The zinc oxide tends to form small homogeneous clusters on the carbon surface. Similar clusters "nanoarchitectures hamburger type" [11], were present in the sample containing zinc, and to a lesser extent when the precursors of calcium and zinc were combined, however they were absent in the sample with the three metal precursors.

The synthesis of carbon with precursors of calcium and zinc, leads to a material that combines the main features of the above materials, still presenting clusters of zinc compounds, and lamellae like structures in the material with calcium.

Considering the mass fraction of precursors used in other studies [3], [4], a material with the three precursors was synthesized according to the composition showed in table 1. The SEM analysis of that sample is shown in part "d" of Figure 3, which shows more defined structures with granules of calcium and magnesium compounds on the surface and very few zinc compounds.

Figure 3. SEM images of the activated carbon after washing with increased 2500. a) carbon with calcium precursor; b) carbon with Zn precursor; c) carbon with calcium and zinc precursors; d) carbon precursor with calcium, magnesium and zinc; e) simple activated carbon.

Comparing the above materials with sample of activated carbon (point "e" in Figure 3), a notorious increase in cavities characteristic of the coals was observed, which leads to a larger surface area in this sample when compared with the other samples.

Specific surface area determination by N_2 adsorption

The textural properties of the materials were analyzed through N_2 adsorption using a Horiba analyzer 9600. The specific areas were determined using the BET model and are shown in Table 3. From these results, we can appreciate that the sample of activated carbon exhibits a greater value of specific surface area when compared with the other samples. However, these values are still small, which indicates that is necessary to improve the activation process and diminish the amount of oxides (CaO, MgO and ZnO). The average particle size for all the samples was smaller than 2 µm after washing.

Regarding the particle size, all samples obtained smaller sizes than 2 µm after washing so as exhausting methodology is not necessary.

Table 3. Specific surface area (SSA).

Sample	Average size particle (µm)	SSA (m^2/g)
Simple Activated Carbon	1.003	68.49
Carbon–$Ca(Ac)_2$	38.270	32.23
Carbon–$Ca(Ac)_2$-washed	1.012	22.40
Carbon–$Zn(Ac)_2$–washed	0.873	30.57
Carbon–$Ca(Ac)_2$–$Zn(Ac)_2$-washed	0.932	34.29
Carbon–$Ca(Ac)_2$–$Mg(Ac)_2$-$Zn(Ac)_2$	23.800	38.31
Carbon–$Ca(Ac)_2$–$Mg(Ac)_2$-$Zn(Ac)_2$ -washed	1.000	59.49

Conclusions

Since hydroxyl groups are present as alkoxy and oxides of magnesium and zinc in the adsorbents, it is likely that most of the adsorption sites on the corresponding materials are of the basic type, which is considered relevant for adsorption, since it may have greater interaction between the acid gases and the surface of the developed materials.

The presence of precursors of Ca, Mg and Zn in the activated carbon, produces a material that combines chemical and structural characteristics of these components, so it could become a highly efficient adsorbent, however, is necessary to improve the surface area of the materials in order to increase the adsorption capacity.

Acknowledgments

We acknowledge the *UMSNH* for the facilities to perform this work, we also thank to *Ma. Remedios Cisneros Magaña* from *UMSNH*, *Lourdes Mondragón Sánchez* and *Ma. Teresita Santoyo Arreguín* from the *ITM* and to the student *Cindy Manríquez Rodríguez* from *UABC Campus Ensenada*.

References

[1] Enviromental Process Development Department, Industrial Machine & Plant Development Center, Ishikawajima-Harima Heavy Industries Co., Ltd., «Adsorption of water vapor and carbon dioxide at carboxilic functional groups on the surface of coal,» *Fuel,* nº 80, pp. 757-764, 2001.

[2] J. Fernández, F. González, C. Pesquera, C. Blanco y M. J. Renedo, «Study of the CO_2/sorbent interaction in sorbents prepared with mesoporous supports and calcium compounds,» *Ind. Eng. Chem. Res.,* nº 49, pp. 2986-2991, 2010.

[3] L. A. Ramírez-Montoya, V. Hernández-Montoya, A. Bonilla-Petriciolet, M. A. Montes-Morán, R. Tovar-Gómez y M. R. Moreno-Virgen, «Preparation, characterization and analyses of carbons with natural and induced calcium compounds for the adsorption of fluoride,» *Journal of Analytical and Applied Pyrolysis,* nº 105, pp. 75-82, 2014.

[4] M. Bhagiyalakshmi, P. Hemalatha, M. Ganesh, P. Mei Mei y H. Tae Jang, «A direct synthesis of mesoporous carbon supported MgO sorbent for CO_2 capture,» *Fuel,* nº 90, pp. 1662-2667, 2011.

[5] O. Haibo, H. Jian Feng, L. Cuiyan, C. Liyum y F. Jie, «Synthesis of carbon doped ZnO with a porous structure and its solar-light photocatalytic properties,» *Materials Letters,* nº 111, pp. 217-220, 2013.

[6] J. Nishino, «Adsorption of water vapor and carbon dioxide at carboxylic functional groups on the surface of coal,» *Fuel,* nº 80, pp. 757-764, 2001.

[7] S. Hosseini, E. Marahel, I. Bayesti, A. Abbasi, L. C. Abdullah y T. S. Y. Choong, «CO_2 adsorption on modified carbon coated monolith: effect of surface modification by using alkaline solutions,» *Applied Surface Science,* nº 324, pp. 569-575, 2015.

[8] M. Fombuena y A. Valentín, «Manual del carbón activado,» Aguapedia, Sevilla, 2010.

[9] L. Li, D. L. King, Z. Nie y C. Howard, «Magnesia-Stabilized calcium oxide absorbents with improved durability for high temperature CO_2 capture,» *Ind. Eng. Chem.,* nº 48, pp. 10604-10613, 2009.

[10] A. Barroso-Bogeat, M. Alexandre-Franco, C. Fernández-González y V. Gómez-Serrano, «Preparation and Microstructural Characterization of Activated Carbon-Metal Oxide Hybrid Catalysts: New Insights into Reaction Paths,» *Journal of Materials Science & Technology,* nº 31, pp. 806-814, 2015.

[11] M. Zhou, X. Gao, Y. Hu, J. Chen y X. Hu, «Uniform hamburger like mesoporous carbon-incorporated ZnO nanoarchitectures: One-pot solvothermal synthesis, high adsorption and visible-light photocatalytic decoloration of dyes,» *Applied Catalysis B: Environmental,* nº 138-139, pp. 1-8, 2013.

[12] A. Zyoud, H. N. Nassar, A. El-Hamouz y H. S. Hilal, «Solid olive waste in environmental cleanup: Enhanced nitrite ion removal by $ZnCl_2$-activated carbon,» *Journal of Environmental Management,* nº 152, pp. 27-35, 2015.

[13] E. Kowsari y B. Bazri, «Synthesis of rose-like ZnO hierarchical nanostructures in the presence of ionic liquid/Mg^{2+} for air purification and their shape-dependent photodegradation of SO_2, NO_X, and CO,» *Applied Catalysis A: General,* nº 475, pp. 325-334, 2014.

[14] K. K. Han, Y. Zhou, Y. Chun y J. H. Zhu, «Efficient MgO-based mesoporous CO_2 trapper and its performance at high temperature,» *Journal of Hazardous Materials,* nº 203-204, pp. 341-347, 2012.

[15] L. K. G. Bhatta, S. Subramanyam, M. D. Chengala, S. Olivera y K. Venkatesh, «Progress in hydrotalcite like compounds and metal-based oxides for CO_2 capture: a review,» *Journal of Cleaner Production,* vol. http://dx.doi.org/10.1016/j.jclepro.2014.12.059, pp. 1-26, 2015.

[16] H. Yang, Z. Xu, M. Fan, R. Gupta, R. B. Slimane, A. E. Bland y I. Wright, «Progress in carbon dioxide separation and capture: A review,» *Journal of Environmental Sciences,* nº 20, pp. 14-27, 2008.

[17] J. Yang, F. N. Gu, H. J. Wang, Y. Zhou, J. Y. Yang, Z. Y. Wu y J. H. Zhu, «The metal-incorporated mesoporous carbon with high performance in capture and degradation of volatile nitrosamines,» *Catalysis Today,* nº 148, pp. 88-96, 2009.

[18] D. I. Jang y S. J. Park, «Influence of nickel oxide on carbon dioxide adsorption behaviors of activated carbons,» *Fuel,* nº 102, pp. 439-444, 2012.

[19] M. A. Béjar-Ubaldo y P. F. Rodríguez-Flores, «Carbón activado vegetal como soporte de un catalizador metálico,» Morelia, 2010.

[20] J. A. Solís-Fuentes, M. Morales-Téllez, R. C. Ayala-Tirado y M. d. C. Durán-de-Bazúa, «Obtención de carbón activado a partir de residuos agroindustriales y su evaluación en la remoción de color del jugo de caña,» *Tecnología, Ciencia, Educación del Instituto Mexicano de Ingenieros Químicos A.C.,* vol. 27, nº 1, pp. 36-48, 2012.

[21] A. Gómez, W. Klose, S. L. Rincón y W. Wiest, «Proceso de producción de carbón activado a partir de cáscaras de palma de aceite en un horno rotatorio y su aplicación en la limpieza de NO,» *Palmas,* vol. 25, nº Especial - Tomo II, pp. 461-471, 2004.

[22] A. Gómez, S. L. Rincón y K. Wolfgang, «Carbón activado de cuesco de palma. Estudio de termogravimetría y esturctura.,» Kassel University Press, Kassel, Alemania, 2010.

[23] E. Schiebold, «Ueber die Kristallstruktur des Periclas,» *Zeitschift fuer Kristallographie, Kristallgeometrie, Kristallphysik, Kristallchemie (-144.1977),* nº 56, pp. 430-430, 1921.

[24] K. Kihara y G. Donnay, «Anharmonic thermal vibrations in ZnO Model: 3-c, at T = 293 K,» *The Canadian Mineralogist,* nº 23, pp. 647-654, 1985.

[25] E. Zolotoyabko, E. N. Capi, J. S. Fieramosca, R. B. Von Dreele, F. Marin, G. Mor, L. Addadi, S. Weiner y Y. Politi, «Differences between Bon Lengths in Biogenic and Geological Calcite,» nº 10(3), p. 1207, 2010.

436654 Toward Understanding the Atmospheric Chromium Chemistry

Abstract:

Trivalent chromium (Cr(III)) and hexavalent chromium (Cr(VI)) are the two common oxidation states of chromium in the environment. Inhalation exposure to Cr (VI) may lead to cancer, nasal damage, asthma and bronchitis. In contrast, Cr(III) is a trace element essential for the proper function of living organisms. Therefore, it is important to accurately discriminate between these two species in atmospheric particulate matter (PM) samples. In the United States, Cr(VI) is one of the criteria pollutants which has been monitored through the National Air Toxics Trends Station (NATTS) Network. Thus, measurement of Cr(VI), is one of the focal points for hazardous air pollutants (HAPs) monitoring in the US.

The focus of my PhD research at Clarkson University was centered around understanding the atmospheric chromium chemistry and improving the sampling of atmospheric Cr(VI). Chromium is available as soluble or insoluble in the atmosphere. When soluble chromium is dissolved in water, the transformation of Cr(VI) is expected to take place in the water associated with particles (aqueous chemistry of chromium). Airborne particles that include soluble salts like ammonium sulfate or ammonium nitrate can spontaneously take up water (deliquescence) from the air and form a solution on the surface of the particle. This liquid layer allows chemistry to take place. I have shown that keeping the sampling particulate matter below the point at which it can take up water provides significant improvements in the amount of Cr(VI) retained in the sample. I have found that Cr(III) is converted to Cr(VI) in the presence of gaseous species like ozone and reactive oxygen species (ROS) at dry condition (gas-solids chemistry of chromium) implying that Cr(III) conversion can produce false positive measurements for Cr(VI) sampling. I have demonstrated that conversion decreases at lower temperature values. Thus, based on my work, it is likely that a sampler that keeps the sample dry and cool will improve the recovery of Cr(VI) that is present in the sample. I built a prototype sampler and deployed 4 of them in a field study in NJ, side-by-side with 4 of the samplers currently used by USEPA in their HAPs monitoring network. These tests showed that the improved samplers did provide better preservation of Cr(VI). Presently, I aim to develop a dynamic model for the simulation of aqueous chromium chemistry in the atmosphere ($1<pH<5$) and during the sampling period (24 hours- pH 9).

Recent studies showed that insoluble Cr(III) and insoluble Cr(VI) dominate atmospheric chromium mass. Regardless of few data in my PhD thesis, no information is available in the literature on the reaction kinetics of insoluble chromium speciation in airborne particles. Therefore my future research will concentrate on insoluble chromium chemistry (gas-solids chemistry of chromium) in the atmosphere (within next five years). My long term research plan (beyond next five years) is to investigate the chemistry/toxicity of soluble and insoluble chromium compounds in human lungs and their impact on lung surfactants. I have proposed three projects on atmospheric chromium. The first project investigates the reaction kinetics of insoluble Cr(III) and insoluble Cr(VI) with reactive oxygen species and volatile organic compounds, respectively (gas-solids reactions). The second project is aimed at designing and field testing a new Cr(VI) sampler to preserve both Cr(VI) and Cr(III) during sampling. The third project focuses on modeling Cr reactions (soluble and insoluble reactions) in the atmosphere and during the sampling.

In addition to my research experiences, I have two years of teaching experiences at Middle East Technical University Northern Cyprus Campus as well as several years of undergraduate research mentoring. The details of my future research plan and potential funding agencies, current research, previous teaching and undergraduate research mentoring experiences are available during the AIChE "Meet the Faculty Candidates" poster session.

436822 Study on Thermal Effects of Natural Gas Adsorption and Desorption in Activated Carbon

Abstract:

In this paper the temperature changes, during adsorption and desorption dynamics of natural gas in activated carbons to quantify their effects on the performance of an Adsorbed Natural Gas System (ANG), are studied. Commercial Norit RB3 activated carbon is used as the adsorbent, which is packed into a 0.47 L volume steel cell. Isothermal storage tests were performed to obtain the corresponding adsorption isotherm of activated carbon at room temperature in the pressure range from atmospheric to 500 psig. It was observed that the amount of gas adsorbed in the coal packed bed plus the compressed gas in the outer porosity is greater than the amount of gas stored in the empty cell at 500 psig.

Dynamic adsorption tests were conducted at the latter pressure using three different load gas rates toward the cell. The results confirm that the increase in temperature during the dynamic adsorption capacity decreases system storage. It was found that the isothermal volumetric storage capacity of carbon Norit RB3 is 80.4 V/V at a storage pressure of 500 psig and that this capacity drops to 16% due to exothermic effect arising from the heat of adsorption. The dynamic desorption is affected mainly by the decrease in temperature accompanied by the drastic drop in pressure during delivery of gas. The delivery capacity was found to be 56.8 V/V. At this stage, the bed undergoes cooling from room temperature (30°C) to -1.7°C.

437181 Accelerating the Development of Green Technologies for Chemical Production through Multiscale Life-Cycle Technology Assessment

Sunday, November 8, 2015
Exhibit Hall 1 (Salt Palace Convention Center)
Yuan Yao, Department of Chemical and Biological Engineering, Northwestern University, Evanston, IL

Evaluating the potential life cycle impacts of emerging technologies is critical for policy makers, manufactures and researchers, because it provides useful references on future technology investment and promotion, and offers valuable information on improvement opportunities and R&D directions. However, it is difficult to conduct such an assessment for new technologies, especially those in early stages, due to lack of process data and flexible frameworks capturing technology changes at different scales.

The focus of my research is to construct a modeling framework evaluating the energy, emissions and economic implications of emerging technologies at both product and economy-wide scales. At product level, this model is able to quantify the net changes of adopting novel technologies through life cycle assessment and chemical process modeling. Regarding innovative technologies that share little similarity of current chemical production process and requires new process design, the model has a conceptual process design module and short-cut models for quickly estimating the energy and emissions of typical unit operations. At economy-wide scale, the model can assess the industrial energy and emission footprint of chemical production in future decades under different economic scenarios. Ethylene production and several technologies related to steam cracker and gas separation are selected as case studies for demonstration. The results of the modeling can shed light on the key drivers of energy and emissions footprint for the petrochemical industry, and highlight promising technology opportunities that can contribute to desirable energy saving and emission mitigations.

Energy and sustainability in the chemical industry will remain the focus of my future research. I will apply and expand my expertise in life cycle assessment and chemical process modeling to enable more robust engineering and policy decisions today to contribute to greatest sustainability benefits tomorrow. In addition to research, as a prospective professor, I would like to incorporate design-to-environment into the undergraduate/graduate curriculum of chemical engineering, and teach students with hands-on experience of chemical process design with consideration of sustainability.

436833 Ballistic Performance Assessment Is a Must for Shelf Life Assessment of Rocket Motors

Abstract:

In the course of rocket motor service life assessment number of different propellants of a certain rocket motor have been examined and analyzed. This paper describes the testing, collects together the results from a number of naturally aged propellant charges to enable reasonable life predictions to be made. The Investigations have been carried out with double based rocket propellants with a view to obtaining predictive data for safe use life of the rocket motors containing these propellants. The examination and analysis procedure, chemically, mechanically and ballistically for accelerated aged propellant charges has been devised to give a comprehensive picture of the future properties of the propellant charge. The chemical and mechanical safe lives of all propellants examined, when stored under normal conditions is more than adequate to meet any sensible requirement. The results for this investigation indicate that a safe life in excess of 30 years is to be expected. The ballistic properties have been thoroughly determined over a temperature range of (-20°, +20°, +50°C) for both test and whole rocket motors. All that can be said is that although the chemical and mechanical properties of these propellant charges are not significantly changing with age and factors connected with them are unlikely to prove life terminating, ballistic properties testing on the other hand is prove life terminating. Taking into consideration this mode of failure, it is evident that the rocket motors under this investigation could have the existing service life of 20-30 years extended subject to satisfactory ballistic properties. The new data increases the confidence that most of the rocket motors will achieve an extended service life, others have to be subjected to periodic investigations.

437502 Performance Comparison of Ethanol and Butanol Production in a Continuous and Closed-Circulating Fermentation System with Membrane Bioreactor

Abstract:

Aerobic ethanol fermentation and anerobic butanol fermentation were conducted in a continuous and closed-circulating fermentation (CCCF) system, with the fermentation duration of 21 and 12 days respectively. Performance of both fermentation and pervaporation were compared, the average cell concentration, glucose consumption rate, cumulated production concentration, production flux and separation factor of ethanol fermentation were 11.45 g/L, 3.70g/L/h, 655.83 g/L, 378.5 $g/m^2/h$, 4.83 respectively, the corresponding parameters of butanol fermentation were 2.19 g/L, 0.61 g/L/h, 28.03 g/L, 58.56 $g/m^2/h$, 10.62. Profiles of fermentation and pervaporation parameters indicated that the intensity and efficiency of ethanol fermentation was higher than butanol fermentation, but the stability and adaptability of butanol fermentation was superior to ethanol fermentation. Although the two fermentation processes had different features, the performance still indicated the application prospect of the ethanol and butanol production by the CCCF system.

437504 A Study on the Liquid Phase Oxidation of Toluene By Pure Oxygen in a Mini-Channel Reactor

Abstract:

Effectiveness and safety are two important and crucial problems in hydrocarbon oxidation. The liquid phase oxidation of toluene by pure oxygen was conducted in a mini-channel reactor in order to improve the operation safety and the reaction effectiveness. The effects of the experimental parameters such as temperature, catalyst, initiator, pressure, gas flow rate and liquid flow rate on the conversion of toluene and the yield of oxidation product were systematically investigated. In the mini-channel reactor, a toluene conversion of approximate 5.7 m% and benzoic acid yield of approximate 4.4 m% were obtained under the conditions: 165 , 0.9 MPa, 0.15 m% of catalyst and 1.0 m% of initiator in the liquid reactants, 0.15 Nl/min of oxygen flow rate and 53.3 ml/h of toluene flow rate. A drift flux model was employed to simulate the flow characteristics of oxygen and toluene in the mini-channel. The results showed that the reaction was greatly influenced by the two-phase flow characteristics.

THE ROLE OF MICROFLUIDIC INTERFACES IN METALS EXTRACTION, SOIL AND FERTILIZERS

Davide CICERI – Department of Materials Science and Engineering, Massachusetts Institute of Technology

➢ ABOUT ME:

2016	Postdoctoral associate	MIT	
2013	Ph.D. Chemical Engineering	University of Melbourne	
2008	M.Sc. Chemistry	University of Milan	
2006	B.Sc. Chemistry	University of Milan	

- Australian Mathematical Sciences Institute (AMSI) industrial internship grant (15,000 AU$)
- International Postgraduate Research Scholarship - The Australian Government
- Melbourne International Research Scholarship - The University of Melbourne
- Erasmus Scholarship - The European Commission
- **Travel awards:** Wilford Gardner fellowship; Melbourne Abroad Travelling Scholarship; Melbourne Research Training Conference Assistance Scheme; Clive Pratt scholarship; ISEC conference attendance scholarship; PFPC travel grant.
- **Industry and Government:** Terrativa (BR), Veolia (AU), EMBRAPA (BR), CEA (FR)

➢ ABOUT MY TEACHING EXPERIENCES:

MIT (current semester)
- Teaching assistant in *Materials Laboratory* (Prof. Donald R. Sadoway)

University of Melbourne (2009; one semester)
- Teaching assistant in *bio/nano-engineering* (Prof. David Dunstan)

University of Milan (2007; one semester)
- Teaching assistant in *Laboratory of General Chemistry* (Prof. Antonio Biondi).

- Kaufman Teaching Certificate Program (KTCP) - MIT
- Supervised the research activity of 3 masters student and 7 undergraduate research students (UROP) from several engineering backgrounds - MIT
- Tutored more than a dozen of high school and university students in a broad range of sciences, particularly mathematics, physics and chemistry (2006-2011).

MICROFLUIDIC INTERFACES

➢ ENERGY (metals)

Metals are at the center of civilization. Their extraction is responsible for significant CO_2 and SO_2 emissions.

My research investigates solvent extraction of metals by manipulating water/oil interfaces in microfluidic systems. Particular emphasis is given to kinetics and improvement of extraction efficiency and throughput.

liquid/liquid

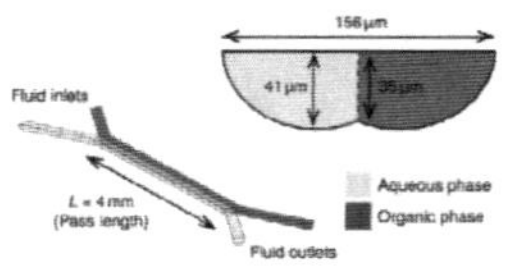

$$Co^{II}_{(aq)} + 2(HL)_{2\,(int)} \rightleftharpoons CoL_2(HL)_{2\,(org)} + 2H^+_{(aq)} \quad \textbf{(a)}$$

$$Fe^{(III)}_{(aq)} + 3(HL)_{2\,(int)} \rightleftharpoons [FeL_3(HL)_{3\,(org)}] + 3H^+_{(aq)} \quad \textbf{(b)}$$

Ciceri *et al.* (**2011**) Microfluid. Nanofluidics 11, 593
Ciceri *et al.* (**2013**) Microfluid. Nanofluidics 14, 213
Ciceri *et al.* (**2014**) Chem. Eng. Res. Des. 92(3), 571

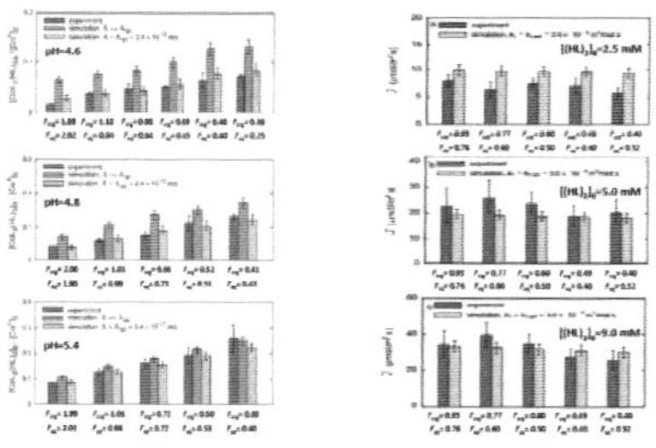

(a) (b)

➢ SUSTAINABILITY (soil)

Soil is a complex composite material at the core of current global challenges (e.g., food security and carbon capture).

My research investigates soil processes such as weathering by using microfluidic devices, which allow to achieve new levels of spatial and temporal resolution.

solid/liquid

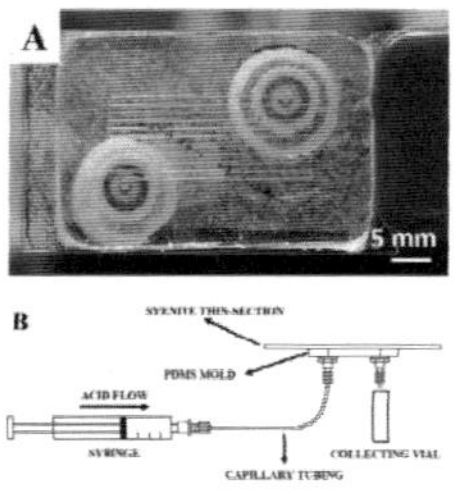

Ciceri and Allanore (**2015**) PLoS ONE 10(10): e0139979

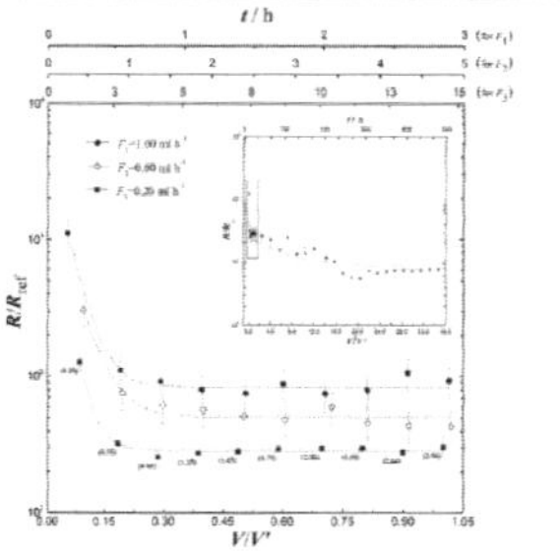

➢ MATERIALS (fertilizers)

Fertilizers (e.g., apatites and potash) are key chemicals for agriculture. Current products do not address the needs of tropical soils.

My research investigates alternative processing routes to the production of fertilizers (e.g., microreactors for hydrothermal treatment of silicates).

solid/fluid

syenite bulk samples

TEM image of hydrothermally altered syenite

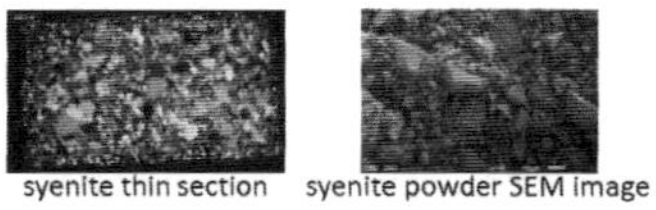

syenite thin section syenite powder SEM image

Ciceri *et al.* (**2015**) Sci. Tot. Env. 502, 590

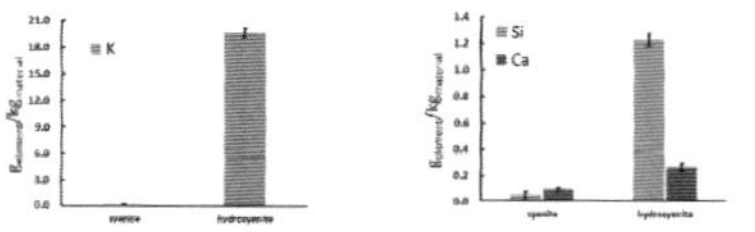

Leaching of hydrothermally altered syenite (pH=5; 24 hrs)

437650 Carbon Capture and Sequestration Technology for Greenhouse Gas Mitigation

Abstract:

'Global warming, Green House Gas emissions, Climate Change' are those terms which are being familiar to every individual on earth since past few decades. Very few were unaware about the three keys of Green House Gas emissions one among them is carbon-dioxide. The relative global contribution of carbon-dioxide in climate change is roughly estimated more than 55% since late 20th century. It is necessary to reduce the carbon-dioxide emissions before it too long. Carbon capture and sequestration (CCS) is a range of technologies that hold the promise of trapping up to 90% of the carbon-dioxide. CCS involves capturing, transporting and burying the carbon-dioxide so that it does not accumulate in the atmosphere and contribute to climate change. The most sensible approach of capturing carbon-dioxide is from large stationary sources, such as fossil fuel power plants. In this technique the challenge is to capture carbon-dioxide at low cost and continuously isolate it securely over long term. In this paper the investigation of capture economics of the three main capture techniques (Pre combustion capture, post combustion capture and oxyfuel) were done. Also the need of CCS and the utilization of captured carbon-dioxide other than putting it in an underground storage (e.g. enhanced oil recovery) too thoroughly studied. After successfully completing the study we come on conclusion that CCS involves the responsible utilization of fossil fuels without endangering the planet. We found that the CCS technology is garnering attention as a bridging technology in the new energy society.

437682 Hydrodynamic Simulation of a Bubbling Fluidized Bed with MP-PIC Method

Abstract:

The gas-solid flows in a bubbling fluidized bed were simulated with MP-PIC (multiphase particle-in-cell) method, which is a newly developed Euler-Lagrangian method that track parcels instead of single particles, to study the hydrodynamic of the bubbling behavior. In the MP-PIC method, a parcel represents several particles with the same diameter, velocity and similar location, thus can significantly speed up the simulations. Different grid and parcel resolutions, the latter of which is the number of particles per parcel, for fluid and solid flows, respectively, were tested to investigate their individual and combined effects on the simulation results. And different drag models, including the homogeneous drag and the newly developed EMMS drags were tested in the simulations. The results show that the EMMS drag can improve the bubbling bed prediction in comparison to that using the homogeneous drag, both qualitatively and quantitatively.

437712 The Effects of Leg Length on the Flow Field and Separation Process of Cyclone Separator

Abstract:

Cyclone is a commonly used device in petrochemical industry. Although effects of operating parameters on cyclone have been studied a lot, the effects of leg length drawed little attention. In this paper, the effects of different leg lengths on flow field and separation processes of cyclone are investigated using numerical simulation and experimental methods. A PV-type cyclone with different leg lengths are simulated with Reynolds Stress turbulence model using Fluent. The results show that with the increase of leg length, the axial velocity curve changed to single peak from Twin Peaks. With further increase, the axial velocity begins to decrease, this leads to the decrease of particle residence time at the central area. After a critical length, the increased dipleg has no effect on the flow field and that critical length should be the most reasonable leg length. Also, as shown in Fig.1, the pressure drop in the cyclone is decreased with the increase of dipleg length and after a critical length, the pressure drop reaches a constant value. Fig.2 further indicates this trend and it also illustrates that the natural vortex length decreases slightly with the increase of leg length. In summary, it's appropriate to extend the length of leg to benefit the separation but it's not simply the longer the better. We should calculate the best leg length of different types of cyclones under their actual operating conditions.

437941 Effect of Filler on Properties Bamboo Fiber Reinforced Epoxy Composites

Abstract:

Polymers have generated a wide interest in the engineering domain, including their tribological applications in view of their good strength and low density as compared to monolithic metal alloys. Being lightweight, they are the most suitable materials for weight sensitive applications, but their high cost becomes a limiting factor for commercial applications. However, by incorporating filler particles into the fiber reinforced composites, higher mechanical strength can be achieved along with the reduced material cost.Filler materials play an important role in determining the properties and behavior of the particulate Composites. In the present work, bamboo (untreated and alkali treated) fiber reinforced epoxy composites are fabricated using Cement By Pass Dust (CBPD) as filler material. Mechanical and physical properties of both categories of the composites (untreated and treated bamboo) are measured and compared. Erosion test has been carried out to study the effect of various operational and material parameters on erosive wear behavior of these composites. In the experimental set-up, an air jet type erosion test rig has been used to carry out both the steady state and unsteady state erosion test. Taguchi's orthogonal arrays are used for analysis of experiential results. It identifies significant control factors influencing the erosion wear and also outlines significant interaction effects. The findings of the experiments indicate that for untreated bamboo composites, impact velocity, stand-off distance, impingement angle and filler content are the most significant factors, while for the treated composites, impact velocity, stand-off distance, impingement angle and CBPD content emerge as the most significant factors in a declining sequence in affecting the wear rate. An optimal parameter combination is determined, which leads to minimization of erosion rate. Analysis of variance (ANOVA) test has also been performed on the measured data and signal to noise (S/N) ratios to find the significance levels of factors.

437979 Sustainability Considerations in Production of Fluids from Shales

Sunday, November 8, 2015
Exhibit Hall 1 (Salt Palace Convention Center)
Palash Panja, Energy & Geoscience Institute, University of Utah, Salt Lake City, UT

Production from unconventional reservoirs such as shales has increased significantly in the recent years due to the technological advancement such as horizontal drilling and hydraulic fracturing. EIA (Energy Information Administration, USA) has reported that 49% of oil and 54% of natural gas in the USA are produced from fractured reservoirs as of February 2015. Current industrial practice of production of liquid hydrocarbons (oil and condensate) and natural gas leads to low recovery and high environmental impact. Only 5-10 % of liquid is recovered from ultra-low permeability (50 – 1000 nD) reservoirs such as shales leaving vast amount of liquids in the subsurface. Hydraulic fracturing is the key technology that has made the enhanced production possible. However, there are potential significant environmental issues related to impact on groundwater, fate of injected water, fugitive emissions and induced seismicity. The challenges to minimize the environmental impacts and to maximize production require understanding the fundamentals of fluid flow in ultra-low permeability porous media in the context of complex geologic and operational controls. In this research, methods to maximize recovery while minimizing some of the environmental considerations are presented.

Reservoir numerical simulation of multiphase fluid flows in tight and ultra-tight reservoirs has become the standard tool to predict production performance from these unconventional resources. Numerical simulation requires accurate modeling of fluid flow, reservoir geometry and spatial distribution of properties. However, the simulations need to be structured in a logical fashion to ascertain accurate results. Even though applicability of Darcy's law for ultra-low permeability porous media is questionable, it is widely used in commercial simulators for flow modeling. More complex fluids and lower permeability reservoirs require better grid resolution. Grid refinement must be applied even more carefully when dealing with the production of condensates. Situations involving multiple, complex parameters need to be handled using specific methods that prioritize the parameter list. Formation permeability is one of the most significant parameters in the production of liquids from shales. Production increases with increase in reservoir permeability but produced gas oil ratio also increases. The oil production is higher with more compressible rock. Produced gas oil ratio is also reduced for higher compressibility rock. Reduced spacing, generally speaking, results in higher recovery. Spacing has the greatest influence up to 500 nD permeability. It is observed that recovery factor is higher in hydraulically fractured reservoirs in primary production than the recovery from the higher permeability conventional reservoirs, assuming fracture spacing of 100 feet or lower. However, production data from unconventional reservoirs do not reflect this fact.

Surrogate reservoir model considering complex interaction of parameters using multivariate regression of simulated results for black oil and condensate from ultra-low permeability reservoirs is an efficient tool. Surrogate models are also utilized to determine the hierarchy of the important factors and to analyze the uncertainty in the production outcomes for given distributions of input factors.

Operating the well at higher flowing bottom hole pressure (FBHP) is preferable for low permeability (100 nD) reservoir and low FBHP for higher permeability (1000 nD) reservoir to recover more liquid such as condensate and volatile oil.

It is known that only about 30% of the water injected during hydraulic fracturing is recovered. It is quite likely that the injected water may be retained in the matrix and is trapped as immovable water saturation. As the wettability of the matrix changes from being water wet to oil wet, the recovery increases by about 15%. Surface equipment and conditions affect subsurface operations and vice-versa. Rapid deployment of resources for liquid shale developments may not always result in optimal choices. It is shown that both from a liquids quality and gas venting perspective, a two-separator arrangement is best when producing moderately volatile oil or a condensate. However, the current practice in most locations is to use single-stage separation.

Measurement of reservoir permeability, thermodynamic properties such as bubble and dew points, gas-oil ratios, etc., are very important. It is thus critically important to understand the impact of the nanoporous structure on the basic properties of shales and fluids in these shales. The current method of measuring permeability in shales consists of crushing the rock, exposing this to pressure and measuring pressure decay. New methods of measurement of rock properties and rock fluid properties such as permeability, relative permeability and porosity should be developed. The fluid sampling technique and re-combination of separator fluids to represent reservoir fluid should be revisited. Determining the suitable location of fracturing in the reservoir is one of the key factors for effective fracturing which will reduce the environmental impact in many ways such as reducing the water requirement for proppant transport. Designing proppant ensuring proper mechanical properties to keep fracture open during production is another key factor. Geomechanical properties such as stress change in the reservoir during fracturing and then production should be studied. The change stress in the field affects many geological properties such as permeability and it also controls the morphology of hydraulic fractures. Injection of gas or water into reservoir is one possibility to exploit more hydrocarbons from unconventional reservoirs. In-fill drilling and refracturing in existing reservoir considering economic constraint are also promising ways to extract more hydrocarbons.

(6cl) Challenges and Progress for Cogeneration of Power and Hydrogen from Nested Carbon-Air/Carbon-Steam Fuel Cells

Stewart, S. M., Stanford University
Mitchell, R. E., Stanford University
Gur, T.,

To address growing energy demands while reducing CO_2 emissions and the environmental footprint of dirty solid fuels such as coal, it is essential to develop advanced technologies with high efficiencies that also facilitate easy carbon capture. To this end, converting a solid fuel to clean hydrogen in a fuel cell with spontaneous and simultaneous generation of electricity while producing capture-ready CO_2 offers attractive opportunities and advantages. The nested air-carbon/steam-carbon fuel cell addresses this need through higher efficiencies, and because the solid fuel is kept separate from the steam and air in the cathode compartments by an ionically conducting electrolyte membrane, the effluent anode product gas is a concentrated CO_2stream that is capture ready. Moreover, due to the downhill chemical potential gradient of oxygen across the electrolyte, steam electrolysis takes place spontaneously under a driving force of > 0.5 V in the steam-carbon fuel cell, which can be supplemented by simultaneous electrical power generation from the air-carbon fuel cell.

Recently our group has published work demonstrating the feasibility of hydrogen production using activated carbon as the solid fuel in a steam-carbon fuel cell.[1] While this work has demonstrated the potential of this technology, the ultimate goal is to convert real fuels such as coal or biomass into clean fuel and electrical energy. Biomass and coal contain a multitude of contaminants, many of which are potent poisons for the Ni-based cermet anodes typically used in these cells. In particular, Ni-cermet anodes, which are commonly employed for solid oxide fuel cells, show susceptibility to H_2S poisoning at concentrations as low as 1 ppm,[2]making their direct use with these fuels unfeasible.

Herein we report our progress on advancing this technology through the use of solid sorbents dispersed throughout a fuel bed of coconut char (Norit®). In particularly, by using sulfur sorbents that are highly dispersed on the solid fuel it is possible to dramatically reduce the H_2S and COS contents of the syngas *in situ* and hence, mitigate the sulfur burden on the catalytic anode material. Indeed, adding group IIA alkali earth–based sorbents to the fuel mitigates the adverse impact of sulfur on fuel cell performance and slows down the degradation process. The rate of current density loss at the constant cell voltage of 0.7 at 900 °C as a function of time for different IIA metal oxide sorbents, is correlated to a reduction in the rate of increase in the kinetic resistance in the anode electrocatalyst as determined from electrochemical impedance spectroscopy (EIS). Elemental depth profiles determined using x-ray photoelectron spectroscopy (XPS) and x-ray diffraction of the anode electrocatalyst of the degraded fuel cells are used to understand the effect of sulfur poisoning on the Ni/YSZ electrocatalyst, and determine if nickel carbide (Ni_3C) also plays a role in increasing the anode kinetic resistance. Finally, unique challenges associated with the use Ni/YSZ in contact with carbon fuels, particularly the formation of Ni_3C, are discussed for direct carbon fuel cells using a sulfur-free carbon source.

The work presented here serves to better understand sulfur mitigation with sulfur sensitive electrocatalyst, and provides an impetus for better understanding of sulfur poisoning mechanisms, development of sulfur tolerant anode materials, and the most effective means of utilizing sulfur sorbents. Through this technology, it will be possible to take dirty fuels containing high concentrations of sulfur and convert them into green energy in the form of hydrogen and electricity.

References:

(1) Alexander, B. R.; Mitchell, R. E.; Gür, T. M. *J. Electrochem. Soc.* **2012**, *159*, F810.

(2) Kurokawa, H.; Sholklapper, T. Z.; Jacobson, C. P.et al. *Electrochem. Solid-State Lett.* **2007**, *10*, B135.

438028 Engineering Therapeutics for Vascular Disease

Sunday, November 8, 2015
Exhibit Hall 1 (Salt Palace Convention Center)
Donny Hanjaya-Putra, Vascular Surgery, Harvard Medical School, Boston, MA; Wyss Institute for Biologically Inspired Engineering, Boston, MA

Engineering Therapeutics for Vascular Disease

Donny Hanjaya-Putra, Ph.D
Vascular Surgery, Harvard Medical School
Wyss Institute for Biologically Inspired Engineering, Boston MA

The circulatory system embodies an exquisite feat of chemical engineering, where blood flow, mass transfer, and reaction kinetic are integrated with vascular and inflammatory responses. Injury to the vasculature of the circulatory system can cause a wide range of life-threatening disease, such as, stroke, hypertension, and diabetes. *My research focuses on using engineering principles and stem cell biology to develop novel tools and therapeutics to detect and treat complications associated with vascular disease.* My background as a chemical engineer, along with my training in stem cell biology and vascular surgery lab, uniquely position myself to contribute substantial progress in the field. During my doctoral work with Dr. Sharon Gerecht at the Johns Hopkins University, I investigated the mechanism of stem cell assembly into functional vascular networks. We developed a synthetic matrix, which properties can be tuned to provide temporal and spatial control for vascular morphogenesis that are useful for therapeutic angiogenesis and tissue regeneration. My postdoctoral training with Dr. Elliot L. Chaikof at Harvard Medical School has further broadened my research horizon in the area of inflammation and thrombosis associated with vascular injury and stem cell transplantation. I led collaborations with engineers and surgical residents to evaluate anti-inflammatory and anti-thrombotic therapeutics in a clinically relevant model of deep vein thrombosis and islet transplantation. Together, my doctoral and postdoctoral work are able to be applied to diverse areas in regenerative medicine, diabetes, stem cell and vascular biology.

Education
B.S., 2007, Chemical and Biomolecular Engineering, University of Notre Dame.
Ph.D., 2012, Chemical and Biomolecular Engineering, Johns Hopkins University.

Awards and Honors
2007, ***B.S. Magna Cum Laude***, University of Notre Dame.
2011, ***U.S. New Investigator Travel Award***, International Society of Thrombosis and Haemostasis (ISTH).
2012, ***Siebel Scholar, Class of 2012,*** Siebel Scholar Foundation.
2014, ***JDRF Postdoctoral Fellowship,*** Juvenile Diabetes Research Foundation.

Selected Publications

1. **Hanjaya-Putra, D.**, Bose, V., Shen, Y.-I., Yee, J., Khetan, S., Fox-Talbot, K., Steenbergen, C., Burdick, J.A., and Gerecht, S. (2011). Controlled activation of morphogenesis to generate a functional human microvasculature in a synthetic matrix. *Blood.*
2. **Hanjaya-Putra, D.**, Wong, K.T., Hirotsu, K., Khetan, S., Burdick, J.A., and Gerecht, S. (2012). Spatial control of cell-mediated degradation to regulate vasculogenesis and angiogenesis in hyaluronan hydrogels. *Biomaterials.*
3. **Hanjaya-Putra, D.**, Shen, Y.-I., Wilson, A., Fox-Talbot, K., Khetan, S., Burdick, J.A., Steenbergen, C., and Gerecht, S. (2013). Integration and Regression of Implanted Engineered Human Vascular Networks During Deep Wound Healing. *Stem Cells Translational Medicine.*
4. Kusuma, S., Shen, Y.-I., **Hanjaya-Putra, D.**, Mali, P., Cheng, L., and Gerecht, S. (2013). Self-organized vascular networks from human pluripotent stem cells in a synthetic matrix. *Proceedings of the National Academy of Sciences.*
5. Krishnamurthy, V.R., Sardar, M.Y.R., Ying, Y., Song, X., Haller, C., Dai, E., Wang, X.,
Hanjaya-Putra, D., Sun, L., Morikis, V., Simon, S.I., Woods, R.J., Cummings, R.D., Chaikof, E.L., (2015). Glycopeptide analogues of PSGL-1 inhibit P-selectinin vitro and in vivo. *Nature Communications.*

438109 A New Approach to Predict the Dynamic Interactions Between an Air Bubble/Drop and a Flat Solid Surface

Abstract:

In this study the Generic Stokes-Reynolds Young-Laplace model (GSRYL model) is proposed to predict the dynamic interaction forces acting between an air bubble and a flat solid surface when they are in relative axisymmetric motion in a Newtonian liquid. In the Generic SRYL model, which is basically similar to the alternative SRYL model, the Stokes-Reynolds equation is combined with the non-linearized second order form of the augmented Young Laplace equation to relate the mean curvature of the fluid interface to the pressure difference across the interface. The non-linearized second order form of the augmented Young Laplace equation built in the GSRYL model, leads to the appearance of the capillary number in the scaled equations.

This new theoretical framework has two important advantages over the alternative SRYL model. First, this model offers an accurate perspective of the bubble geometry at the initial separation and eliminates the approximation of the elliptical bubble geometry used as the initial separation in the alternative SRYL model. The second advantage of this model is that in spite of the SRYL model in which the scaled equations of system have a universal nature which predicts the bubble behaves nearly in a universal way over all ranges of capillary numbers, the scaled equations of the GSRYL model do not have a universal nature and depend on the physical parameters of system via capillary number.

The accuracy of this model is tested with the recent experimental data reported in the literature and confirms that this model can be successfully applied to predict the non-equilibrium interactions between an air bubble and a flat solid surface. This new model is worth further investigation, as we can take into account the bubble deformation, the original shape of the bubble at initial separation and all the physical parameters of system for this prediction.

438128 Orders of Magnitude of Sudden Increases of X-Ray Diffraction Intensity in Surfactant-Based Liquid Crystals Triggered By Co-Self-Assembly

Sunday, November 8, 2015
Exhibit Hall 1 (Salt Palace Convention Center)

Yoon Seob Lee, University of Dayton, Dayton, OH and James Rathman, The Ohio State University, Columbus, OH

Orders of Magnitude of Sudden Increases of X-ray Diffraction Intensity in Surfactant-Based Liquid Crystals Triggered by Co-Self-Assembly

Yoon Seob Lee*

Department of Chemistry, The University of Dayton, Dayton, OH 45469

James F. Rathman

Department of Chemical and Biomolecular Engineering, The Ohio State University, Columbus, OH 43210

* To whom correspondence should be addressed. E-mail: ylee01@udayton.edu

Orders of magnitude of sudden increases of x-ray diffraction intensity have been observed in cationic surfactant-based silicate-containing liquid crystals with different structures and symmetries. These increases have been triggered by the co-self-assembly of one particular co-surfactant. A direct comparison reveals that the x-ray intensity and number of peaks that can be indexed are matched with those of solid single crystals. To the best of our knowledge, this kind of observations has never been reported before, and we here present its early results. The liquid crystals have been prepared in cetyltrimethylammonium chloride (CTAC) micelle solutions with the concentration ranges from 1.0 to 24.8 wt.% with decyl alcohol (DeOH) as a co-surfactant. Tetraethylorthosilicate (TESO) has been used as a silicate source. Once all of the components are mixed with a pre-determined order, the liquid crystals begin to form within 1-7 minutes at room temperature. They have been further aged for 20 hours – 3 days, and collected from the solutions through filtration and characterized. The mixing order of the components is one of the critical factors to have the desired liquid crystals, and aging in some cases further improves the crystallinity of the liquid crystals. X-ray powder diffraction (XRD) patterns were taken on Scintag PAD-V diffractometer with Cu Ka radiation of 1.54060 Å wavelength at 20 mA, 45 kV at room temperature. To ensure the direct comparison, the same silicon holder, the same scanning condition, and the same slit set-ups have been used throughout the entire measurements. The changes in the x-ray intensity have been compared with those liquid crystals obtained from the same or similar self-assembly systems but without the co-surfactant, and those liquid crystals that are obtained with other additives which have a potential to act as a co-surfactant. The additives that have been tested in this study include simple organic molecules (benzene, phenol, benzyl alcohol, cyclohexane, n-hexane, cyclohexanol, 1-methyl-cyclohexanol, 3-methyl-cyclohexanol, 1-hexanol, 2-hexanol, and 3-hexanol) and surfactants (dodecyldimethylamine oxide and perfluoro-1-octanesulfonyl fluoride). One such example that shows this dramatic increase of the x-ray intensity is shown in the figure below. The one on the right is the XRD pattern of the bicontinuous cubic liquid crystal obtained without decyl alcohol, and the one on the left is the bicontinuous cubic liquid crystal obtained with decyl alcohol. The x-ray intensity of the highest peak has been increased over 50 times at the same condition and there are 18 peaks that are clear enough to be indexed (shown is a part of them at low 2q region and typical solid single crystals with cubic structure show 12 – 20 peaks). We have identified a region where the most of the sharp-jumps like this take place for this particular self-assembly system and the jump is often more than 100 times. This not only drastically improves the quality of the existing liquid crystals but creates the liquid crystals with new symmetries as well.

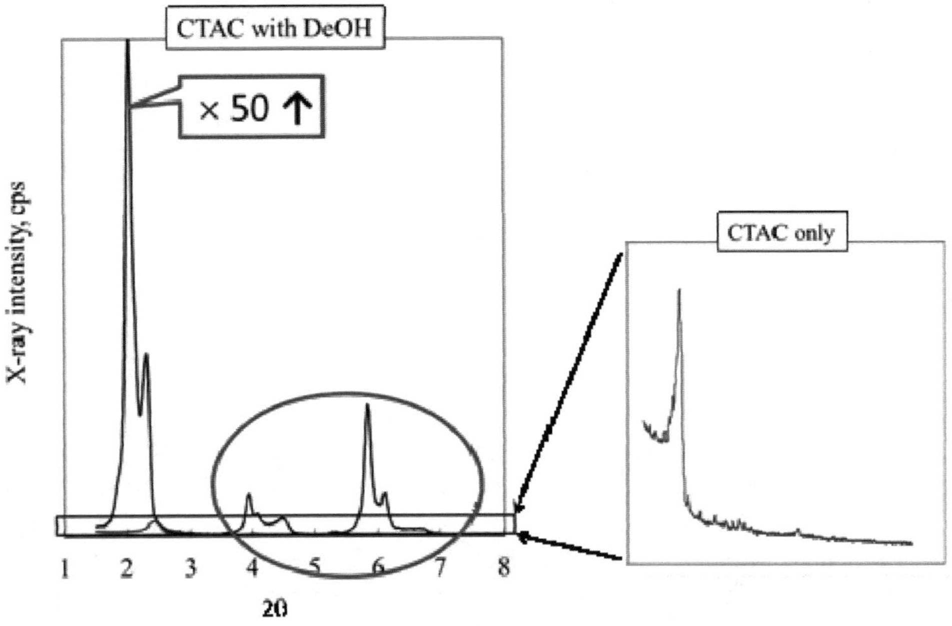

The kinetic studies that have been performed along with the XRD measurements show that there is a clear correlation between the increases of the x-ray intensity and the rate of the initial reaction. As the degree of the co-self-assembly increases, this rate is getting faster, and results in the higher increases of the x-ray intensity. Decyl alcohol, once co-self-assembled with the cationic micelles, reduces the distance between the adjacent cationic binding sites, which provides the geometrically-matching environment for the silicate anions to be condensed evenly and faster on the surface of the micelles, which increases the rate of the inter-micellar condensation. This uniform condensation of the silicate anions, each of which bound with each of the cationic head-group of the surfactant molecule, reduces the defects and mismatches within the self-assembling liquid crystals, thus making the final liquid crystals to have much higher degree of the long-range order, possibly over the macroscopic scale. Nature uses a very similar strategy for her biomineralization, in which the proteins provide the geometrically-matching binding sites for inorganic precursor ions. The difference is that those condensing inorganic ions on the surface of proteins do not re-self-assemble the whole structures of the proteins.

These very unusual phenomena can logically be explained using the concept of force balance for self-assembly (Lee, Y. S. *Self-Assembly and Nanotechnology: A Force Balance Approach*, Wiley, **2008**; Lee, Y. S. *Self-Assembly and Nanotechnology Systems: Design, Characterization, and Applications*, Wiley, **2012**), and can be expanded to the viewpoint of the materials discovery by self-assembly. The concept of force balance states that all possible self-assembly building units including atomic, molecular, polymeric, colloidal, bio-mimetic, and biological are composed of up to five distinctive structural segments. What determines the process (critical and hierarchical) and outcome (structural diversity and emergent properties) of self-assembly is how these segments interact and balance each other through intermolecular and colloidal forces by following five distinctive rules. Those rules are: *Rule 1*. To have self-assembly, self-assembly building units must have at least one segment that can generate an attractive force between them, *Rule 2*. They also must have at least one segment that can generate a repulsive force which can counterbalance the attractive force, *Rule 3*. Certain segments such as hydrogen-bonding group make self-assembly to grow in one direction, *Rule 4*. Asymmetric packing segments such as multiple hydrogen bonding site and chiral carbon make self-assembly to assemble in an uneven manner, and *Rule 5*. Functional segments such as ferrocenyl group and azo-benzene group make self-assembly system to communicate with the environment.

How far can we push chemical self-assembly? was identified as one of the top 25 big questions for the next quarter-century by Science (*Science*, **2005**, *309*, 95). Self-assembly offers a powerful way to fabricate nanoscale materials and systems with controlled complexity and hierarchy. The problem however has been how we use self-assembly which is seemingly chaotic in a logical way while maintaining a good overview over its promises. Using the results that are being reported here, a discussion will be made how to use self-assembly to control nanoscale features at the level of individual building unit, which is to show how to design the bottom-up building of nanoscale materials and systems for a given purpose. It strongly suggests that there might be a family of similar liquid crystal materials to be discovered from other self-assembly systems, which of course could lead us to find new structures and new properties. We anticipate that the properties of these highly-ordered liquid crystals might be quite different from those of the known liquid crystals whose structural order extends usually only up to the microscopic domains. These properties include birefringence, viscoelasticity, the effect of chiral dopants, the responses to light (with light harvesters) and temperature, and the effects of electric and magnetic fields.

438160 Genome-Scale Models for Systems Biology and Combinatorial Drug Discovery

Abstract:

Research Overview:

In my current position as a Harvard Junior Fellow, I have established an independent research program focused on developing systems biology approaches for multi-omic data integration and drug discovery. As a faculty, I propose to lead a research lab that develops genome-scale approaches that enable model-driven combination therapy and genome engineering – two challenging problems that are high dimensional in nature and have a large search-space of solutions, necessitating a computational approach. Yet solving them can have a tremendous global impact by addressing drug resistance in cancer and infectious diseases, and can enable rational genome design for biotechnology applications. My proposed research builds on my strength in developing innovative methods that reconstruct and model biological networks, such as PROM, ASTRIX, INDIGO and GEMINI. My accomplishments, described below, highlight my ability to successfully secure research funding, execute innovative research projects, establish collaborations with academic labs and biotech startups, and mentor scientists. (web: http://scholar.harvard.edu/sriram)

Positions:

1. 2008 – 2013 PhD (Systems Biology), University of Illinois, Urbana (Advisor: Nathan Price)
2. 2011 – 2013 Visiting Fellow, Institute for Systems Biology, Seattle
3. 2013 – Junior Fellow and Principal Investigator, Harvard University, Cambridge (Mentor: James J. Collins, MIT)

4. 2013 – Visiting Fellow, Broad Institute of MIT and Harvard, Cambridge

Successful Proposals:

1. William Milton Fund, Harvard Medical School (PI)
2. Harvard Junior Fellowship from the Harvard Society of Fellows
3. Robert Emerson Fellowship
4. Howard Hughes Medical Institute (HHMI) International Predoctoral Fellowship

Current and Past Research Accomplishments:

A. Genome-scale Modeling and Drug-discovery:

1. Developed the first integrated framework called PROM (Probabilistic Regulation of Metabolism), for automated integrated modeling of metabolic and regulatory networks at the genome-scale (Chandrasekaran and Price, PNAS 2010).

2. Created the GEMINI (Gene Expression and Metabolism Integrated for Network Inference) algorithm, that integrates data across many published experiments to iteratively test and curate network models to improve their predictive ability (Chandrasekaran and Price, PLOS Computational Biology, 2013).

3. Developed a new approach called INDIGO (Inferring Nonlinear Drug Interactions using a chemo-Genomic Omnibus), which predicts novel antibiotic combinations with enhanced potency from chemogenomic data (Chandrasekaran et al, in revision)

B. Reverse-engineering and Multi-Omic Data Integration:

1. Designed a new reverse engineering pipeline called ASTRIX (Analyzing Subsets of Transcriptional Regulators Influencing eXpression), to reconstruct the first genome-scale model of a gene regulatory network operating in the brain, and identified regulatory genes that control behavior states in the brain (Chandrasekaran et al PNAS 2011).

2. Integrated analysis of transcriptomic and metabolomics data led to the discovery of a novel metabolic state in the brain of aggressive honeybees that is strikingly similar to a metabolic alteration that occurs in cancer cells called the Warburg Effect. (Chandrasekaran et al, Genes Brain and Behavior, 2015).

3. Currently developing new metabolic models that integrate transcriptomic, epigenomic and metabolomics data to investigate normal and aberrant metabolic states, such as the Warburg Effect, in diverse cancers and in brain disorders (supported by seed funding from the William Milton Fund to S.C). This can lead to metabolic drug targets or adjuvants for combination therapy for cancer and neurological disorders

Peer-reviewed Publications

$^{+}$- Corresponding Author, * - Co-first author

Research Articles:

1. **Chandrasekaran S^{+}**, Cokol-Cakmak M, Sahin N, Yilancioglu K, Kazan H, Collins J.J^{+} and Cokol M^{+}, "Chemical-genetic Inference of Antibiotic Interactions for Combination Therapies" (in revision)
2. **Chandrasekaran S***, Rittschof C*, Djukovic D, Gu H, Raftery D, Price N.D, and Robinson G.E, "Aggression is Associated with Aerobic Glycolysis in the Honey Bee Brain", Genes Brain and Behavior, 2015
3. **Chandrasekaran S** and Price N.D. "Metabolic Constraint-based Refinement of Transcriptional Regulatory Networks", Plos Computational Biology, 2013. Highlighted in Nature Reviews Genetics
4. **Chandrasekaran S,** Ament S.A, Eddy J.A, Rodriguez-Zas S.R, Schatz B.R, Price N.D, and Robinson G.E,

"Behavior-specific changes in transcriptional modules lead to distinct and predictable neurogenomic states", ***PNAS***, 2011 (Cover Article). Highlighted in PNAS

5. **Chandrasekaran S** and Price N.D. **,** "Probabilistic integrative modeling of genome-scale metabolic and regulatory networks in Escherichia coli and Mycobacterium tuberculosis,**"** ***PNAS***, 2010
6. Brooks A.N, Reiss D.J, Allard A, Wu W, Salvanha D.M, Plaisier C, **Chandrasekaran S**, Pan M, Kaur A, Baliga N.S. "A system level model for the microbial regulatory genome", Molecular Systems Biology, 2014

Review Articles:

1. **Chandrasekaran S**, "Predicting Phenotype from Genotype through Reconstruction and Integrative Modeling of Metabolic and Regulatory Networks", Systems and Synthetic Biology: A Systematic Approach, 2014
2. Simeonidis E, **Chandrasekaran S**, Price N.D. "A guide to integrating transcriptional regulatory and metabolic networks using PROM (Probabilistic Regulation of Metabolism)", *Methods in Molecular Biology: Systems Metabolic Engineering*, 2012.
3. Sung J, Wang Y, **Chandrasekaran S,** Witten DM, and Price N.D, "Molecular signatures from omics data: from chaos to consensus"**.** *Biotechnology Journal*, 2012.
4. L.B. Edelman**, Chandrasekaran S,** and Price N.D. "Systems biology of embryogenesis"**,** *Reproduction, Fertility, and Development*, 2010.

Honors and Activities:

1. Elected to the Harvard Society of Fellows, 2013
2. Lemelson MIT Illinois Student Award for innovation - finalist, 2012
3. HHMI International Predoctoral Fellowship, 2011
4. Outstanding Oral Presentation Award, CMBTG Symposium, UIUC, 2010
5. Biophysics Travel Grant Award, 2010
6. Q-Bio Conference Travel Fellowship, 2010
7. Reviewer for PNAS, Plos One, BMC Systems Biology, Molecular Biosystems, Journal of Theoretical Biology

438210 Modeling the Impact of Bubbling Bed Hydrodynamic Oscillations on the Yield of Biomass Fast Pyrolysis Oil

Sunday, November 8, 2015
Exhibit Hall 1 (Salt Palace Convention Center)
Qingang Xiong, Computer Science and Mathematics Division, Oak Ridge National Laboratory, Oak Ridge, TN

Modeling the impact of bubbling bed hydrodynamic oscillations on the yield of biomass fast pyrolysis oil

Qingang Xiong

Oak Ridge National Laboratory, Oak Ridge, Tennessee

Email: xiongq@ornl.gov

Abstract Fast pyrolysis of biomass, in combination with catalytic upgrading of the resulting oils, has become a leading candidate technology for producing renewable transportation fuels and high-value organic chemicals from biomass. Bubbling fluidized bed reactors are widely used for biomass pyrolysis because of their excellent heat and mass transfer, but their complex nonlinear dynamics pose a significant challenge to achieving the needed degree of chemical reaction control. As a part of our participation in DOE's Computational Pyrolysis Consortium (www.cpcbiomass.org), we are conducting numerical simulations of a laboratory-scale bubbling fluidized-bed pyrolysis reactor to investigate the effects of bubbling bed hydrodynamics on the temporal variations in bio-oil yield and composition. The yield and composition of the raw pyrolysis oil is critical to the success of further upgrading steps, such as vapor-phase or liquid-phase deoxygenation and hydrogenation. In the simulated pyrolysis reactor, all phases were modeled as interpenetrating continua through the so-called multi-fluid model within the framework of the open-source code MFIX. A global multi-component multi-step mechanism proposed in the literature for biomass was chosen for simulating the fast pyrolysis reactions. The instantaneous product yields over time were predicted at the reactor exit and their overall statistics characterized. The initial bed height, fluidization intensity, and biomass injection rate were changed to shift the hydrodynamic state in order to observe the resulting temporal and statistical variations in the product yield and composition. Based on these results, we recommend general limits on the operating conditions for biomass bubbling bed pyrolyzers and suggest specific directions for future experimental and computational studies of biomass fast pyrolysis.

438242 Interface Engineering for Sustainability and Health Care

Sunday, November 8, 2015
Exhibit Hall 1 (Salt Palace Convention Center)
Rong Yang, Harvard Medical School, Boston, MA

Energy, clean water, and medicine are the three most prominent global problems for sustainability. Control of surface interactions via organic thin films is critical to saving energy, providing clean water, and engineering better medicines and diagnostics. My research program will have a strong emphasis on multidisciplinary collaborations with biologists, chemists, and mechanical engineers. I will initially focus on the following topics:

1) At the meso- to macro-scale, controlling biofilm formation and cell assembly on a surface to direct tissue repair and fight fouling, chronical infection and antibiotic resistance;

2) At the mesoscale, controlling the transport of clusters of molecules for new membrane design;
3) At the molecular scale, directing the assembly of functional molecules at the interface to fabricate inexpensive and rapid diagnostic devices for personalized medicine.

This focus will require a highly interdisciplinary approach, a deep understanding of surface science and engineering, polymer and material science, nanotechnology and a good grasp of the current technology in the chemical and biomedical engineering areas. My Ph.D. training in fouling mitigation of seawater desalination membranes and post-doctoral research in biomaterials in clinical applications bring together experience in each of these fields.

438249 Engineering Interleukin-2 Antibodies to Shape Immune Homeostasis

Abstract:

Interleukin-2 (IL-2) is a multifunctional cytokine that is essential to the differentiation, growth, and activity of both stimulatory and suppressive cells of the immune system. Its key role in regulating immune homeostasis renders IL-2 an attractive therapeutic target for a wide range of immune-linked disorders such as autoimmune disease, cancer, and infectious disease. Unfortunately, efforts to develop IL-2 as a therapeutic have been limited by its concurrent promotion of immune effector cells (such as memory-phenotype [MP] $CD8^+$ T and natural killer [NK] cells) and regulatory T (T_{Reg}) cells, which leads to undesirable off-target effects and dose-limiting toxicity. It would therefore be of tremendous therapeutic value to isolate and selectively modulate the immunostimulatory and immunosuppressive effects of IL-2 in order to bias immune homeostasis for particular disease applications. We have implemented a structure-guided antibody engineering approach to achieve selective bias of IL-2 activity through engagement of particular cell subsets based on their surface receptor expression patterns. Our work represents a major advance in IL-2 therapeutic development and provides a molecular blueprint for the design of other cytokine-targeted drugs.

Whereas most cytokines signal through engagement of only two receptor chains, IL-2 signals through formation of either a high-affinity quaternary complex (K_d≈10 pM) with the interleukin-2 receptor-α (IL-2Rα, also known as CD25), IL-2Rβ, and IL-2Rγ (also known as common gamma, γ_c) chains, or an intermediate-affinity ternary complex (K_d≈1 nM) with only the IL-2Rβ and γ_c chains. Consequently, expression of the non-signaling IL-2Rα subunit regulates cytokine sensitivity whereas the IL-2Rβ and γ_c chains mediate signaling. IL-2Rα is robustly expressed on T_{Reg} cells but virtually absent from naïve effector cells, resulting in differential IL-2 responsiveness of these subsets. It was recently noted that immunocomplexes consisting of mouse IL-2 (mIL-2) and certain anti-cytokine antibodies preferentially stimulate particular immune cell subsets, presenting an enticing strategy for selective modulation of immune homeostasis. Immunocomplexes comprised of IL-2 and the JES6-1 antibody actively induce proliferation of IL-2Rα^{High} T_{Reg} cells but not IL-2Rα^{Low} effector cells. In contrast, immunocomplexes comprised of IL-2 and the S4B6 antibody potentiate growth of all immune cells, but particularly favor IL-2Rα^{Low}/IL-2Rβ^{High}effector cells. Subsequent work has validated a vast array of therapeutic applications for these two IL-2-targeted antibodies: JES6-1 immunocomplexes promote graft tolerance and show efficacy in preclinical models of autoimmune disease and S4B6 immunocomplexes exhibit potent anti-tumor activity without inducing toxicity. However, in the absence of structural or molecular characterization, the mechanistic basis for selective cytokine potentiation was unknown.

We combined crystallographic, biophysical, and functional data to elucidate the molecular rationale for antibody-induced bias of cytokine activity. Through determination of the high-resolution molecular structures of both the IL-2:JES6-1 and IL-2:S4B6 immunocomplexes and implementation of surface plasmon resonance-based competition studies, we found that JES6-1 and S4B6 exert complex steric and allosteric effects on IL-2 to alter its molecular properties and, consequently, its functional behavior. JES6-1 sterically blocks binding of IL-2Rβ and γ_c and also allosterically impedes binding of IL-2Rα to the cytokine. Allosteric disruption of the IL-2:IL-2Rα interaction results in exquisite IL-2 sensitivity of IL-2Rα^{High} T_{Reg} cells but not IL-2Rα^{Low} effector cells, as high amounts of IL-2Rα expression are required to overcome this disruption to allow for IL-2 complex formation and signal activation. In vivo immune cell subset expansion studies revealed that selective proliferation of T_{Reg} cells is perpetuated by a transcriptional feedback loop. IL-2:JES6-1 immunocomplex treatment increases immune cell surface expression of IL-2Rα through two concurrent effects: (1) Increased IL-2Rα transcription, which has been shown to occur in response to IL-2 signaling, and (2) Selective proliferation of cells with particularly high surface densities of IL-2Rα. This elevated IL-2Rα expression in turn results in enhanced sensitivity to the IL-2:JES6-1 immunocomplex, further heightening selectivity for IL-2Rα^{High} T_{Reg} cell expansion. In the case of the IL-2:S4B6 immunocomplex, we found that S4B6 sterically occludes IL-2Rα and mildly obstructs IL-2Rβ binding to the cytokine but it also allosterically strengthens the IL-2:IL-2Rβ interaction by inducing an affinity-enhancing conformational change in IL-2. The net result of these structural effects is desensitization to IL-2Rα expression as S4B6-bound IL-2 signals equivalently through the IL-2 ternary complex on all IL-2 responsive immune cells, and susceptibility to IL-2 signaling is instead governed by IL-2Rβ expression. Allosteric enhancement of IL-2:IL-2Rβ interaction by S4B6 leads to increased stimulatory activity that particularly favors IL-2Rβ^{High}effector cells, resulting in the observed bias toward immunostimulatory effects following IL-2:S4B6 immunocomplex treatment. These insights provide, for the first time, a direct link between the molecular interactions of cytokine-antibody complexes and their unique functional outcomes, establishing a new paradigm for antibody-mediated modulation of IL-2 behavior.

We harnessed our structural insights to engineer variants of the JES6-1 antibody with enhanced functional properties by combining molecular engineering tools with immunological characterization techniques. We pursued two objectives: (1) Development of antibodies that accentuate the T_{Reg} cell-biased activity of JES6-1 and (2) Development of a JES6-1 variant that cross-reacts with the human IL-2 (hIL-2) cytokine. Currently, JES6-1 is the gold standard antibody used ubiquitously to expand T_{Reg} cells, but there is significant room for improvement as IL-2:JES6-1 complexes are effective prophylactically in mouse models of autoimmune disease, but actually *accelerate* pathogenesis if administered after disease onset. This results from IL-2:JES6-1 immunocomplex-induced upregulation of IL-2Rα on activated effector cells, which heightens their sensitivity to IL-2 signaling and counteracts the T_{Reg} cell bias. Furthermore, JES6-1 is presently limited to mouse studies as it does not interact with hIL-2. We leveraged our exclusive structural insights to design yeast surface displayed libraries of JES6-1 that mutagenized residues in the cytokine-interacting epitope, and evolved these libraries to fulfill our two objectives. To develop antibodies with enhanced T_{Reg} cell selectivity, we evolved JES6-1 variants that bind IL-2 with a range of affinities and inhibit the IL-2:IL-2Rα interaction to varying extents. We then directly evaluated the relative in vivo T_{Reg}:effector cell expansion ratio elicited by these variants to determine the optimal IL-2:JES6-1 interaction affinity for T_{Reg}cell-biased growth. To isolate a JES6-1 variant that interacts with hIL-2, we selected and functionally characterized antibody clones that interact with both the mouse and human cytokines.

Collectively, our work establishes structure-based engineering as a general approach toward developing effective cytokine-biasing antibodies and introduces novel therapeutic molecules with immediate clinical relevance. Furthermore, our insights and methodologies can readily be applied to other systems to evolve cytokine- or growth factor-directed antibodies with optimized receptor binding and competitive properties to address a host of immunotherapeutic objectives.

438258 Seeing Is Believing - Macrophage-Targeted Theranosis

Sunday, November 8, 2015
Exhibit Hall 1 (Salt Palace Convention Center)
Rahul Keswani, University of Michigan, Ann Arbor, MI

While advanced genetic technologies, precise drug synthesis chemistry and controlled release systems have resulted in remarkable progress in the field of personalized medicine, gene and cell therapy; critical barriers – untargeted therapy, immune system defense & active non-invasive therapy monitoring – continue to hamper the realization of superlative medical treatments. Perhaps, the optimal way to design therapeutics would be to seek out our natural defense mechanisms and empower them with the potential to either transport the therapeutic or treat the disorder themselves while allowing the active monitoring of the therapy. My research interests lie in the development of such intelligent therapeutics via self-assembly and depot formulations of drug or gene theranostics (therapy + diagnostics).

The immune system is remarkably efficient in defending the human body from unwanted pathogens. Unfortunately, the same defense mechanisms disallow most current therapeutics from performing to their full potential due to ill-serving pharmacology leading to inefficient biodistribution and targeting. Pharmaceutical research has advanced to the point wherein the application of current advanced bioinspired materials such as synthetic lipids, biopolymers, polymersomes, cell-derived supramolecular structures such as exosomes combined with functional moieties such as active pharmaceutical ingredients, genetic constructs, hormones and peptides can be used to generate super-therapeutics and advanced synthetic viruses that can deliver the desired therapy. My research focus is to use the aforementioned toolkit and add immune cell targeting via specific linkers and targeting moieties and theranostic elements – such as dye molecules, contrast agents, activated peptides, acoustic probes amongst others. This would enable the design of a library of smart theranostics with a range of pharmacological activity and detection capabilities.

My first plan of action would be to target ***macrophages***– capable systemic immune cells that can transport and distribute foreign supramolecular assemblies. Macrophages are immune cells with unparalleled capabilities to reach diseased locations such as tumors, tissue inflammation, infections and physical injury. Targeting macrophages could have a dual purpose – one, to either transport the therapy to their intended site of action or two, provide the necessary therapy themselves via promotion of their natural defense mechanisms. For example, such a strategy could be used to target synovial macrophages for arthritic theranostic therapy or peritoneal macrophages for theranostic therapy of systemic inflammation and bacterial infections.

Additional design elements would come from ligands or peptides that would modulate the pharmacology of intracellular interactions or binding to the intended site. This library of theranostic treatments could be further expanded to include new epigenetic and gene editing/regulating based therapies such as DNA methyltransferase (DNMT) inhibitors and CRISPR-associated genetic switches as well as the inclusion of more complex contrast agents which can be detected via PET, CT scans and microwave medical imaging. Overall, these design principles represent a strong leap forward in the field of pharmaceutics design and theranosis that can lead to significant advancement in global medicinal practice and personalized medicine.

(6cx) Mechanical Work Makes Important Contributions to Surface Chemistry

Francis, M., EPFL

Surface chemistry is thought to be controlled exclusively by electronic factors. In this work, we show that in addition to the established electronic factors a mechanical work term arises upon the application of stress (*Nat Comm* **6**, 2015). This mechanical work term is due to the interaction of the applied stress field with the relaxation (strain field) induced by the presence of a bound molecule. The mechanical work term has dependencies different from established electronic parameters and allows for cases of stronger (weaker) binding upon the application of compression (tension), opposite to expectations from electronic parameters. It is shown explicitly for the case of methanation catalysis that stress induced chemistry changes may shift the activity of Ni to that of Ru and Co; an activity shift comparable to moving one element across the periodic table with compositional or structural changes. When designed in using core-shell particles a stress field may be used to tune the activity of a catalyst to hit peak activity precisely adding a new technique for catalyst design.

NUMERICAL SIMULATION OF A DELAYED COKING REACTOR

Fabián A. Diaz[1,*], *Arlex Chaves*[1], *Maria Maradei*[1], *David Fuentes*[1], *Alexander Guzmán*[2], *Hector Picón*[2].

1. *Universidad Industrial de Santander, Bucaramanga, Colombia.*
2. *Instituto Colombiano del Petróleo, Bucaramanga, Colombia.*

**. First author email: fabiandiazmateus@hotmail.com*

Abstract

A numerical unsteady simulation of a pilot plant delayed coking reactor was performed, it was modelled as a three-phase system being the vacuum residue the liquid phase, the distillables the gas phase and the coke the solid phase. The dynamics of the liquid and the gaseous phase were described by continuity, momentum and energy balances; the solid phase is described only by continuity and energy balances since it is defined static and remains still where it is formed. The finite volume method was employed to carry out the discretization of the governing equations; a modified SIMPLEC algorithm achieved the coupling of the pressure-momentum equations which were solved on a 2D axisymmetric grid. A 4 hours run of the coking reactor was simulated; as this investigation is still at early stage, the only experimental parameter available is the size of the coke bed, the results of the simulation present good agreement when compared with this experimental data.

Keywords: CFD, computational fluid dynamics, vacuum residue, coke, mathematical modelling.

1. Introduction

The delayed coking process is a purely thermal process where a heavy residual oil is converted into coke, liquids and gases at temperatures above 400°C. At industrial level, the residue is heated in furnaces and rapidly transferred to a drum, the word "delayed" comes from the attempt to delay the formation of coke inside the furnace and take it only inside the drum. The process is semi-continuous because when one drum is full of coke the feed switches to a second drum while the first one is discharged and so on.

The delayed coking process is the preferred choice for refiners for the thermal conversion of petroleum residua due to the flexibility to process different types of residua (Rana *et al.*, 2007) and the superior economic benefits when compared to other processes for thermal conversion of heavy oil (Vartivariam & Andrawis, 2006).

A detailed description of the process for coke formation inside a drum has been presented by Ellis & Hardin (1993), few attempts to numerically simulate this unsteady process have been presented in the literature as most of the work focuses on the coking furnace, such as Nunhez *et al.* (2006) who performed, with their own programming code, a numerical simulation where the coke is deposited inside the pipe walls. Also, Zhang et al. (2015) performed a numerical simulation with CFD (Computational Fluid Dynamics) commercial code of a coking furnace with deposition of coke inside the pipe walls. However, numerous simulators based on correlations or statistical models have been presented in the literature, some of the most recognized are Gary & Handwerk (1975); Castiglioni (1983); Maples (2000); Zambrano (2001); Volk, Wisecarver & Sheppard (2002), Pushpalayari (2004). These models are based on the assumption that the process is in steady state and they calculate the yield of coke, liquids and gases according to the properties of the feed and the operating conditions. An interesting comparison of these correlation models for delayed coking simulation can be found in Ancheyta *et al.* (2013).

Simulation models based merely on chemical kinetics can also be found in the literature such as Bozzano & Dente (2005) and Zhou *et al.* (2007). These models also calculate the yield of coke, liquids and gases according to the properties of the feed and the operating conditions, however, the composition of the liquid is relatively more defined as it is divided into pseudo-components. An even more detailed simulator is the one presented by Tian, Shen & Liu (2012a, 2012b) based on structured oriented lumping (SOL), with this lumping the feed and the products are characterized at molecular level offering up to 7000 pseudo-components. The simulation is also based on the assumption that the process is in steady state.

In this work, a numerical unsteady simulation of a pilot plant delayed coking reactor from ICP (Colombian Institute of Petroleum) is performed; the mathematical modelling is based on unsteady continuity, momentum and energy balances for three phases (gas, liquid and solid) which are solved using CFD techniques. The results are presented throughout the run time and provide a changing coke bed which resembles the process for coke formation inside a drum presented by Ellis & Hardin (1993).

2. *The Coking reactor*

A scheme of the coking reactor from the delayed coking pilot plant at ICP is presented on Figure 1. The reactor is basically a pipe with walls that are heated by electrical resistances, given the geometry, a 2D axisymmetric grid can simulate the entire physical geometry of the reactor. The only experimental data available is the size of the coke bed, which was measured after its removal from the reactor. The complete set of operating conditions used in the experimental procedure and the simulation is on Table 1.

Table 1. Operating conditions for the delayed coking reactor

Temperature of the walls	510°C
Temperature of the feed	370°C
Feed mass rate	5g/min
Reactor diameter	2.5 NPS, SCH 40
Reactor length	569.2 mm
Operating pressure	1 atm.

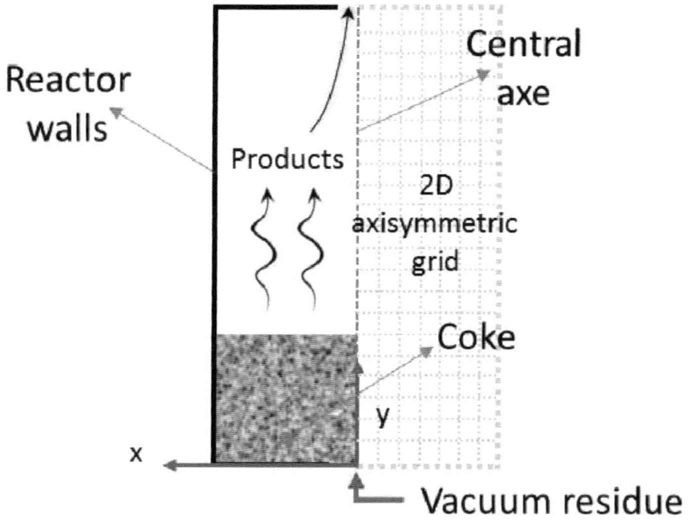

Figure 1. Scheme of the delayed coking reactor at ICP

3. *Kinetic model*

The kinetic model used in the simulation of the delayed coking reactor is taken from Diaz *et al.* (2015). It consists in a simple kinetic model where the vacuum residue decomposes into coke and distillables, in this model the reaction mechanism changes from first to second order depending on the temperature. The complete characterization of the vacuum residua used in this work can also be found on Diaz *et al.* (2015) and it corresponds to the number 3 vacuum residue.

4. *Mathematical model*

For the development of the governing equations, a series of assumptions were made in order to simplify the modelling without significantly affecting the results, these assumptions are:

- The model is 2D axisymmetric orbiting the central axe.
- The system is composed by three phases: Liquid (vacuum residue) and gas (distillables) at laminar flow and solid (coke).
- The liquid is the primary phase and the gas is the secondary.
- The chemical reaction can be model by the thermal decomposition (pyrolysis) of vacuum residue into distillables and coke.
- The coke does not have movement, it remains still at the place where it is formed.
- The momentum loss through the coke bed can be calculated as a loss through a packed bed.
- The porous medium (coke bed) is isotropic (homogenous).
- The effects of pressure and viscous heating on temperature are discarded.

4.1 Continuity equations

Continuity equation for the residue (R) and distillable (D) phase:

$$\frac{\partial(\gamma\alpha_n\rho_n)}{\partial t} + \nabla\cdot(\gamma\alpha_n\rho_n\vec{v}_n) = -\gamma\alpha_n\rho_n\Gamma_n \qquad \textit{Eq. 1}$$

Where $n = R, D$

Continuity equation for the coke phase:

$$\frac{\partial(\alpha_C\rho_C)}{\partial t} = \gamma\alpha_R\rho_R\Gamma_C \qquad Eq.\ 2$$

Continuity equations can also be called volume fraction equations. The porosity of the coke bed is calculated as:

$$\gamma = 1 - \alpha_C \qquad Eq.\ 3$$

4.2 Momentum equations

Momentum equation for the residue (R) and distillable (D) phase:

$$\frac{\partial(\gamma\alpha_n\rho_n\vec{v}_n)}{\partial t} + \nabla\cdot(\gamma\alpha_n\rho_n\vec{v}_n\vec{v}_n) = -\gamma\alpha_n\nabla\cdot\mathrm{P} + \nabla^2\cdot(\gamma\alpha_n\mu_n\vec{v}_n) + \gamma\alpha_n\rho_n\vec{g} + \gamma K_{mn}(\vec{v}_m - \vec{v}_n) - \alpha_n\vec{v}_n\left(\frac{\mu_n}{\varepsilon} + \frac{C_2\rho_n|\vec{v}_n|}{2}\right) \qquad Eq.\ 4$$

Where $n, m = R, D$ and $m \neq n$

The coke phase does not have momentum equation.

4.3 Energy equations

Energy equation for the residue (R) and distillable (D) phase:

$$\frac{\partial(\gamma\alpha_n\rho_n C_{P,n}T_n)}{\partial t} + \nabla\cdot(\gamma\alpha_n\rho_n C_{P,n}\vec{v}_n T_n) = \nabla^2\cdot(\gamma\alpha_n\kappa_n T_n) + \gamma\alpha_n\rho_n\Gamma_n\Delta\dot{H}_n + \gamma H_{mn}(T_m - T_n) + \alpha_n H_{Cn}(T_C - T_n) \qquad Eq.\ 5$$

Energy equation for the coke (C) phase:

$$\frac{\partial(\alpha_C\rho_C C_{P,C}T_C)}{\partial t} = \nabla^2\cdot(\gamma\alpha_C\kappa_C T_C) + \alpha_R H_{CR}(T_R - T_C) + \alpha_D H_{CD}(T_D - T_C) \qquad Eq.\ 6$$

4.4 Parameters

The momentum exchange coefficients (K_{mn}) are calculated with the symmetric model (See ANSYS (2014)), the symmetric model is recommended where a secondary phase becomes primary phase in different places of the domain, as it happens in this simulation. The permeability (ε) and the inertial resistance factor (C_2) are calculated with the correlations from Ergun (1952). The fluid-fluid heat transfer coefficients (H_{mn}) are calculated with Tomiyama´s correlation (Tomiyama (1998)) which is appropriate for low Reynolds flow. The fluid-solid heat transfer coefficients (H_{CR}, H_{CD}) are calculated with the correlations from Wakao and Kaguei (1982) also recommended for low Reynolds flow.

5. *Solution of the equations*

The governing equations were solved using CFD techniques, the equations were discretized by the finite volume method and the pressure-velocity coupling was achieved by means of a SIMPLEC algorithm (Van Doormal & Raithby (1984)). The procedure proposed by Patankar & Spalding (1972), to obtain the pressure correction equation through an arrangement of the continuity and momentum equations, was used in this work.

As the momentum exchange (Drag) between the phases gas and liquid is considered in the momentum equations, the partial elimination algorithm (PEA) (Spalding & Markatos (1983)) was used to avoid convergence issues caused by the coupling of the momentum equations. In order to avoid the famous "pressure checker-board" or unrealistic pressure oscillations caused by the interpolation of the pressure, a staggered grid is used for the solution of the momentum equations.

Continuity, momentum and energy equations were solved with a Gauss-Seidel procedure. This method performs a sweeping through all the cells of the grid solving an equation of the form of equation 7 to obtain the value of the variable in the cell. This method provides an advantage in multiphase models, if the volume fraction value of certain component is zero in the cell, the value of other parameters, such as velocities, take the value of zero immediately in the cell, in this manner, the solution of the equations is more precise and faster than other solution methods.

$$\phi_P = \frac{\sum a_{NB}\phi_{NB} + \hat{S}_P}{a_P} \qquad \textit{Eq. 7}$$

The pressure correction equation was solved with a SOR (Successive over relaxation) method, this method is practically the same Gaus-Seidel method but over-relaxed with a relaxation factor between 1 – 2 that accelerates convergence. The simulator was programmed in C++ code and compiled in Linux with an Intel compiler; with a time step size of 0.01s, the simulation of the 4 hours run took about 6 days.

6. *Results*

The results of the simulation are presented according to the simulation time, phase velocities are presented as a magnitude not as a vector. Results are presented as colored contours, the colormap at the right side of the graphs represent the range of the values, volume fractions and temperatures of the phases are grouped to provide a single colormap which makes it more clear.

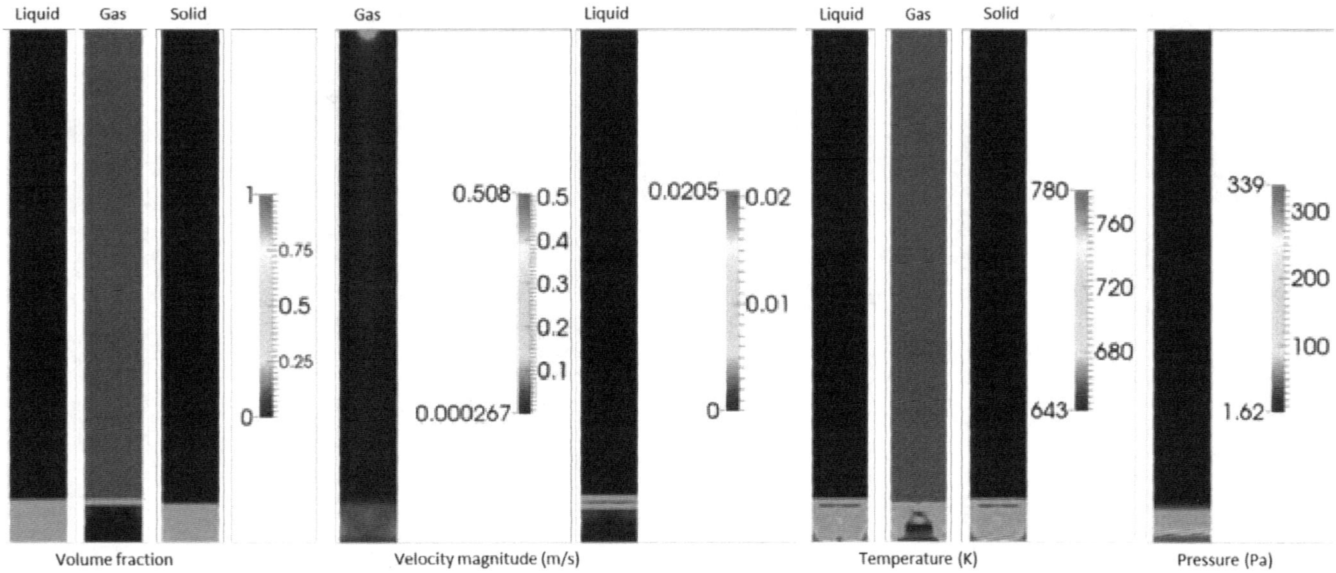

Figure 2. Results of the coking reactor simulation at a simulation time of 1h.

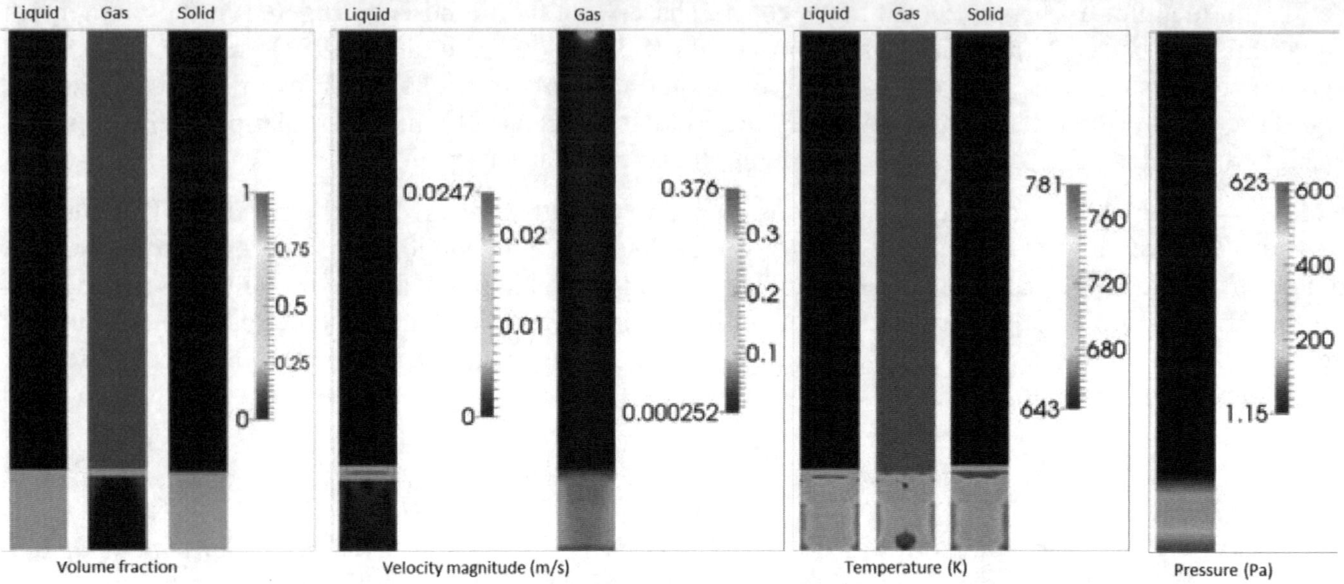

Figure 3. Results of the coking reactor simulation at a simulation time of 2h.

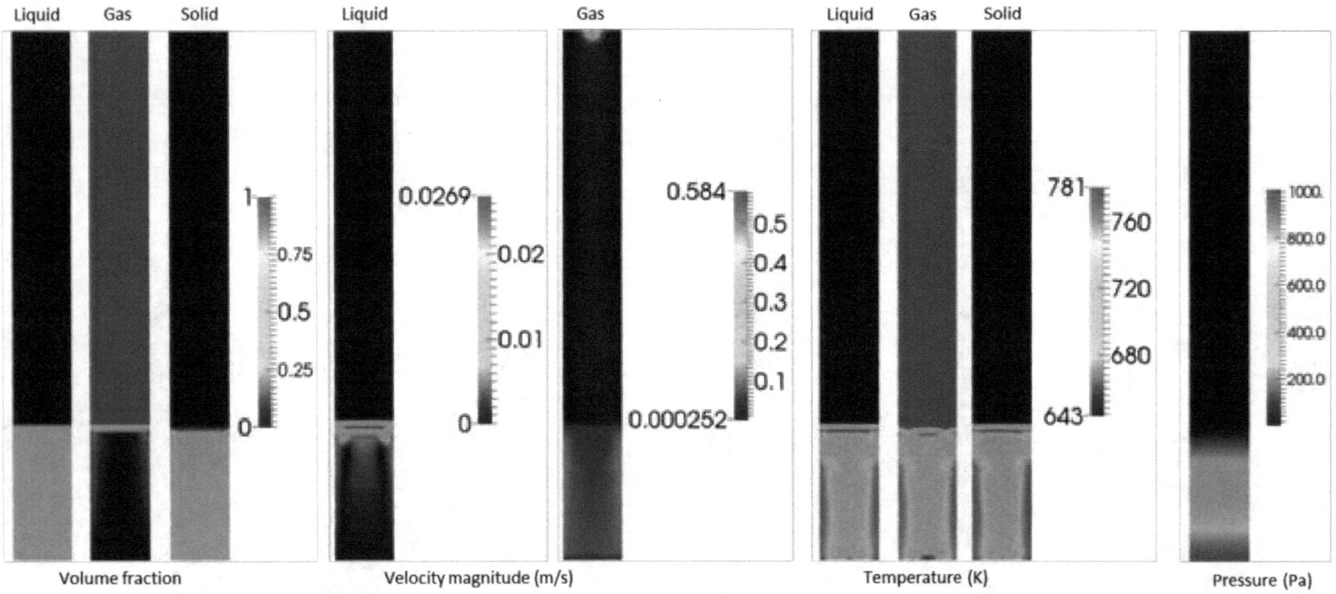

Figure 4. Results of the coking reactor simulation at a simulation time of 3h.

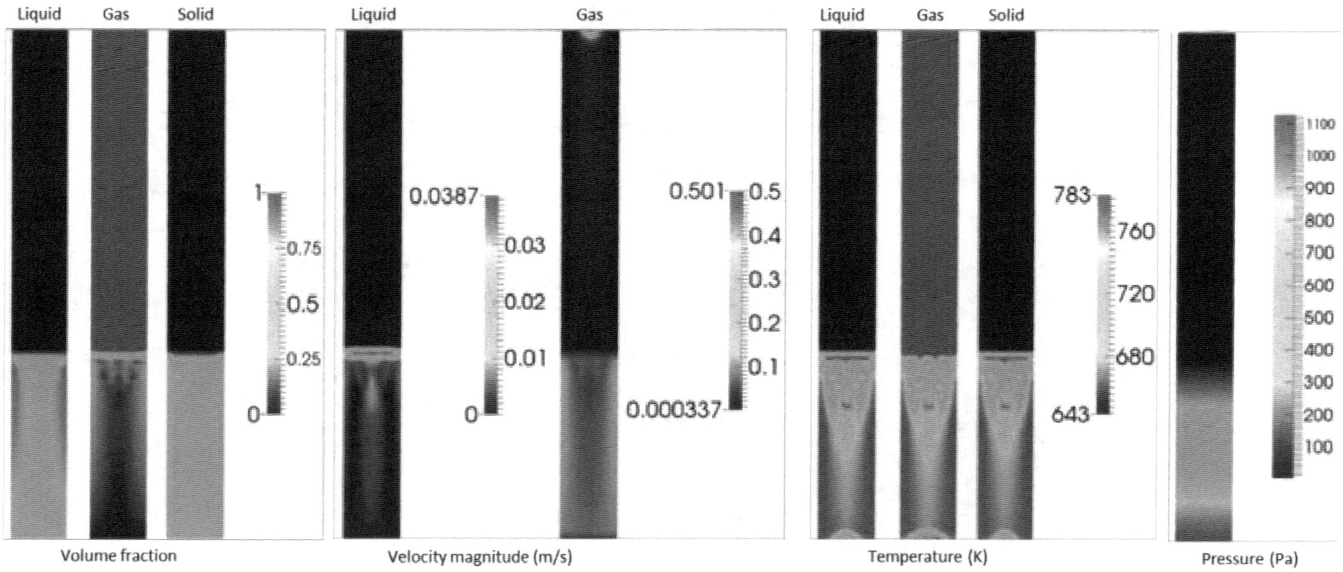

Figure 5. Results of the coking reactor simulation at a simulation time of 4h.

In figures 2 – 5 it can be seen the volume fractions, velocity magnitudes, temperatures and pressure drop calculated at 1, 2, 3 and 4 hours of run. The graphs of temperature for the liquid and solid phase in the upper part of the reactor shows a temperature of 643°C, but in the volume fraction graph it is seen that the volume fraction is zero in the same part. This occurs because the simulator is programmed to assign the lowest temperature of the system to cells with zero volume fraction of any phase. A temperature of zero degrees could not be assigned to cells with zero volume fraction because it generated an error in the simulation.

Table 2. Size of the coke bed at the end of the simulation

Size of the coke bed (calculated)	Size of the coke bed (measured)
20.78 cm	20.5 cm

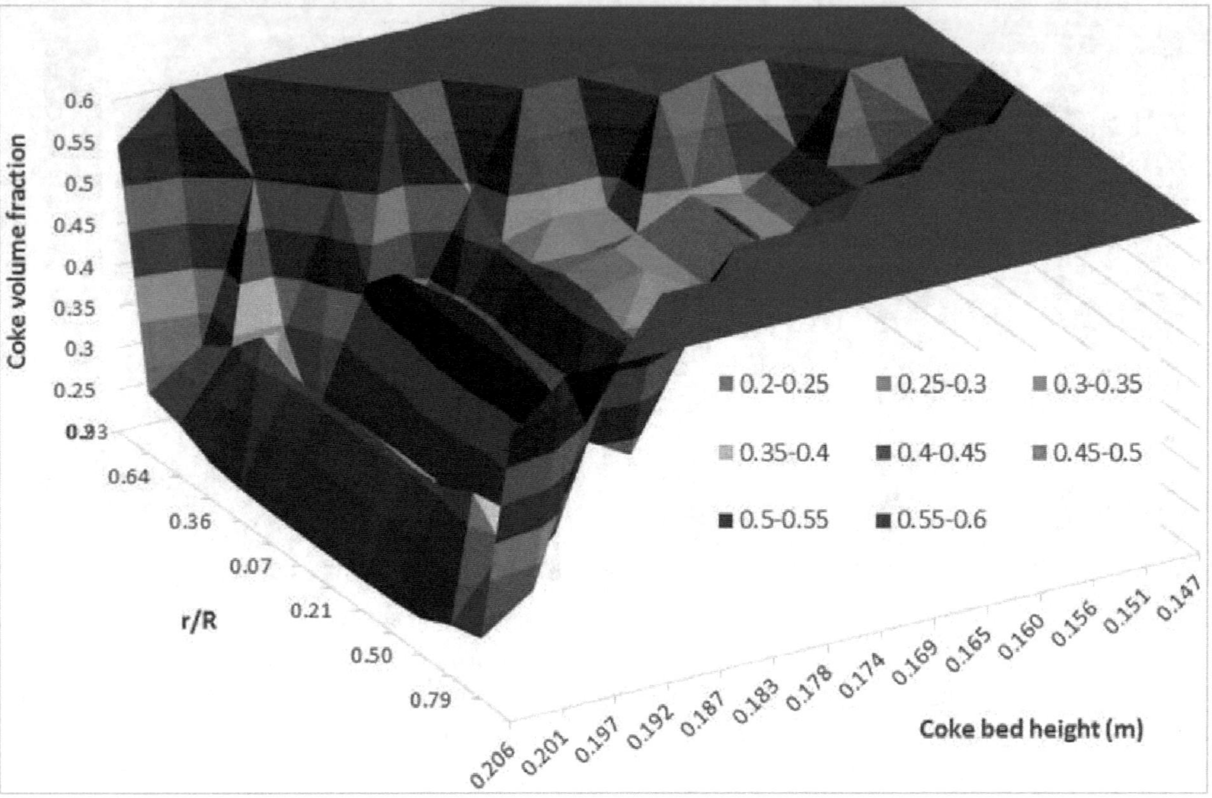

Figure 6. Shape of the upper part of the coke bed at the end of the simulation.

7. *Discussion*

The predicted and measured size of the coke bed, presented on Table 2, are practically equal. The volume fraction of coke calculated by the simulator shows a tendency to gather at the reactor walls, which means the porosity of the coke bed is lower at the walls and higher at the center of the reactor, it also provides an inverted cone shape in the upper part of the coke bed, as it can be seen on Figure 7. This observation was confirmed by visual inspection of the samples of coke extracted from the reactor. Due the small diameter of the reactor, the coke bed tends to channel towards the center.

The temperature of the three phases are practically equal (See Figures 2 – 5) which suggests that the phases are in thermal equilibrium, this is probably caused by the small diameter of the reactor. Figure 5 shows how the temperature is higher at the walls and decreases around 40°C at the top of the coke bed due to the heat absorbed by the pyrolysis reaction. Unfortunately, there is not experimental data to confirm these results yet they are consistent with what is expected in this process.

8. *Conclusions*

The size of the coke bed calculated by the numerical simulation of a delayed coking reactor presented in this work shows good agreement when compared with experimental data, even though most of the results could not be compared with experimental data, they are consistent with what is expected in this process. The results of the simulation present a porosity of the coke bed that decreases from the walls to the center of the reactor and a coke bed with the shape of an inverted cone. These observations were confirmed by visual inspection of the samples of coke extracted from the reactor, however, porosity values have not been measured to confirm the values found in the simulation.

Parameters such as temperatures, velocities and pressure drop still have to be measured to confirm the validity of the mathematical model proposed to simulate the delayed coking process.

Nomenclature

$\vec{v}$	Velocity vector	a	Convection-diffusion coefficient
t	Time	ϕ	Random variable
P	Pressure	S	Source term
T	Temperature	γ	Porosity of the coke bed
$\vec{g}$	Gravitational acceleration vector	α	Volume fraction
K	Momentum exchange coefficient	ρ	Density
C_P	Heat capacity	Γ	Reaction rate
$\Delta\dot{H}$	Heat of reaction	C_2	Inertial resistance factor
H	Heat transfer coefficient	ε	Permeability of the coke bed
μ	Viscosity	κ	Thermal conductivity

Subscripts

n	Residual phase or distillable phase	C	Coke phase
R	Residual phase	NB	Neighbored cell
D	Distillable phase	P	Core cell

References

1. Rana, Mohan S., Samano, V., Ancheyta, J. & Diaz, J.A.I. (2007). A review of recent advances on process technologies for upgrading of heavy oils and residua. *Fuel*, 86 (9), pp. 1216 – 1231.
2. Varvitariam, D. & Andrawis, H. (2006). Delayed coking schemes are most economical for heavy-oil upgrading, *Oil & Gas Journal*, 104 (6), pp. 52 – 56.
3. Ellis, P. & Hardin, E. (1993). How petroleum delayed coke forms in a drum, *Light Metals*, pp. 509 – 515.
4. Souza, B., Matos, E., Guirardello, R. & Nunhez, J. (2006). Predicting coke formation due to thermal cracking inside tubes of petrochemical fired heaters using a fast CFD formulation, *J. Pet. Sci. Eng.* 51(1-2), pp. 138 – 148.
5. Li, X., Zhang, R., Suna, Y., Jiang, B., Luo, M., Li, X. & Zhang, L. (2015). CFD modeling of phase change and coke formation in petroleum refining heaters, *Fuel Process. Technol.* 134, pp. 18 – 25.
6. Castiglioni, B. P. (1983). How to Predict Coker Yield, Hydrocarbon Processing, 62(9), pp. 77 – 79.
7. Gary, J. H. & Handwerk, G. E. (1975). Petroleum Refining – Technology and Economics, Fourth Edition, Marcel Dekker, New York, USA.
8. Maples, R. E. (2000). Petroleum Refinery Process Economics. Penn Well Publishing Co., Tulsa, USA.
9. Zambrano, T. A. (2001). Micro-correlations and data analysis, M.Sc. Thesis dissertation, Department of Chemical Engineering, University of Tulsa, Tulsa, USA.
10. Volk, M., Wisecarver, K. & Sheppard, C. (2002). Fundamentals of delayed coking joint industry project, Department of chemical engineering, University of Tulsa, Tulsa, USA.
11. Pushpalayari, A. G. (2004). Micro-Coker Continuation Studies, M.Sc. Thesis dissertation, Department of Chemical Engineering, University of Tulsa, Tulsa, USA.

12. Muñoz, J., Aguilar, R., Castañeda, L. & Ancheyta, J. (2013). Comparison of Correlations for Estimating Product Yields from Delayed Coking, *Energy & Fuels*, *27* (11), pp. 7179 – 7190.
13. Bozzano, G. & Dente, M. (2005). A Mechanistic Approach to Delayed Coking Modelling, European Symposium on Computer Aided Process Engineering – 15, 20, pp. 529–534.
14. Zhou, X. L., Chen, S. Z. & Li, C. L. (2007). A predictive kinetic model for delayed coking, *Petroleum Science and Technology,* 25 (12), pp. 1539 – 1548.
15. Tian, L., Shen, B. & Liu, J. (2012a). A delayed coking model built using the structure-oriented lumping method, *Energy and Fuels*, 26 (3), pp. 1715 – 1724.
16. Tian, L., Shen, B. & Liu, J. (2012b). Building and Application of Delayed Coking Structure-Oriented Lumping Model, *Ind. Eng. Chem. Res.*, 51 (10), pp. 3923 – 3931.
17. Diaz, F. A., Chavez, A., Fuentes, D., Maradei, M., Guzman, A. & Picon, H. (2015). Kinetic analysis of the thermal decomposition of colombian vacuum residua by thermogravimetry, *Ingenieria e investigacion*.
18. ANSYS Fluent 14.0 User Manual, in: ANSYS Inc.
19. Ergun, S. (1952). Fluid Flow through Packed Columns, *Chem. Eng. Prog.*, 48(2), pp. 89 – 94.
20. Tomiyama, A. (1998). Struggle with computational bubble dynamics. Third International Conference on Multiphase Flow, Lyon, France.
21. Wakao, N., Kaguei, S. (1982). Heat and Mass Transfer in Packed Beds, Gordon and Breach Science Publishers, New York, USA.
22. Patankar, S. & Spalding, D. (1972). A Calculation Procedure for Heat, Mass and Momentum Transfer in Three-Dimensional Parabolic Flows, *Int. J. Heat Mass Transfer*, 15(10), pp. 1787 – 1806.
23. Van Doormal, J. & Raithby, G. (1984). Enhancements of the SIMPLE Method for Predicting Incompressible Fluid Flows, *Numer. Heat Transfer*, 7, pp. 147 – 163.
24. Spalding, D. & Markatos, N. (1983). Computer Simulation of Multi-Phase Flows: A Course of Lectures and Computer Workshops, Report CFD/83/4, Mech. Eng., Imperial College Sci. Tech., London, UK.

437833 Leveraging Supramolecular Interactions for Therapeutics

Sunday, November 8, 2015
Exhibit Hall 1 (Salt Palace Convention Center)
Matthew Webber, Massachusetts Institute of Technology, Cambridge, MA

Pressures arising from the increasing incidence of disease, coupled with a focus on improving the quality of life for patients, inspires the development of creative healthcare solutions. The next generation of "smart" therapies must include strategies for autonomous sensing of disease state and disease site, routes for high affinity and high efficiency therapeutic targeting, strategies that address the complications that arise from the immune system, and materials that provide both structural *and* functional recapitulation of native tissue properties, among many others. In this regard, therapies based on supramolecular principles could have broad impact. Supramolecular chemistry, defined as "chemistry beyond the molecule", is based on rational design of specific, directional, tunable, reversible, non-covalent molecular recognition motifs that exploit combinations of hydrogen bonding, metal chelation, hydrophobic interactions, pi-pi interactions, and/or van der Waals interactions. Through these motifs, materials can be created with properties that follow from the dynamic nature of their constituents. Leveraging these interactions may contribute distinct and useful properties that have been unrealized thus far by materials used in medicine. Though non-covalent interactions are individually weak in comparison to a covalent bond, the summation and directionality of these interactions leads to materials with mechanical properties approaching those of covalent systems, with the added benefit that these interactions are reversible, highly tunable, dynamic, and modular. My own training has included work with peptide amphiphile supramolecular polymers during my PhD in Prof. Samuel Stupp's lab, along with postdoctoral training in the laboratories of Profs. Robert Langer and Daniel Anderson developing new "smart" strategies for insulin delivery. The Webber laboratory, with a basis in my own personal training, will be built on the use of a diverse toolbox of supramolecular systems to build highly functional next-generation disease therapeutics. A specific emphasis will be placed on therapeutic applications for diabetes, cancer, and inflammatory diseases. Students within this training environment will have the opportunity to cross-train within various disciplines, as developing new supramolecular therapeutics will encompass areas of chemistry, materials science, biology, and biomedical engineering. The interdisciplinary nature of the group will also necessitate strong collaborations, particularly with clinicians, that will further enrich this training environment.

438287 Novel Nanostructured Coatings for Use in Transport Applications

Abstract:

Shipping accounts for at least 90% of the global trade of goods. In terms of distance travelled, shipping is by far the most efficient method of transporting goods. The biggest problem in this method of transportation is the biofouling on ships hulls. Marine biofouling is caused by the adhesion of barnacles, macroalgae and microbial slimes. On ships hulls, biofouling results in an increase in roughness, which in turn leads to an increase in hydrodynamic drag as the vessel moves through water. Enhanced fuel consumption, hull cleaning, paint removal and repainting, and associated environmental compliance measures all contribute to the costs of biofouling. New, effective, and environmentally compatible options are needed to control fouling. One critical point in the development of such paints is that the outer surface has to exhibit certain properties in order to repel fouling microorganisms, such as superhydrophobic behavior and reduced drag force.

The product introduced in this work represents a key enabling technology for manufacturers to maximize profit and gain competitive advantages. Silicone based coatings were prepared with multi-wall nanotube (MWCNTs) of extremely low bulk density that enables efficient dispersion in resins. The goal was to create superhydrophobic surfaces that can also exhibit fouling release properties. Besides the use of pristine nanotubes, chemical functionalization of MWCNTs also took place for comparison purposes as well as graphene-based additives. Different protocols for the nanostructured coatings preparation were also investigated.

The obtained surfaces were fully characterized in terms of hydrophobicity as determined by contact angle measurements. Higher elongation and remarkably improved mechanical strength were concurrently achieved, with respect to the pure resins, thus expecting to lead to increased coating lifetime. The mechanical properties of the pristine MWCNTs were more enhanced than those of functionalized nanotubes, indicating that the production of the novel coatings can be cost effective if the synthesis of MWCNTs has been properly tailored.

The coatings with the most promising results (contact angle > 100°) were immersed in static seawaters to visualize any incurred fouling phenomena as a function of time. Results on the MWCNTs-silicone formulations that resulted in the best fouling release performance are discussed and compared with results from previous related work using other resin matrices. It should be pointed out that the developed coatings do not contain any harmful for the marine ecosystem biocides as the 2nd generation antifouling coatings available in the market. The fouling release properties of the coatings result from the needle-like structure of nanostructured carbon that can facilitate release of microorganisms responsible for biofouling.

438346 Vine Copula-Based Dependence Description for Multivariate Multimode Process Monitoring

Abstract:

For complex chemical processes, process variables with high dimensionality, nonlinearity and non-Gaussian properties are frequently encountered. To maintain process safety and product quality of chemical processes, in recent years, a host of data-based methods have been successfully applied in chemical process monitoring. The existing methods mainly focus on dimensionality reduction or variable decoupling processes, which more or less, will result in information lost or distortion for the process data, thus leading to poor monitoring performance.

As an efficient tool in dependence modeling, vine copula aims to simplify multivariate dependence problem into dealing with optimization problems for a series of bivariate copulas listed in a sparse matrix. It somewhat prevents the problem called "curse of dimensionality" that exists for traditional multivariate copula. In addition, due to its flexibility structure and the rank correlation information preserved, vine copula is capable of handling data reflecting both nonlinear and non-Gaussian properties.

In this work, a novel vine copula-based method is proposed for multimode process monitoring. C-vine copula is initially introduced to establish the joint distribution of normal samples for each mode. The offline modeling procedure includes (a) pair copula construction, (b) conditional distribution calculation, (c) selection and optimization of vine copula models and (d) goodness-of-fit testing. Herein, the statistical information of each mode can be obtained via sampling methods (e.g., Markov Chain Monte Carlo Method). To achieve online monitoring for multimode processes, the global Bayesian Inference-based probability (BIP) index proposed by Yu and Qin is used. Note that Mahalanobis distance is no longer efficient on measuring the distance of monitored data from each non-Gaussian mode, a generalized local probability (GLP) index is then defined. Because the traditional numerical integration method appears rather time-consuming in estimating GLP, density quantile approach is introduced. By analyzing the one-dimensional distribution of probability density function values of each mode, the corresponding density quantile table can be constructed. Therefore, the GLP index as well as the BIP index can immediately be updated in real time by just checking the static table created offline.

The proposed vine copula-based dependence description (VCDD) approach is successfully applied to a numerical example and TE benchmark process. Compared with the FGMM-based approach, the proposed method can achieve higher detection rates, lower false alarm rates and shorter time delays especially to those non-Gaussian processes with tail dependence. Moreover, It is demonstrated that both offline modeling and online monitoring of this approach achieves low computation load, which will make this approach more practical.

438460 Visulization of Transport Dynamics in Complex Fluids

Abstract:

2nd year Post-doctoral Fellow:

Research Interests:

Complex fluids are ubiquitous in nature and the industry. Interesting behaviors arise from these fluids in ways that are difficult to predict directly from their single constituents. This sometimes makes complex fluids useful in some processes that simple materials cannot be used for. It is clear that the complex nature of these fluids is directly connected to their microstructure . Moreover, experimental observations of transport processes in these systems help to elucidate the microscopic nature of their complex behavior. This in turn enables us to tailor the properties of complex fluids for a specific target function. For example, visualization of flow of liquid foams in simple geometries reveals how the interaction between microscopic constituents (i.e. bubbles) affect the macroscopic behavior of material. In another example, simple visualization of the transport of a solid object in a shear banding polymeric fluid reveals how strong and/or flexible the polymer chains are.

Proposals: NSERC Discovery Grant, Canadian Foundation for Innovation, National Science Foundation and American Chemical Society-Petroleum Research Fund.

Postdoctoral Project: "Instability of Shear Banding Fluids in Taylor-Couette Geometry"

Under supervision of Susan J. Muller , Chemical and Biomolecular Engineering, University of California, Berkeley.

PhD Dissertation: "Dynamics and Rheology of Sheared Liquid Foams".

Under supervision of James J. Feng, Chemical and Biological Engineering, University of British Columbia.

Research Experience:

Over the past few years, I have been investigating few projects from both engineering and scientific perspectives. I have been formally trained as chemical engineer with specialty in transport processes in complex systems such as Liquid foams, Viscoelastic, and shear banding fluids. My approach to tackle these pressing problems is mostly through experimentation. Moreover, I have extended my knowledge and expertise by collaborating with multi-disciplinary team of engineers and scientists at UBC and UC Berkeley on couple of transport projects in micro-fluidic devices. As a result of my training and collaborations, I have been proficient in designing, and developing new methods to study the complex behavior and gained a deep understanding of the physics of these systems by directly visualizing their transport process. This in turn greatly helped illuminating the nature of the complex interactions between microscopic elements of these materials which serves as a crucial step to design a smart material or process.

Teaching Experience:

Teaching is an important component of the academic profession. I strongly believe that teaching and research go hand-in-hand. I am committed to the teaching as evidenced from my teaching experiences. I have been teaching assistant for multiple courses in Faculty of Applied Science at University of British Columbia and often taught the course materials as guest instructor. I have instructed undergraduate students in variety of chemical engineering courses such as, Transport Phenomena, Thermodynamics and also in Numerical Computations for Engineering and Science. In addition, I have supervised two senior undergraduate students at UC Berkeley for their projects. My main interest lies in teaching of transport phenomena that includes heat, mass and momentum transfer as well as Thermodynamics. I am well experienced in teaching thermo-fluids for undergraduate students. During my teaching experience at UBC, I realized that teaching is very rewarding to me. This led me to take an intense teaching course offered by learning and education centre at UBC. In this course, I was instructed by well known world class teachers on several teaching techniques. Those techniques were adapted to scientific courses at the university level, and at the end of program I was issued a teaching certificate from UBC teaching and learning centre.

Vision for Future direction:

As a faculty member, I would like to continue applying my experimentations and analytical skills in studying the complex fluids that are applicable in areas from biotechnology to food, and oil-gas industries. My overall approach is to make the investigation of these systems more tractable by using simple and well-developed devices. This will be accomplished by fabricating simple devices and incorporating several techniques such as, particle image velocimetry, and viscometry in experiments. I would also like to continue collaborating with scientist from all other related fields which will at the end, bring a solid and concrete understanding of the nature of the complex behavior.

Selected Publications and Submissions:

- **Mohammadigoushki, H.**, and Muller, S. J., " Interface dynamics in a wormlike micellar fluid sheared in a Taylor-Couette cell: A study on interface travel " Submitted to *Soft Matter*, (2015).
- **Mohammadigoushki, H.**, and Feng, J. J., "Temporal evolution of microstructure and rheology of sheared two-dimensional foam" *J. Non-Newtonian Fluid Mech.* 223, 1-8 (2015).
- **Mohammadigoushki, H.**, Yue, P. and Feng, J. J., "Bubble migration in two-dimensional foam sheared in a wide-gap Couette device: effects of non-Newtonian rheology". *J. Rheol.* **58**, 1809-1827 (2014).
- **Mohammadigoushki, H** and Feng, J. J., "Size segregation in sheared two-dimensional polydisperse foam." *Langmuir* 29, 1370-1378 (2013).
- **Mohammadigoushki, H.** & Feng, J. J., "Size-differentiated lateral migration of bubbles in Couette flow of two-dimensional foam". *Phys. Rev. Lett.* **109**, 084502 (2012).

NIST
National Institute of
Standards and Technology
U.S. Department of Commerce

John R. Royer
National Institute of Standards and Technology
Gaithersburg, MD 20899
(443)-852-1682
jroyer@gmail.com

Extended abstract - *Rheology and dynamics of colloidal superballs*

I am currently at the National Institute of Standards and Technology (NIST), working in the Polymers and Complex Fluids group within the Materials Science and Engineering Division. At NIST I am working on the properties of suspensions of anisotropic colloidal particles, covering both particle synthesis and, in collaboration with Georgetown University, their dynamics under shear. This project is motivated by issues in protein rheology and is tied to the NIST Bio-manufacturing initiative.

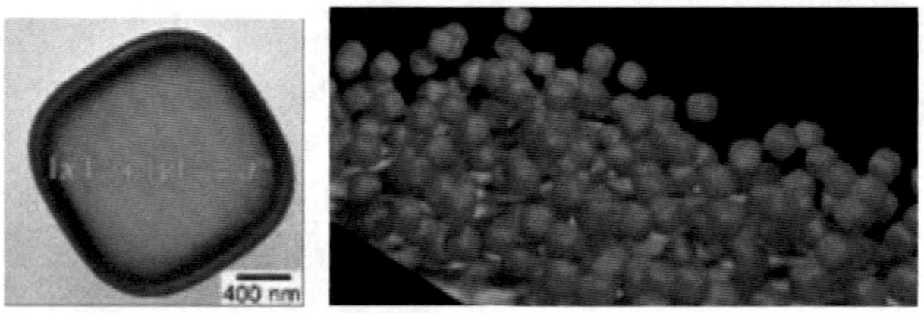

Colloidal superball suspensions

Here I present work detailing particle shape effects in suspensions of cube-like ‘superball’ shaped silica colloids. These colloidal superballs can be synthesized in bulk quantities with well-controlled particle sizes and shapes, making them an excellent experimental model system to study shape effects. We explore how shape-mediated interactions alter the suspension microstructure, particle dynamics and bulk rheology. We find that the shear thickening onset stress is dramatically reduced compared to otherwise identical silica sphere suspensions. This suggests that particle shape could provide both a new window into origins of shear thickening, as well as a new control parameter for industrial applications.

Though my academic background is primarily in physics, my interests have led me to research topics which bridge departmental boundaries between physics and engineering and directly relate to industrial applications. Prior to my work at NIST, I obtained my Ph.D. at the University of Chicago then worked as a post-doc at NYU. For my Ph.D. I worked with Heinrich Jaeger studying granular flows, pioneering novel imaging techniques and material characterization methods to resolve previously unexplained features of granular impact dynamics and stream breakup. As a post-doc at NYU with Paul Chaikin, I worked on both Molecular Dynamics simulations of granular materials and experiments on self-assembly of DNA coated colloidal particles.

My proposed research focuses on the link between the microstructure and macroscopic dynamics in soft, disordered systems far from equilibrium. I will develop a robust research program focused on exploring the role of particle shape and directional interactions in suspensions and granular materials. Initially I intend to focus on three related, yet distinct topics: the role of grain shape in granular clustering, shear thickening in suspensions of anisotropic colloids, and shear-induced aggregation in colloids with directional interactions. This work will bring together elements of chemical engineering, materials science and physics. Throughout my career as a graduate student and post-doctoral researcher, I have had the opportunity to closely interact with engineers in a variety of industries, including Particulate Solid Research, Inc. (Chicago, IL), ExxonMobile Corporate Strategic Research (Clinton, NJ) and Medimmune (Gaithersburg, MD). My proposed research is directly motivated by current industrial needs, and as a faculty member in a chemical engineering department, I plan to maintain and strengthen these connections to industry.

438483 Discovering Novel Catalysts for Production of Renewable Energy and Fuels

Abstract:

The eventual exhaustion of fossil fuels and the increasing environmental impacts of greenhouse gas emissions have generated tremendous interest in renewable energy resources and clean energy technologies. The success of alternative technologies to address these issues in many cases relies on advances in new catalyst materials and catalytic processes, for example in fuel cells and electrolyzers, solar fuels by photocatalysis, biomass conversion, and CO_2remediation. To meet the challenges in new catalyst development, my research is primarily focused on understanding high performance catalysts and discovery of new catalysts for energy and environment-related catalysis.

My previous research extensively explored Pt-monolayer and novel non-Pt electrocatalysts for proton exchange membrane (PEM) fuel cells and electrolyzers. This work sought insight into reaction mechanisms and discovery of new materials by establishing clear structure-activity relationships for electrocatalysts. My studies combined synthesis and characterization of single crystal surfaces in ultrahigh vacuum prepared with well-defined surface structure and evaluation of their catalytic performance under electrochemical conditions. Such samples serve as ideal, model surfaces to study fundamental behavior and properties for functional materials. Supplemented with theoretical simulations, this approach creates an excellent platform to develop new insights into relationships between catalytic performance and surface structure/electronic properties of electrocatalysts.

My recent research focuses on developing high performance heterogeneous catalysts for alkene epoxidation and CO_2 conversion. Direct propylene epoxidation by O_2 is a challenging reaction because of the strong inherent tendency toward complete combustion. Carbon dioxide emission has been associated with adverse global climate changes. Instead of utilizing capture and sequestration of CO_2, a promising way to alleviate this problem is to utilize CO_2 as an inexpensive feedstock to produce fuels and possibly other chemicals. My research addresses key scientific issues by combining theoretical and experimental studies on both model surfaces and practical powder catalysts. Advancing our understanding of fundamental principles that can be used to increase selectivity and activity will enable the design of better catalysts. This approach is proving to be valuable. Our discovery of the exceptional catalytic properties of highly mixed oxides is a breakthrough that could lead to developing a new process for propylene epoxidation. In addition, by synthesizing nanocatalysts and decoding the nature of the active sites, I have identified an unprecedented highly active catalyst for converting CO_2/CH_4 into methanol. This finding could potentially lead to the design of high performance catalysts for CO_2conversion and novel fuel production.

My future work targets grand challenges involving catalytic processes for developing alternative fuels, new energy technologies, and sustainable green chemistry. Through the integration of both mechanistic studies and practical catalyst development, my research aims to achieve both in-depth understanding of the origins of enhanced catalysts performance and the ability to tune this performance for important applications.

438495 New Frontiers: Membrane Fouling Remediation for Sustainable Water & Industrial Wastewater Treatment Technologies

Abstract:

One of the key issues that keep coming up in current research trends is the sustainable water use and water re-usability. The low-pressure membrane applications are considered to be the most effective and sustainable methods of addressing environmental problems when it comes to treated water and wastewater that meets or exceed stringent environmental standards. Nevertheless, membrane fouling is one of the primary operational concerns that is currently hindering a more widespread application of ultrafiltration (UF) with a variety of contaminants. Membrane fouling leads to higher operating costs, excessive energy demand, reduced membrane life time, and increased cleaning frequency. As a consequence, an efficient and well-planned UF process is becoming a necessity for consistent and long-term monetary returns. Examining the source and mechanisms of foulant attachment to the membranes surface is critical when it comes to the research of membrane fouling and its potential practical implementation. A mathematical model was developed for homogeneous and heterogeneous membranes in order to predict the amount of fouling and the increase in transmembrane pressure in UF of simulated latex effluent based on the analyses of particle attachments. A power consumption model, incorporating fouling attachments, as well as chemical and physical parameters involved in membrane fouling, was developed in order to ensure accurate prediction and the scale-up applications. Innovative remediation techniques were likewise developed and applied so as to minimize membrane fouling, enhance membrane performance, and save energy. Based on these achievements, it is essential to direct current research trends toward sustainable water and wastewater treatment technologies that can solve the existing industrial and environmental issues, especially when it comes to solutions that can be successfully commercialized on the global scale.

438498 Influence of Cu-Cr Layered Double Hydroxide (LDH) on the Rheological Properties and Thermal Degradation Kinetics of PMMA Nanocomposites

Abstract:

Abstract: The objective of present work is to study the effect of Cu-Cr LDH on the rheological properties and thermal degradation kinetics of PMMA/Cu-Cr LDH nanocomposites. The PMMA nanocomposites with Cu-Cr LDH (1-5 wt%) were prepared by melt intercalation method. The rheological properties of PMMA/Cu-Cr LDH nanocomposites were characterized by parallel-plate rheometer at a fixed temperature of 200 ^{o}C. Viscoelastic parameters such as storage modulus, loss modulus and complex viscosity exhibit the change from primarily viscous to elastic behaviour as the temperature increased. The experiment results are interpreted by using a theoretical prediction of storage and loss modulus values developed by Einstein and Guth. The thermal degradation activation energy of the nanocomposites is determined by Coats-Redfern method. The enhanced thermal stability of PMMA nanocomposites is also confirmed by integral decomposition temperature (IPDT) compared with neat PMMA. Criado method is finally used to determine the degradation reaction mechanism.

438504 Thermochemical Cycles for the Production of Essential Chemicals

Sunday, November 8, 2015
Exhibit Hall 1 (Salt Palace Convention Center)

Timothy Davenport, Materials Science, California Institute of Technology, Pasadena, CA

Solar-driven thermochemical water splitting has received significant recent attention for solar energy storage because of its potential for high solar-to-fuel conversion efficiency and, when deployed using nonstoichiometric oxides, its chemical simplicity. One particularly attractive approach uses ceria as the reactive medium due to its rapid reaction kinetics, which has been attributed to its nonstoichiometric behavior and fluorite structure. My research program will focus on the development of novel energy efficient thermochemical cycles based on nonstoichiometric oxides to synthesize other chemicals essential to modern society, such as ammonia and ethylene. Research in the laboratory will emphasize on the synthesis and characterization of new materials, catalysis, and process analysis, combining the fields of materials science, chemistry, and chemical engineering.

Postdoctoral Research:

Advised by Sossina Haile, Department of Applied Physics and Materials Science and Department of Chemical Engineering, California Institute of Technology

Optimization of Fuel Production by Ceria-Based Isothermal Pressure-Swing Cycles under Gas-Flow Limited Kinetics

In isothermal pressure-swing cycles (in which water-splitting is driven by switching between an inert, reducing gas and an oxidizing gas, such as steam, at a constant temperature) material kinetics of ceria were determined to be so high that the reaction kinetics were limited only by the rate of flow of reducing or oxidizing gas through the reactor. With knowledge of the gas flow limited kinetics, an integrated kinetic and thermodynamic model was developed with which the optimal fuel production rate could be determined for any possible operating conditions.

Determination of Thermodynamic Parameters in Doped Ceria Materials

The enthalpy and entropy of reduction as a function of nonstoichiometry was determined for Pr-doped ceria (for 5, 10, 20, 40, and 60% Pr content) and Tb-doped ceria (for 10% Tb content) by thermogravimetric analysis under varying atmospheres.

Graduate Research:

Advised by T. Don Tilley, Department of Chemistry, University of California, Berkeley

Dissertation: "Development of Dinuclear Metal Complexes for Catalytic Transformations in Artificial Photosynthesis"

Dinucleating ligands were designed to coordinate two metal centers with a specific geometry. This geometry was then tested for catalytic activity towards water oxidation.

Undergraduate Degrees – Massachusetts Institute of Technology

S.B. Chemical Engineering

S.B. Chemistry

Physics - Minor

438519 Thermochemical Cycles for the Production of Essential Chemicals

Abstract:

Solar-driven thermochemical water splitting has received significant recent attention for solar energy storage because of its potential for high solar-to-fuel conversion efficiency and, when deployed using nonstoichiometric oxides, its chemical simplicity. One particularly attractive approach uses ceria as the reactive medium due to its rapid reaction kinetics, which has been attributed to its nonstoichiometric behavior and fluorite structure. My research program will focus on the development of novel energy efficient thermochemical cycles based on nonstoichiometric oxides to synthesize other chemicals essential to modern society, such as ammonia and ethylene. Research in the laboratory will emphasize on the synthesis and characterization of new materials, catalysis, and process analysis, combining the fields of materials science, chemistry, and chemical engineering.

Postdoctoral Research:

Advised by Sossina Haile, Department of Applied Physics and Materials Science and Department of Chemical Engineering, California Institute of Technology

Optimization of Fuel Production by Ceria-Based Isothermal Pressure-Swing Cycles under Gas-Flow Limited Kinetics

In isothermal pressure-swing cycles (in which water-splitting is driven by switching between an inert, reducing gas and an oxidizing gas, such as steam, at a constant temperature) material kinetics of ceria were determined to be so high that the reaction kinetics were limited only by the rate of flow of reducing or oxidizing gas through the reactor. With knowledge of the gas flow limited kinetics, an integrated kinetic and thermodynamic model was developed with which the optimal fuel production rate could be determined for any possible operating conditions.

Determination of Thermodynamic Parameters in Doped Ceria Materials

The enthalpy and entropy of reduction as a function of nonstoichiometry was determined for Pr-doped ceria (for 5, 10, 20, 40, and 60% Pr content) and Tb-doped ceria (for 10% Tb content) by thermogravimetric analysis under varying atmospheres.

Graduate Research:

Advised by T. Don Tilley, Department of Chemistry, University of California, Berkeley

Dissertation: "Development of Dinuclear Metal Complexes for Catalytic Transformations in Artificial Photosynthesis"

Dinucleating ligands were designed to coordinate two metal centers with a specific geometry. This geometry was then tested for catalytic activity towards water oxidation.

Undergraduate Degrees – Massachusetts Institute of Technology

S.B. Chemical Engineering

S.B. Chemistry

Physics - Minor

438522 Carbon Materials As High-Efficiency and Metal-Free Catalyst for the Synthesis of Cyclic Carbonates from CO2 and Epoxides

Abstract:

Carbon materials as high-efficiency and metal-free catalyst for the synthesis of cyclic carbonates from CO_2 and epoxides

Dong-Hui Lan[1], Lang Chen[1], Chak-Tong Au[1,2], Shuang-Feng Yin[1] *

[1] *State Key Laboratory of Chemo/Biosensing and Chemometrics, College of Chemistry and Chemical Engineering, Hunan University, Changsha 410082,*

Hunan, China

[2] *Department of Chemistry, Hong Kong Baptist University, Kowloon Tong,*

Hong Kong

E-mail: sf_yin@hnu.edu.cn

From the viewpoints of "green chemistry" and "atom economy", it is attractive to utilize CO_2 to replace harmful reactants such as CO and phosgene for the generation of cyclic carbonates. Ionic liquids (ILs) as metal-free catalysts for the cycloaddition of CO_2 to epoxides are mostly poor in activity. For good performance, the use of ILs is commonly accompanied with a co-catalyst [1].

Recently we combined Bu_4NBr with graphene oxide (GO) and found that the binary system is highly efficient towards the cycloaddition of CO_2 to propylene oxide (PO) at room temperature and atmospheric pressure [2]. The synergetic action between a Lewis acid and nucleophilic anion results in enhanced catalytic activity. With the realization of the effects of basic sites on CO_2 adsorption and activation, we designed a Bu_4NBr/Acid-Base bi-functionalized C_3N_4 system to catalyze the cycloaddition reaction. The catalytic activity can be improved by increasing the number of acid sites on C_3N_4 through the control of phosphorus amount.

Since the recycle of Bu_4NBr is complex, it is desirable to immobilize the quaternary ammonium salt on a carbon material. By silylanization of GO with chlorine-terminal silanes followed by subsequent nucleophilic substitution reaction with N,N'-dimethylethylenediamine in a one-pot approach, we prepared GO functionalized with silanol group, amine and quaternary ammonium salts [3]. The non-toxic, metal-free, water-tolerant, and high-efficiency catalyst shows a TOF of 46.4 h^{-1} under mild condition, which is the highest among the reported CN-based heterogeneous catalysts [4]. The excellent performance is attributed to the synergetic effect of silanol group and halide anion for epoxide ring opening as well as the ability of amine for the adsorption and activation of CO_2. The catalyst can be easily separated and reused for at least five times without significant loss of activity.

[1] W.Y. Gao, Y. Chen, S.Q. Ma, et al. *Angew. Chem. Int. Ed.*, 2014, 53, 2615-2619.

[2] D.H. Lan, C.T. Au, S.F. Yin, et al. *Carbon*, 2014, 73, 351-360.

[3] D.H. Lan, L. Chen, S.F. Yin, et al. *Carbon*, 2015, 93, 22-31.

[4] J. Xu, F. Wu, Y.X. Li, et al. *Catal. Sci. Technol.*, 2015, 5, 447-454.

438542 Programming 3D Energy-Efficient Nano-Electronics at 2-Nm Resolution

Sunday, November 8, 2015
Exhibit Hall 1 (Salt Palace Convention Center)
Wei Sun, Department of Systems Biology, Harvard Medical School, Boston, MA; Wyss Institute for Biologically Inspired Engineering at Harvard University, Boston, MA

Background:

Recent development of portable and wearable electronics inspires diverse application potentials in health care, communications, and aerospace. Meanwhile, the growing need for fast data processing in wearable electronics also challenges the material and device design for efficient energy generation, storage and consumptions. Using non-Si nano-materials in future electronics/optoelectronics is expected to promote data processing speed while dramatically lower energy consumption by 1 or 2 magnitudes. However, even though energy-efficient prototypical devices are demonstrated with non-Si materials, they still suffer from low manufacturing resolution/precision, limited manufacturing scalability, and simple architectures complexity. As results, non-Si devices still cannot compete with existing Si devices.

Vision:

I envision that the missing link bridging current synthetic foundation for functional nano-materials to the emerging challenges in energy-efficient 3D electronic/optoelectronic is a cross-scale patterning technique that precisely positions functional non-Si nano-materials into programmable 3D architectures at both high spatial resolution (sub-10 nm) and lithographic precision (sub-5 nm). Such assembled architectures should be designed *de novo* and fabricated in a scalable fashion to direct heterogeneous nano-materials with high spatial positioning resolution and balance the competing needs from high device performance and low energy consumption.

I believe that recent development of bio-molecular self-assembly, in particular DNA self-assembly, will promise an innovative strategy to potentially address this emerging challenge, through interfacing with functional materials. Based on my previous experiences in programming DNA self-assembly and DNA-directed material fabrication for electronics and plasmonics, I propose to develop *a scalable framework using DNA brick crystals as templates for cross-scale positioning of functional nano-materials, such as carbon nanotubes and nanoparticles, into energy-efficient 3D electronic/optoelectronic devices at 2-nm spatial positioning resolution.* **Micron-scale mono-dispersed 3D DNA brick crystals** will be firstly designed and assembled from millions heterogeneous building blocks at 2-nm spatial resolution, following *de novo* developed growth pathway. Next, **spatially confined chemical recognitions** between DNA brick crystal templates and selected **non-Si nano-materials** direct the high-precision positioning of nano-entities exclusively at pre-designed positions with desired orientations, forming arbitrary prescribed 3D architectures tailored to device architecture at lithographic precision. **Cross-scale integration** of the micron-scale architectures into millimeter scale, in collaboration with soft lithography (imprint lithography and contact printing) and dry etching, constructs the designed energy-efficient 3D electronic/optoelectronic devices from novel non-Si materials, with resolution and precision beyond the scope of existing photolithography and block-copolymer lithography.

Towards this framework, I will lead the transformative researches addressing several scientific and technical challenges, and demonstrate the innovative 3D electronic/optoelectronic products with unique energy-efficient features. Scientifically, I will explore the **crystallization principles for complex self-assembly systems** with highly distinct local interactions and diverse conflict growth pathways dominated by entropy and enthalpy effects. Technically, I will develop **spatial-confined chemical patterning at single-molecule resolution**, which simplifies orientation control during scaling nano-entities and enables lithographic precision (down to 1 nm) for complex nano-entities within prescribed irregular architectures. **High-precision reproducible pattern transfer** from DNA brick crystal template to functional relevant substrates expands the device dimension from individual micron array to millimeter scale integrated circuit, and finalizes the scalable production of target device with 10 to 100 times lower energy consumption than Si analogues.

Research experience:

The general theme of my postdoctoral research focuses on using self-assembled DNA nanostructures for digital fabrication and assembly of inorganic nano-materials, such as metals, carbon, and Si, under solution synthesis and lithographic conditions. By transferring geometrical information from the DNA templates into inorganic nano-materials, nano-materials with complex prescribed shapes can be rationally designed and reliably fabricated at high spatial resolution. To expand the DNA self-assembly into micron scale architectures, we also develop a new framework, DNA brick crystal, which is a 2D DNA film with complex prescribed 3D morphologies assembled from modular DNA bricks at 2-nm spatial resolution. Using these DNA brick crystals as structural templates for scaling up nano-materials into larger prescribed architectures, we further demonstrate the periodicity scaling of carbon nanotubes arrays and Si trench arrays down to 10-12 nm ranges, beyond the scope of existing photolithography and block copolymer lithography.

Specifically, my postdoc work includes: (1) **DNA nano-foundry**. We have developed the synthetic foundation towards a digital fabrication framework using DNA nanostructure as templates: Casting to confine 3D metal nanoparticle growth using DNA mold (*Science 2014*); Etching to use metallized DNA nanostructure as lithographic masks for graphene patterning (*Nature Commun. 2013*); Welding to fuse nanoparticles patterned on DNA template into continuous metal nanostructure, and Coating to grow conformal inorganic oxides on DNA substrates (*J. Am. Chem. Soc. 2013*). This synthetic foundation enables us to design a complex 2D/3D inorganic shape *de novo* and fabricate it with high-resolution. (2) **DNA brick crystal with prescribed depths**. Using DNA bricks, we invented a simple, robust and general approach to engineer complex micron scale two-dimensional crystals (lattice structures with repetitive structural units) with prescribed depths of up to 80 nm (*Nature Chem. 2014*). These self-assembled DNA brick crystals exhibit sophisticated user-specified nanometer-scale three-dimensional features, including intricate cavities, channels, and tunnels. During crystal growth, assembly and disassembly occur via relatively weak intermolecular interactions involving the addition or subtraction of a single short strand at a time. (3) **Scaling-up with DNA brick crystals**. Self-assembled micron scale DNA brick crystals provide new platforms for scaling-up the integration of nano-materials into larger *in silico* designed architectures. For example, metal nanoparticles can be selectively docked onto DNA brick crystals to form the prescribed 2D and 3D nanoparticle arrays with precise inter-particle spacing control. These DNA brick crystals could also be used as lithographic masks to create parallel Si lines on substrate with 10-12-nm periodicity or for spatially confining carbon nanotubes with programmable periodicity from 24 nm to 8 nm at lithographic precision.

AIChE Annual Meeting
Salt Lake City, Utah
8-13 Nov, 2015
ID: 438549

Numerical experiments of density driven CO_2 saturated brine migration in heterogeneous geologic fabric materials

Akand Islam, Tip Meckel, Alexander Sun, Prasanna Krishnamurthy
Bureau of Economic Geology, The University of Texas at Austin, TX 78758, USA

Introduction

What is "Carbon capture and sequestration"?

Carbon dioxide capture and sequestration is a set of technologies that can greatly reduce CO_2 emissions from origin sources (mainly coal- and gas-fired power plants). This consists of following three step processes:

- Capture from power plants or industries.
- Transport of the captured and compressed CO_2.
- Injection into underground rock formations.

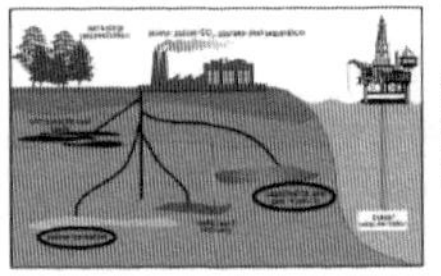

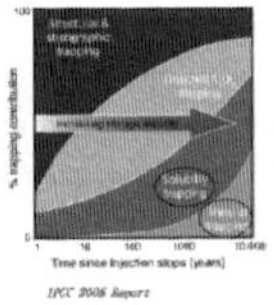

http://eurasiangeologic.blogs.com/GeoLogic/CO2Sequestration.html IPCC 2005 Report

Primary objective

Our aim here, based on simulation results, is to estimate the storage efficiency (capacity) of subsurface geologic formations, and to find out which facies exhibit higher convection rate and therefore are suitable candidates for CO_2 sequestration.

The Reservoir Model and Fabric Formations

Reservoir Model

- Two-dimensional rectangular domain filled with brine saturated porous medium.
- Single phase saturated brine with dissolved CO_2 at the interface.
- Permeability varies spatially (i.e., k=$k(x,z)$).
- The Dirichlet boundary conditions are, $\psi(x,z)$=0, $c(x\neq 0,z)=0$.

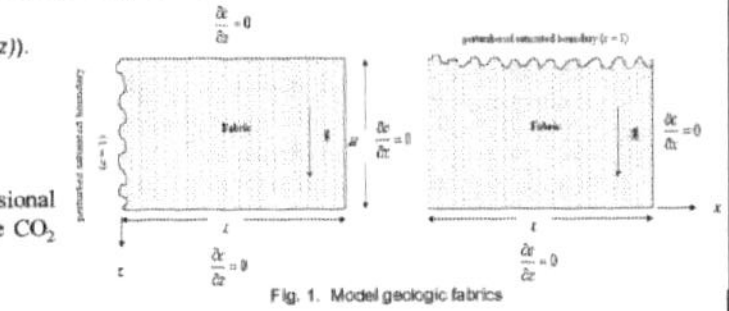

Fig. 1. Model geologic fabrics

Governing equations

- Following final forms of non-dimensional equations are to be solved to compute CO_2 dissolution through free convection:

Stream Equation: $k_z\frac{\partial^2\psi}{\partial x^2}+k_x\frac{\partial^2\psi}{\partial z^2}=Ra\frac{\partial c}{\partial x}$,

Mass Conservation Equation: $\frac{\partial c}{\partial t}+u_x\frac{\partial c}{\partial x}+u_z\frac{\partial c}{\partial z}=\frac{\partial^2 c}{\partial x^2}+\frac{\partial^2 c}{\partial z^2}$.

where, ψ :stream function

Ra : Rayleigh number $=\frac{\Delta\rho gkH}{\phi D\mu}$

c :Concentration

Above equations are solved to compute ψ and c.

❖ **Assumptions:**

- Pressure change due to dissolution is not accounted.
- Capillary transition between gas and brine phases is ignored and therefore only the liquid phase is modeled.
- The velocity induced dispersion effect is ignored.
- Boussinesq and Darcy flow model are followed.
- Isothermal flow condition is kept valid throughout the dissolution process.

Quantification of CO_2 deposition

We quantify CO_2 deposition owing to dissolution by the formula below

$$\bar{c}=\frac{\int_0^x\int_0^z c(x,z,t)dxdz}{\int_0^1\int_0^1 dxdz}.$$

Fabric Formations

The 2D geostatistical models used here are built using a stochastic simulation technique that utilizes Fast Fourier Transforms (FFT) to generate heterogeneous property fields that we name as "fabrics". This program known as FFTSIM, generates log-normally distributed geological property fields with the desired mean, variance and correlation lengths. While the correlation lengths of the generated fabrics are controlled by the user input, the grain sizes of the fabrics are populated based on a variety of rock/facies types with defined median grain size and sorting (grain size histogram), characteristically described by geologists. The nine facies used in the models are displayed in Fig. 2. The rows have fixed sorting (standard deviation) while the columns have fixed median grain size values. Moving down the rows, sorting becomes worse, indicating wider grain size distributions and greater heterogeneity.

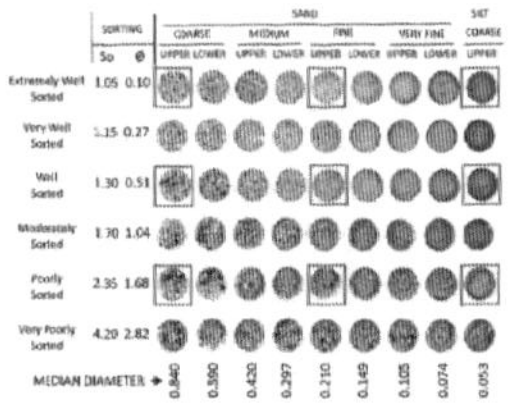

Fig 2. The 9 selected facies (highlighted in red) from the library of 54 sandstone facies used in this work. These facies are characteristic of a wide range of depositional processes and environments. Sorting is given in terms of Trask coefficient S_o and on the phi scale. Median grain diameters at the bottom are in millimeters. Wet packed porosities for these materials range from 43.1% (upper coarse extremely well sorted sand; upper left) to 32.6% (lower very fine sand; lower right). We assume porosities to be in the similar range as that of lower fine sand.

Results and Discussions

Permeability distributions

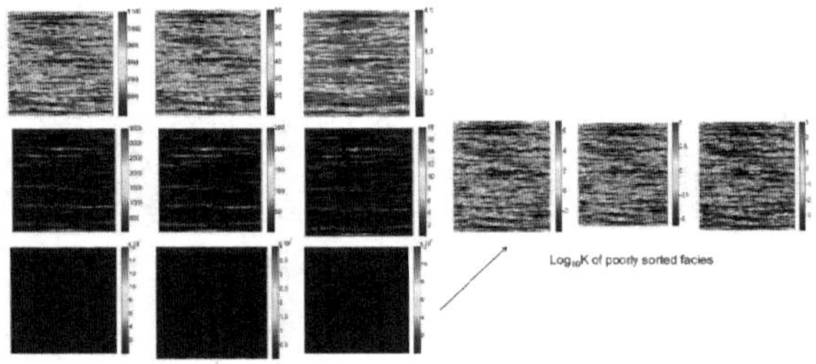

Fig. 3. Fabric 1 (50:1). Colorbars are in mD

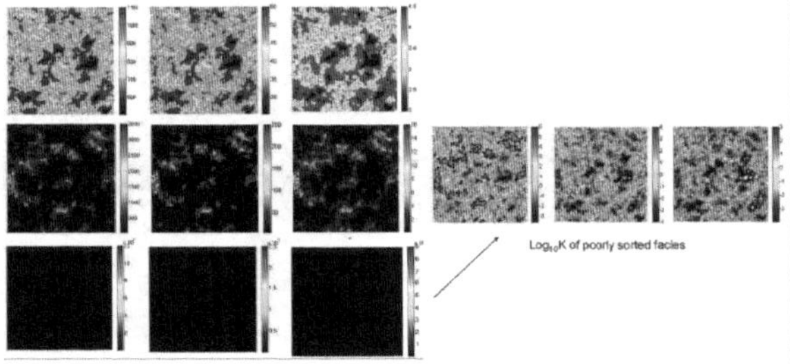

Fig. 4. Fabric 3 (1:1). Colorbars are in mD

❖ **Key Points**

- **The level of heterogeneity varies as low as by 2 times in EWUSCSa to maximum 7 orders-of-magnitude in PSUCSa.**
- **Because of extreme variations the permeability contrasts are not visible in absolute scales for the case of poorly sorted facies. The distributions are shown in logarithmic scales.**
- **The least permeable formation is EWUSi of which mean permeability 2.5 mD.**
- **In the case of well and poorly sorted facies the heterogeneity ranges are very wide. The permeability variations are far away than normal distributions.**
- **In PSUCSa of Fabric 2 mean permeability is 4400 mD where the median is only 86 mD. In PSUSi mean and median permeability are 340 and 0.67 mD, respectively.**

Propagation of CO_2 fronts

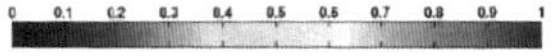

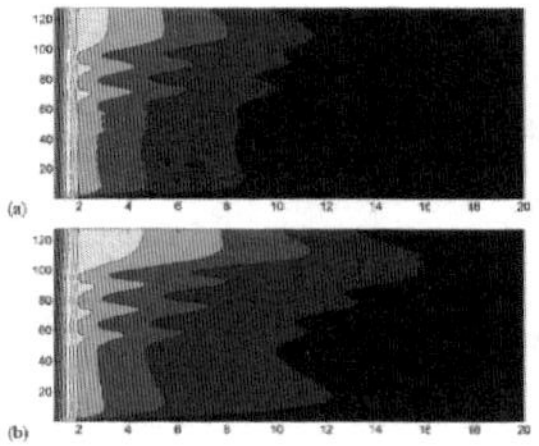

Fig 5. Concentration profiles of EWUCSa of Fabric 1 at (a) 1 and (b) 12 months

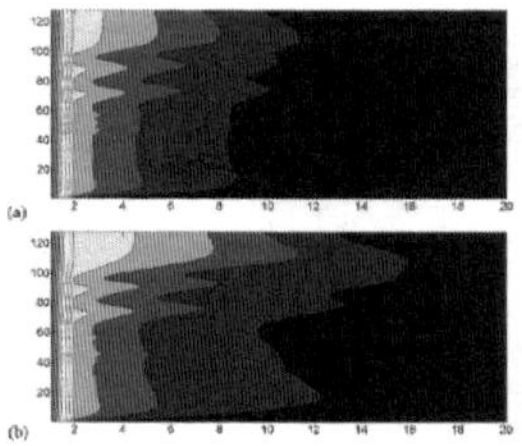

Fig 6. Concentration profiles of EWUCSa of Fabric 3 at (a) 1 and (b) 12 months

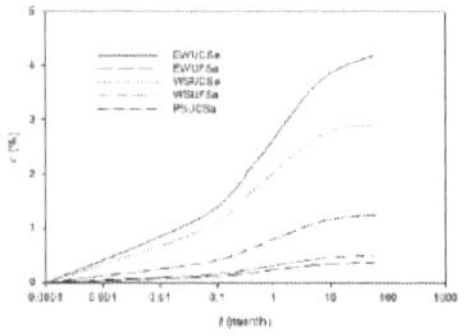

Fig. 7. CO_2 concentrations over time for Fabric 1

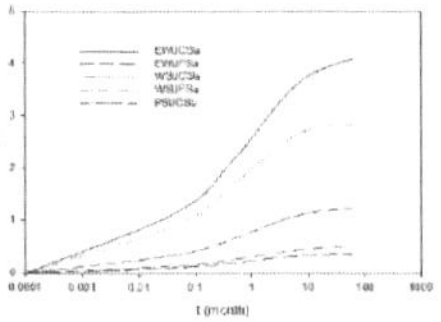

Fig. 8. CO_2 concentrations over time for Fabric 2

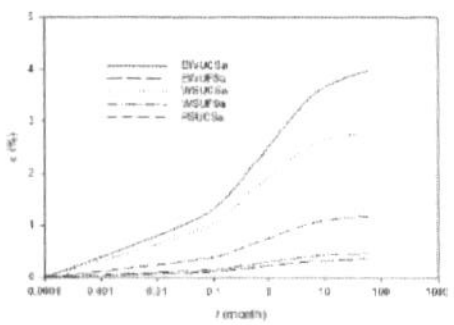

Fig. 9. CO_2 concentrations over time for Fabric 3

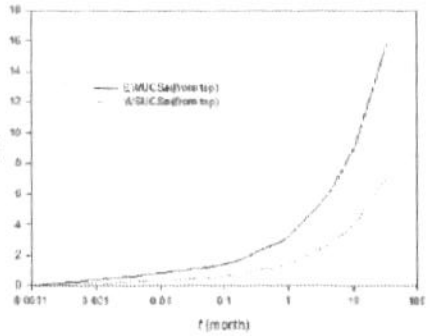

Fig. 10. CO_2 concentrations over time for Fabric 1. Saturated boundary is at the top.

❖ **Key Points**

- **We have found following descending order after sorting out the suitability of sequestration based on the amount of CO_2 can be trapped through dissolution and natural convection processes,**
  EWUCSa → WSUCSa → PSUCSa → WSUFSa → EWUFSa.
- **The coarse formations are relatively suitable for sequestration than the fine facies.**
- **In silt formations the plumes are arrested from the very beginning resulting in no free convection.**
- **When CO_2 fronts spread from the top boundary saturation results are similar to the lateral cases of side boundary. However, because of gravity effect quantitatively concentrations are higher by few times.**

Effects of correlation lengths

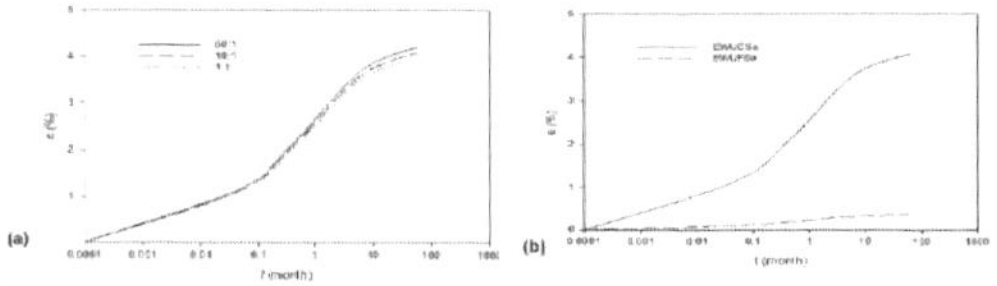

Fig. 11. CO_2 concentrations over time (a) for EWUCSa in different correlation lengths (b) for two different facies at the same correlation length of 10:1.

Summary and Conclusions

- **The coarse sand materials are better candidates for increased CO_2 dissolution while the silts are the least preferable because of their low permeability. The facies can be ranked with respect to dissolution as EWUCSa → WSUCSa → PSUCSa → WSUFSa → EWUFSa. The others do not meet critical conditions to drive any convection. In these cases the plumes are blocked from very beginning.**
- **In the same fabric with specific correlation length the convection rates can vary significantly in different facies because of wide range in grain sizes. This reveals that the heterogeneity has very strong effect in CO_2 dissolution.**
- **In different facies with the same heterogeneity results do not vary much. This is due to the weak effect of correlation lengths. Dissolution magnitudes and rates are loosely dependent on the correlation length of sedimentary fabrics.**
- **The vertical line source cases result in higher saturation than the lateral flows because of gravity effect.**

Acknowledgment

The authors acknowledge computational support from Texas Advanced Computing Center (TACC). This work was funded by the US Department of Energy (DE-FE0012231).

438550 Systems Biology Approaches to Develop Precision Therapeutic Approaches to Overcome Drug Resistance in Cancer

Sunday, November 8, 2015
Exhibit Hall 1 (Salt Palace Convention Center)

Mohammad Fallahi-Sichani, Systems Biology, Harvard Medical School, Boston, MA

My goal is to direct a research program that combines multiplex biochemical measurement, single-cell analysis and novel multi-scale and network-level computational modeling to address fundamental and clinically significant challenges in cancer biology: (i) relating drug action at the molecular scale to phenotypes at the cellular and tissue (tumor) levels, (ii) understanding signaling mechanisms involved in drug-induced adaptive responses that lead to submaximal cell killing and ultimately drug resistance in patients, (iii) developing predictive mathematical models that guide development of safe, effective and personalized therapeutic approaches to overcome drug resistance. My PhD background in multi-scale computational modeling and signal transduction, postdoctoral training in cancer cell signaling and systems pharmacology, strong publication record and successful fellowship and grant applications have all prepared me to successfully accomplish the goals outlined above.

Education:

(2012) PhD, Chemical Engineering, University of Michigan, Ann Arbor, MI

(2009) MSc, Chemical Engineering, University of Michigan, Ann Arbor, MI

(2007) BSc/MSc (Integrated), Biotechnology, University of Tehran, Iran

Research/Work Experience:

(2012-Present) LSRF Postdoctoral Fellow in Systems Biology, Harvard Medical School (Advisor: Dr. Peter Sorger)

(2007-2012) Graduate Student Research Assistant, Chemical Engineering, University of Michigan, Ann Arbor (Advisors: Dr. Jennifer Linderman and Dr. Denise Kirschner)

Grants and Honors:

(2015) NIH K99 Pathway to Independence Award, NCI (selected for funding in Aug 2015)

(2013-Present) Life Sciences Research Foundation (LSRF) Postdoctoral Fellowship

(2012) Distinguished Young Scholars Seminar Speaker, University of Washington, Seattle

(2011) Richard and Eleanor Towner Prize for Outstanding PhD Research Award, University of Michigan

(2011) Rackham Predoctoral Fellowship, University of Michigan

(2010) Phi Kappa Phi Graduate Award

(2009) Rackham International Student Fellowship, University of Michigan

Conference Awards:

(2015) American Association for Cancer Research (AACR) Scholar-in-Training Award

(2014) Outstanding Poster Award, FEBS-EMBO 2014 Conference, Paris, France

(2013) Systems Biology of Human Disease Conference Travel Award, Germany, Heidelberg

(2011) 5thq-bio Conference in Cellular Information Processing Travel Award, Santa Fen, NM

Selected Publications:

1) Lin JR, **Fallahi-Sichani M**, Sorger PK. Highly multiplexed high-throughput imaging of single-cell using CycIF cyclic immunofluorescence (***in revision at Nature Communications***).

2) **Fallahi-Sichani M**, Moerke NJ, Niepel M, Zhang T, Gray NS, Sorger PK. Systematic analysis of BRAF(V600E) melanomas reveals a role for JNK/c-Jun pathway in adaptive resistance to drug-induced apoptosis, ***Molecular Systems Biology*** (2015) 11:797.

3) **Fallahi-Sichani M**, Honarnejad S, Heiser LM, Gray JW, Sorger PK. Metrics other than potency reveal systematic variation in responses to cancer drugs, ***Nature Chemical Biology*** (2013) 9:708-14.

4) Kirschner DE, Hunt CA, Marino S, **Fallahi-Sichani M**, Linderman JJ. Tunable resolution as a systems biology approach for multi-scale, multi-compartment computational models, ***WIREs Systems Biology and Medicine*** (2014) 6:289-309.

5) **Fallahi-Sichani M**, Kirschner DE, Linderman JJ. NF-kappaB signaling dynamics play a key role in infection control in tuberculosis, ***Frontiers in Physiology*** (2012) 3:170.

6) **Fallahi-Sichani M**, Flynn JL, Linderman JJ, Kirschner DE. Differential risk of tuberculosis reactivation among anti-TNF therapies is due to drug binding kinetics and permeability, ***Journal of Immunology*** (2012) 188:3169-78.

7) **Fallahi-Sichani M**, El-Kebir M, Marino S, Kirschner DE, Linderman JJ. Multiscale computational modeling reveals a critical role for TNF-a receptor 1 dynamics in tuberculosis granuloma formation, ***Journal of Immunology*** (2011) 186:3472-83.

8) **Fallahi-Sichani M**, Schaller MA, Kirschner DE, Kunkel SL, Linderman JJ. Identification of key processes that control tumor necrosis factor availability in a tuberculosis granuloma, ***PLoS Computational Biology*** (2010) 6:e1000778.

9) **Fallahi-Sichani M**, Linderman JJ, Lipid raft-mediated regulation of G-protein coupled receptor signaling by ligands which influence receptor dimerization: A computational study, ***PLoS ONE*** (2009) 4:e6604.

438552 Integrating Computational Chemistry, Molecular Simulation, and Chemical Engineering

Abstract:

An overview of my postdoctoral and graduate research is presented. Computational chemistry and molecular simulation are employed to solve complex challenges in materials science, catalysis, and biocatalysis. Tools include density functional theory, ab initio molecular dynamics, classical molecular dynamics, accelerated sampling techniques, kinetic modeling, and predictive reduced models. Currently at the Pacific Northwest National Laboratory my postdoctoral research is directed toward the molecular design of carbon capture solvents. Computational chemistry calculations are done for parameters to perform accurate molecular simulation-based prediction of properties, and to develop reduced models for quick screening of solvent libraries. Additional projects include establishing detailed reaction pathways and kinetic modeling of metal organic framework synthesis, dynamics of water on metal oxide surfaces, and catalysis in liquid-solid systems. My dissertation work at Iowa State University, as part of the Center for Biorenewable Chemicals, focused on enzymes. Statistics and bioinformatics were used to find target enzymes with desired activities among the many ones existing in nature, and mixed quantum mechanics/molecular mechanics simulations were done to understand in detail the mechanism of enzymatic catalysis.

438554 Understanding and Harnessing Nature's Synthetic Potential to Advance Modern Drug Development

Abstract:

Research Experience:

2nd Postdoctoral Project (2013-present): Utilizing Synthetic Biology to Understand and Engineer the Biosynthesis of Plant Natural Products in Yeast. Under supervision of Dr. Christina D. Smolke, Bioengineering, Stanford University

1st Postdoctoral Project (2012-2013): Investigating Microbial Interactions on Microfluidic Devices. Under supervision of Dr. Rustem Ismagilov, Chemistry & Chemical Engineering, California Institute of Technology

PhD Dissertation (2008-2012): "Engineered Biosynthesis of Aromatic Polyketides". Under supervision of Dr. Yi Tang, Chemical and Biomolecular Engineering, University of California, Los Angeles

Nature endows organisms including microbes, plants and animals with enormous power to synthesize complex functional molecules from simple building blocks. These complex molecules from natural sources and of diverse bioactivities are so called natural products. Natural products play an indispensable role in modern drug discovery and development. Comprehensive understanding of the biosynthesis of these complex molecules is essential for better harnessing nature's synthetic potential to address the remaining challenges in the sourcing of many natural products, and to advance natural product based drug discovery and development. My research training has focused on utilizing engineering approaches (such as synthetic biology, metabolic engineering, and protein engineering) to understand the biosynthesis of important natural products (including bacterial and fungal aromatic polyketides, plant benzylisoquinoline alkaloids); and utilizing this enhanced understanding of nature's synthetic power to advance engineering approaches to biosynthesis. Engineered biosynthesis systems can include engineering native hosts for higher accumulation of key natural products, heterologous reconstitution of valuable natural product biosynthesis pathways in fermentable yeast or *E. coli* as potential alternative to traditional native host or chemical synthesis for industrial production, and engineering the biosynthesis for novel derivatives as potential lead compounds for drug discovery.

Future Research Interests:

In recent decades, there has been a decline in natural product-based drug development. This is in part due to a number of challenges associated with decreased frequency of identifying novel NP structures with traditional methods, obtaining sufficient supply of natural product lead compounds through the drug development pipeline, and modifying the highly complex natural product structures for lead optimization. The genomic information obtained through advances in sequencing and bioinformatics suggests that the natural products isolated thus far represent only a small portion of nature's true synthetic potential, which has been largely underestimated and unused by human society. My future research program will leverage advances in genome sequencing, bioinformatics, and synthetic biology to study and engineer novel natural product and biosynthesis from the less explored sources, so as to expand our understanding of nature's synthetic potential, and to address these key challenges in natural product discovery and related drug development.

Selected Publications:

Li, Y., Smolke, C., "Elucidating and Engineering Biosynthesis of the Anticancer Alkaloid Noscapine in Yeast", *Nat. Chem. Biol.*, in revision.

Li, Y., Chooi, Y.-H., Sheng, Y., Valentine, J. S., Tang, Y.*, "Comparative Characterization of Fungal Anthracenone and Naphthacenedione Biosynthetic Pathways Reveals an α-Hydroxylation-Dependent Claisen-Like Cyclization Catalyzed by a Dimanganese Thioesterase.", *J. Am. Chem. Soc.* **2011**, 133, 15773-15785.

Li, Y., Xu, W., Tang, Y.*, "Classification, Prediction and Verification of the Regioselectivity of Fungal Polyketide Synthase Product Template Domains." *J. Biol. Chem.* **2010**, 285, 22762-22771. (*Highlighted by Nat. Chem. Biol.*)

Zhou, H.†, **Li, Y.†** (*equal contribution*), Tang, Y.*, "Cyclization of Aromatic Polyketides from Bacteria and Fungi."*Nat. Prod. Rep.* **2010**, 27, 839-868.

Zhang, W., **Li, Y.**, Tang, Y.*, "Engineered Biosynthesis of Bacterial Aromatic Polyketides in Escherichia coli."*Proc. Natl. Acad. Sci. U. S. A.* **2008**, 105, 20683-20685.

438557 Material Preparation and Kinetic Study of Catalysts

Abstract:

Catalytic reaction has been an essential process for wide industrial application, and it will support the future science and industry to establish the sustainable society. Although it has been placed more expectations on catalyst, many aspects of this field need to be clarified. The preparation procedures, the factor of each materials (surface structure, active sites, and so on), and the catalytic activity are closely connected. Therefore, having a broad understanding of this field is essential for developing a novel catalytic processes. In my research career, I have focused on this field from a materials synthesis to a catalytic reaction.

The first part of my poster will focus on the catalyst synthesis, especially on microporous materials synthesis. During my Ph.D. work (with Prof. Tatsuya Okubo in the University of Tokyo, Japan) followed by one and half years of postdoctoral research (with Prof. Raul F. Lobo in University of Delaware), we developed novel synthesis methods of zeolites. Zeolites, which is a crystalline microporous material, has been widely used in industrial catalytic fields as ion-exchanger, catalyst, adsorbent, and so on. The synthesis of zeolite with controlled framework structure and compositions, however, still relies on a lot of try-and-error trials. We developed several synthesis methods of zeolites; 1) a novel synthesis pathway of silica-sodalite via 2-D crystalline precursor without hydrothermal treatment, 2) a method to use a highly amphiphilic molecule as a structure-directing agent in a pure-silica zeolite synthesis, 3) a seeding method to reduce the use of an organic structure-directing agent in the synthesis of UZM-4 zeolite, and 4) a general methodology to synthesize high-silica zeolites in fluoride media.

The second part of my poster will focus on the catalytic C-C bond formation reaction, which is my current work (with Prof. David W. Flaherty in University of Illinois, Urbana-Champaign). Ethanol derived from the fermentation of biomass can be catalytically converted into higher molecular weight hydrocarbons and oxygenates to produce advanced biofuels and lubricants. The Guerbet reaction is one pathway for this process. The catalyst hydroxyapatite (HAP) exposes both acid and base sites, and therefore, HAP gives better selectivities. Cascades of primary and secondary Guerbet reactions give broad distributions of heavier products; however, little is known about how the acid-base properties of catalysts affect the distribution of carbon numbers (C_n) or branching of the products. We use steady-state rate measurements to study the reaction network for ethanol Guerbet on HAP materials that produces C_4-C_{12} alcohols via cascades of coupling steps.

438558 Hierarchical Nanostructured and Polymeric Materials for Energy Storage and Conversion

Abstract:

Hierarchical Nanostructured and Polymeric Materials for Energy Storage and Conversion

Zheng Chen, Postdoctoral Fellow

Stanford University

Research Background:

Facing fossil-fuel shortage and ecological deterioration, humankind has been diligently seeking for clean, safe and renewable energy sources to offset the diminishing availability or to take place of fossil fuels. At the same time, new strategies to increase energy-use efficiency, reduce fossil-fuel consumption and decrease CO_2 emission, is highly demanded. These essential needs have made energy storage and conversion a critical component in the creation of sustainable society. At the forefront of such endeavor is to develop high-performance materials to store charges, catalyze reactions and convert energy. My research interests focus on designing and synthesizing new materials based on fundamental principles of these processes. The goal is to enhance the capability and efficiency of materials in various energy-related applications.

PhD Dissertation: *"Rational Material Architecture Design for Better Energy Storage"*

Advisor: Prof. Yunfeng Lu, Chemical and Biomolecular Engineering, **University of California, Los Angeles**

Postdoctoral Project: *"Novel Polymeric Materials for High-Performance Energy Storage"*

Advisors: 1) Prof. Zhenan Bao, Chemical Engineering, Stanford University; 2) Prof. Yi Cui, Materials Science and Engineering, **Stanford University**

Research Experience:

My PhD research focused on designing and fabricating multifunctional nanoarchitectures by integrating distinct material structures and properties to address the kinetics limitation of traditional materials. Different types of energy storage architectures were investigated and compared with conventional structures to demonstrate such design concepts. As model systems, "hierarchically porous spherical carbon particles" with graphitized structures, "interpenetrating oxide nanowire/CNT network", "self-assembled ultrafine nanocrystals/CNTs" were designed and fabricated using various approaches. The microstructures were carefully analyzed and electrochemical reaction kinetics was systematically studied. It was demonstrated that thick electrodes with high charge capacity, high rate performance and cycling stability rely on functional architecture that simultaneously provides high electronic conductivity, easy ion diffusion, abundant surface actives sites and robust structure and interfaces.

My postdoc research is to understand and utilize novel properties of polymer materials for energy storage devices. Polymeric materials can offer a variety of functions such as good electronic and ionic conductivity, mechanical flexibility, stretchability and self-healing capability. These properties are of great interest for electronic and energy-related applications. In next-generation high-energy lithium-ion batteries, their electrode active materials (e.g., silicon, sulfur) suffer from mechanical and electronic degradation, which limits the battery capacity and cycling life time. In my research, novel functional polymers were used to solve some of the most critical issues related to high-energy materials. Specifically, a dynamic hydrogen-bonding based self-healing polymer is developed to effectively coat and bond onto silicon particle surface, which can maintain good electronic conductivity and mechanical integrity in electrodes over repeatedly charging/discharging. In addition, conducting polymer hydrogels were designed to make flexible electrodes and produce graphitic carbon with record-high surface area. Thermal responsive polymers were used to improve the battery's safe property. Our polymer approach provides novel platform to renovate electrode structure and improve battery performance by tuning their structures and compositions.

With the synthetic capability and fundamental understanding of charge transport of energy materials, my future research will focus on materials development for advanced energy applications: 1) Precise synthesis of new materials for large scale energy storage, CO_2 conversion and membrane, and 2) Understanding charge transport behavior on electrochemical interfaces.

Selected Publications:

1. John W. F. To,† **Zheng Chen**,† HongbinYao, Jiajun He, Kwanpyo Kim, Ho-Hsiu Chou, Lijia Pan, Jennifer Wilcox, Yi Cui and Zhenan Bao,* Ultra-High Surface Area Three-Dimensional Porous Graphitic Carbon from Conjugated Polymeric Molecular Framework, ***ACS Central Science***, 2015, in press (**DOI:**10.1021/acscentsci.5b00149). (***co-first author***)
2. **Zheng Chen**, Chao Wang, Jeffrey Lopez, Zhenda Lu, Yi Cui*, Zhenan Bao*, High-Areal-Capacity Silicon Electrodes with Low-Cost Silicon Particles Based on Spatial Control of Self-Healing Binder, ***Advanced Energy Materials***, 2015, DOI:10.1002/aenm.201401826.
3. **Zheng Chen**, Yin Yuan, Huihui Zhou, Xiaolei Wang, Zhihua Gan*, Fosong Wang* and Yunfeng Lu*, 3D Nanocomposite Architectures from Carbon-Nanotube-Threaded Nanocrystals for High-Performance Electrochemical Energy Storage, ***Advanced Materials***, 2014, 26, 339-345.
4. **Zheng Chen**, John W. F. To, Chao Wang, Zhenda Lu, Nan Liu, Alex Chortos, Lijia Pan, Fei Wei, Yi Cui* and Zhenan Bao*, A Three-Dimensionally Interconnected Carbon Nanotube-Conducting Polymer Hydrogel Network for High-Performance Flexible Battery Electrodes, 2014, ***Advanced Energy Materials***, 2014, 4: 1400207.
5. Yanhua Cheng,† **Zheng Chen**,† Meifang Zhu*, Yunfeng Lu*, Polymer-Assisted Assembly of Oxide Particles and Carbon Nanotubes for High-Performance Flexible Battery Anode, ***Advanced Energy Materials***, 2014, DOI: 10.1002/aenm.201401207. (***co-first author***)
6. Chao Wang, Hui Wu, **Zheng Chen**, Matthew T. McDowell, Yi Cui*, and Zhenan Bao*, Enabling Stable

Operation for Silicon Microparticle Anodes for High Energy Lithium Ion Batteries using Self-healing Chemistry, ***Nature Chemistry***, 2013, 5, 1042-1048.

7. **Zheng Chen**, Dieqing Zhang, Xiaolei Wang, Xilai Jia, Fei Wei, Hexing Li* and Yunfeng Lu*, High-Performance Energy Storage Architectures from Carbon Nanotubes and Nanocrystals Building Blocks, ***Advanced Materials***, 2012, 24, 2030-2036.
8. **Zheng Chen**, Veronica Augustyn, Xilai Jia, Qiangfeng Xiao, Bruce Dunn* and Yunfeng Lu*, High-Performance Sodium-ion Pseudocapacitors Based on Hierarchically Porous Nanowire Composites, ***ACS Nano***, 2012, 6(5), 4319-4327.
9. Xilai Jia,† **Zheng Chen**,† Arnold Suwarnasarn, Benjamin M. Wu, Fei Wei* and Yunfeng Lu*, High-Performance Flexible Lithium-Ion Electrodes Based on Robust Network Architecture, ***Energy Environ. Sci.***, 2012, 5, 6845-6849. (***co-first author***)
10. **Zheng Chen**, Jing Wen, Chunzhu Yan, Lynn Rice, Hiesang Sohn, Mei Cai*, Bruce Dunn and Yunfeng Lu*, High-Performance Supercapacitors Based on Hierarchically Porous Graphite Particles, ***Advanced Energy Materials***, 2011,1, 551-556.
11. **Zheng Chen**, Veronica Augustyn, Jing Wen, Yuewei Zhang, Meiqing Shen*, Bruce Dunn* and Yunfeng Lu*, High-Performance Supercapacitors Based on Intertwined CNT/V_2O_5 Nanowire Nanocomposites, ***Advanced Materials***, 2011, 23, 791-795.
12. **Zheng Chen**, Yaochun Qin, Ding Weng, Qiangfeng Xiao, Yiting Peng, Xiaolei Wang, Hexing Li*, Fei Wei* and Yunfeng Lu*, Design and Synthesis of Hierarchical Nanowire Composites for Electrochemical Energy Storage, ***Advanced Functional Materials***, 2009, 19(21), 3420-3426.

438571 Site-Specific Techniques for Identification of Active Sites of Supported Transition Metal Oxide and Late Transition Metal Catalysts

Sunday, November 8, 2015
Exhibit Hall 1 (Salt Palace Convention Center)

Kunlun Ding and Peter C. Stair, Chemistry, Northwestern University, Evanston, IL

Identification and characterization of catalytic active sites are the prerequisites for an atomic level understanding of the catalytic mechanism and rational design of high-performance heterogeneous catalysts. Supported transition metal oxide and late transition metal catalysts represent two groups of most studied and widely used catalysts for many industrial applications. One common feature of these catalysts is the structural diversity of their surface species. In many cases, only a limited portion of the surface species determines overall catalytic performances. For instance, it has been well known that only less than 1-2% of the total metal atoms in supported molybdena catalysts contribute to olefin metathesis reactions. Identifying the structure of such a small portion of active species is an extremely challenging task. Most techniques employed by researchers either provide statistically limited information such as microscopy, or sample-averaged information such as X-ray absorption spectroscopy. From this point of view, site-specific techniques that are capable of providing statistically sufficient information on site identification and quantification as well as the activity evaluation of specific sites are highly desired.

Here we show our recent efforts in developing site-specific spectroscopy to identify active sites in several heterogeneous catalytic reactions. In the first case, we combine UV resonance Raman spectroscopy and electron microscopy measurements to unambiguously identify the monomeric nature of the active sites for MoO_3/SiO_2 catalysts in olefin metathesis. In another case, we show that infrared spectroscopy, using CO as a probe molecule, can distinguish and quantify supported Pt single atoms and nanoparticles. Further, we observe that only the CO molecules adsorbed on Pt nanoparticles can react at low temperatures upon O_2 or H_2O exposure, showing that the active sites in CO oxidation and water-gas shift reactions are present on the nanoparticles but not on single atoms. The atomic-level understanding of the active site structure will advance the rational design of high-performance heterogeneous catalysts.

438581 Increasing Global Access to Diagnostic Testing Using Low-Cost, Non-Instrumented Paper-Based Microfluidics

Sunday, November 8, 2015
Exhibit Hall 1 (Salt Palace Convention Center)
Bhushan J. Toley, Bioengineering, University of Washington, Seattle, WA

INCREASING GLOBAL ACCESS TO DIAGNOSTIC TESTING USING LOW-COST, NON-INSTRUMENTED PAPER-BASED MICROFLUIDICS

Bhushan J. Toley

4th year Postdoctoral Fellow

Bioengineering Department, University of Washington

Doctoral Thesis

Title: Microtechnologies for mimicking tumor-imposed transport limitations and developing targeted cancer therapies

Funding: NIH, NSF

Advisor: Prof. Neil Forbes

Postdoctoral Project

Title: Development of Multiplexable Autonomous Disposables for Nucleic Acid Amplification Tests (MAD NAAT)

Funding: DARPA, NIH

Advisor: Prof. Paul Yager

Research Interests

The primary aim of my research is to reduce the cost of, and increase access to state-of-the-art medical diagnostic testing by bringing it to the point-of-care. Through my research, I aim to develop portable and low-cost diagnostic tests that are deliverable to low resource settings and are operable by untrained users, and that emulate the abilities of a high-tech lab. The core technology I plan to use is paper-based microfluidics, which enables controlled transport of microliter volumes of fluids without the need of pumps or electricity.

Research Experience

The common theme of my research projects till date has been the design of devices that enable fluid manipulation at sub-millimeter length scales, the use of such (microfluidic) devices for conducting biological processes, and quantitatively analyzing these biological processes by applying core chemical engineering principles of transport, thermodynamics, and chemical kinetics.

The aim of my doctoral research was to develop methods that would enable quantification of drug and nutrient gradients in tumor tissues in vitro. To this end, I developed a microfluidic in-vitro model of continuously perfused tumor tissue that mimics the region of a tumor adjacent to a blood vessel. I used this device for a) assessing the anti-cancer therapeutic potential of several bacterial strains, and b) estimating the transport parameters of two standard chemotherapeutic drugs – doxorubicin and Doxil. By applying chemical transport models to these systems, I estimated key transport parameters of the therapeutic agents and showed that a) motility governs the location of bacterial accumulation in tumor tissue, and b) ideal chemotherapeutic drugs must have an optimal combination of diffusion, binding, and cytotoxicity to be effective. In another study, I quantified the effect of surface charge of gold nanoparticles carrying doxorubicin on the rate and location of their tumor penetration. This work was published in *Nature Nanotechnology*and has been cited over 150 times.

The aim of my postdoctoral work is to develop a paper-based platform device that can conduct a sample-to-answer nucleic acid amplification test (NAAT). While paper-based microfluidic devices enable fluid flow without pumps, they have relatively limited flow control and valving capabilities compared to conventional microfluidic systems. To bridge this gap, I developed several new paper-microfluidic valving techniques. In my first publication, I demonstrated that relative differences in properties of porous materials can be used to shunt fluids from the main fluidic path to generate controlled delay lines. In my second publication, I presented a valving toolkit based on expandable elements that can conduct many types of fluidic operations, like on-switches, off-switches, and flow-diversion switches. A combination of these valves can be used to conduct arbitrarily complex, multi-step fluidic operations in paper-based devices. In addition to designing fluidic systems, I conducted extensive experiments to optimize the performance of isothermal strand displacement amplification (iSDA) – a new method of DNA and RNA amplification – in paper substrates. Simultaneously, I developed kinetic mathematical models of the iSDA reaction network that helped us gain key insights into the molecular mechanisms of iSDA. My research has led to the development of a prototype paper-based microfluidic device that can detect as low as 100 copies of *Staph. aureus*bacteria from a nasal swab by conducting an automatic multi-step molecular assay, making it the first device of its kind.

Mentoring and Teaching Experience

Over the course of my career, I have mentored 14 undergraduate students and 2 high school students; 7 out of these mentees have coauthored papers in peer-reviewed journals with me. I have served as teaching assistant for core chemical engineering undergraduate courses taken by juniors: i) Separation Processes – for 3 semesters and ii) Process Control and Dynamics – for 1 semester.

Future Research Directions

The aim of my future research is to develop patentable and commercializable technologies that will be used in devices that can provide the quality of testing available in fully equipped diagnostic labs at locations where access to such labs may be limited or non-existent. My initial projects will leverage my expertise in paper-based microfluidics in developing portable non-instrumented systems for conducting a) protein detection tests, and b) nucleic acid amplification tests. At the same time, I want to better understand how fluids flow in porous matrices like paper. I am interested in applying concepts developed in the field of geology to model fluid flow in paper-based materials. Ultimately, I would like to develop a paper microfluidic design tool that will guide researchers in the field to easily design paper-based microfluidic devices.

Protein Detection: Enzyme-linked immunosorbent assay (ELISA) is a well-established analytical lab technique used to quantify the amount of a substance in a sample. The ELISA procedure requires several hours to complete and requires a trained technician and lab instruments to conduct the multiple reagent dispensing and washing steps. Automation of ELISAs has traditionally required an instrumented system for pumping fluids. I am interested in developing a paper-based microfluidic platform device that will store dried ELISA reagents for at least 6 months and conduct the ELISA at the push of a button after a sample is introduced into the device. Such a device will be completely electricity and battery-free and can be disposed-off after a single use. Because all ELISAs require the same fundamental fluidic operations, this platform can be adopted to conduct any ELISA, even in remote geographical locations.

Molecular detection of DNA/RNA: I am also interested in developing a platform for conducting automatic sample-to-answer NAATs based on human sputum as the sample. Such a platform will have tremendous utility in diagnosis of TB. Recently, the Cepheid GeneXpert instrument revolutionized TB diagnostics; the GeneXpert automates the multiple steps in the molecular detection of TB from sputum, thus providing sophisticated diagnosis at the point-of-care. However, over the last few years, it has been realized that thc cost of the instrument and disposable cartridges is prohibitive for the people who have the most need for it, i.e. those in the developing world. The paper-based sample-to-answer TB test that I will develop will operate on a single battery but will be instrument-free and significantly cheaper than the GeneXpert.

List of Peer-Reviewed Publications

1.**Toley BJ,** Wang JA, Gupta M, Buser JR, Lafleur LK, Lutz BR, Fu E, Yager P, "A versatile valving toolkit for automating fluidic operations in paper microfluidic devices", *Lab on a Chip*, 2015, 15, 1432-44. *(Featured on back cover)*

2.Panpradist N, **Toley BJ**, Zhang X, Byrnes S, Buser JR, Englund JA, Lutz BR, "Swab sample transfer for point-of-care diagnostics: characterization of swab types and manual agitations methods", *PLoS ONE*, 2014, 9(9), e105786

3.**Toley BJ**, McKenzie B, Yager P, Fu E et al, "Tunable delay shunts for paper microfluidic devices", *Analytical Chemistry*, 2013, 85(23), 11545-11552.

4.Spicar-Mihalic P*, **Toley BJ***, Houghtaling J, Liang T, Yager P, Fu E; "CO_2 laser cutting and ablative etching for the fabrication of paper-based devices", *Journal of Micromechanics and Microengineering*, 2013, 23(6) (*equal contribution)

5. **Toley BJ**, Tropeano-Lovatt Z, Harrington J, McGarry MJ, Forbes NS; "Transport-reaction-coupled pharmacokinetic pharmacodynamics model captures key mechanisms of Doxorubicin and Doxil therapeutic efficacy", *Integrative Biology*, 2013, 5, 1184-1196

6.Dai Y, **Toley BJ**, Swofford C, Forbes NS; "Construction of an inducible cell-communication system that amplifies Salmonella gene expression in tumor tissue", *Biotechnology and Bioengineering,* 2013, 110(6), 1769-81

7.**Toley BJ**, Forbes NS; "Motility is critical for effective distribution and accumulation of bacteria in tumor tissue in vitro", *Integrative Biology*, 2012, 4, 165-176

8. **Toley BJ**, Park J, Kim BJ, Venkatasubramanian R, Maharbiz MM, Forbes NS; "Micrometer-scale oxygen delivery rearranges cells and prevents necrosis in tumor tissue in vitro", *Biotechnology Progress*, 2012, 28(2), 515-525

9.**Toley BJ*,** Ganz DE*, Walsh CL, Forbes NS; "Microfluidic device for recreating a tumor microenvironment in vitro", *Journal of Visualized Experiments*, 2011, 57, DOI: 10.3791/2425 (* equal contribution)

10. Kim BJ, Han G, **Toley BJ**, Kim C, Rotello VM, Forbes NS; "Tuning payload delivery in tumor cylindroids using gold nanoparticles", *Nature Nanotechnology* 2010 Jun, 5(6), 465-72

438582 Rational Design of Electrochemical Interfaces for Control over Separation and Catalytic Processes

Abstract:

Although it is a well-established field in its own, electrochemistry has not yet been widely adopted as a control strategy for separation and catalytic processes, both of which are of central importance to chemical engineering. My research aims to establish design principles for efficient electrochemical interfaces, and develop novel electrochemical modulation strategies for separation and catalysis applications. An electrochemical interface generally consists of three key components: the electrode phase, the electrolyte phase, and the functional layer between these two phases. Rational design of an electrochemical interface for a specific application requires careful consideration of these three key constituents and the simultaneous electron/ion transport process across the interface. Identification of the critical factors that govern the efficiencies of electrochemical interfaces is also important for other future technologies of relevance to energy, the environment and health care.

First I will discuss several generalizable approaches to engineering the three key constituents for an efficient electrochemical interface: (i) electrode design with focus on tuning of electronic structure, together with theoretical calculations to elucidate the origin of anomalously high heterogeneous electron transfer (HET) kinetics obtained by modulating electronic properties of electrodes; (ii) molecular engineering of electrolytes with focus on synthesis of novel surfactant-like ionic liquids (SILs), combined with molecular dynamic (MD) simulations to decipher the unique structuring phenomenon of SILs in the bulk and near charged surfaces; (iii) surface functionalization of electrodes by redox-responsive materials with focus on improvement of accessibility of redox sites by creating nanostructured hybrid systems, and on elucidation of the charge transport mechanism in redox thin films.

Next I will discuss electrochemistry-based control strategies for several exemplary separation and catalytic processes, using highly efficient electrochemical interfaces with optimized structural, electronic, and chemical properties. First, I will show a new electrochemical mediation strategy for separation of organics-water mixtures using materials that exhibit redox-tunable hydrophobicity, and for separation of ion-ion mixtures using materials that bear stimuli-controlled charge densities. Second, I will introduce a catalysis control concept, electrochemically responsive heterogeneous catalysis (ERHC), that allows for temporal and spatial control of reaction kinetics and product selectivity in both batch and flow systems. In these examples, special attention is paid to (i) interrogating the relationship between electrochemical activities of material systems and the overall efficacies of separation or catalytic processes, (ii) comparing electrochemical modulation approaches to conventional methods in terms of energy consumption, and (iii) identifying general descriptors for high selectivity in separation and catalysis based on molecular-level interactions between target species and redox moieties.

With the new insights obtained into electrochemistry on electrode/functionalities/electrolyte systems, and the novel methodologies developed, my future research will expand on the concept of electrochemical modulation in order to achieve molecular precision in separation processes and synthetic chemistry, for environmental, energy and pharmaceutical applications. I will describe a few proposed projects consistent with this plan, such as (i) conductive networks with potential-dependent pore structures for highly selective one phase (liquid or gas) separation, (ii) biomimetic membranes with tunable multiphase transport properties using electrochemical gating mechanisms, and (iii) electrochemical control over radical-based chemical transformations to achieve high predictability and selectivity for synthesis of key pharmacophores. I expect that the materials and technologies that my group develops may be applied to a broad range of problems such as water treatment, protein separation, CO_2 capture and transformation, environmental remediation, and drug discovery.

I aspire to introduce an unprecedented level of control in critical processes in chemical industries by developing new technology platforms based on “designer” electrochemical interfaces, elucidate the charge/electron transport mechanisms in complicated hybrid systems that are key to enhancing electrochemical activities, and advance the fundamental understanding of molecular-level interactions between target chemicals and the electrochemical interfaces using theoretical approaches. The long-term trajectory of my group will be defined by bringing the high precision, controllability, and energy efficiency of electrochemical modulation strategies to technologically important separation and catalytic processes at large scale.

438587 A New Hybrid Modeling Strategy: Data-Driven Models with First-Principle Constraint

Abstract:

Chemical Process Systems (CPSs) is complicated nonlinear processes. In the cognitive process of a system, two kinds of knowledge can be gotten that is reaction mechanism and sampled data information, causing two different modeling methods. Based on the reaction mechanism, first-principle methods are developed to establish a rough description model via mass energy equations, the reaction kinetics equations and other mechanisms. With adding complexity of modern CPSs and the improving controlling accuracy, first-principle models are often difficult to describe systems precisely, sometimes fails because of the unknown reaction knowledge. Meanwhile, data-driven methods, based on the historical or operational sampled data and requiring no detailed information about systems, are rapidly developing and widely used in CPSs' modeling and optimization process. Distinctly, neither modeling method makes full use of all information. Drawing on each other's strength, hybrid modeling becomes a hot topic in the modeling field.

Common structure of a hybrid model is parallel or serial structures. Parallel structures need clear reaction mechanism, which is very difficult for a complex CPS. Serial structures build each separate model and combine models by superposition or multiplication, i.e. the weighted method. Unfortunately, one serial sub-model may violate another, especially that the data-driven model with noise samples may violate the rough first-principle model. There is no verification between sampled data and reaction mechanism.

In order to fully utilize all information and verify two kinds of information mutually, a new hybrid modeling strategy to model a nonlinear industrial process is proposed. This approach is based on data-driven models combining with constraint from the reaction mechanism analysis. The main contribution is to construct a unified framework for modeling process by integrating data-driven model, knowledge detection, constraint handling method, and evolutionary algorithm. Initially, a data-driven model, e.g. artificial neural network model, is chosen to describe the input-output relationship. In the preparation stage, constraint conditions are extracted through the reaction mechanism analysis, for example, response restrictions, restricting response derivatives, the initial or extensional values, equal or unequal functions. Thus, the training process changes to a constraint optimization problem (COP) to solve optimal model parameters matching reaction mechanism constraints. Intelligent algorithm combining constraint handling method is applied to solve such a COP. Finally, the model parameters are solved and taken into the original data-driven model.

Moreover, both function simulation and the soft-sensor application to estimate average particle size of ZrO_2-TiO_2 composite colloidal sols indicate that the novel hybrid modeling approach can avoid the over-fitting problem to some extent. In addition, the Bayesian inference gives an explanation for the effectiveness of this method theoretically.

(6z) Ultrasound-Responsive Nanoparticles for Drug Delivery

Kwan, J. J., Columbia University
Myers, R.,
Graham, S.,
Coviello, C.,
Carlisle, R.,
Stride, E.,
Coussios, C.,

The potential of acoustic inertial cavitation, i.e., the radial collapse of a bubble due to ultrasound exposure, has been widely recognised for enhancing both transport and uptake of drugs in cancer treatment. For example, it has been proposed to provide a triggering mechanism for drug release from liposomes that extravasate into tumours due to the enhanced permeability and retention (EPR) effect and to increase the permeation of drugs throughout the tumour volume. In the past decade, much research has been devoted to initiating and controlling inertial cavitation using microbubbles. However for tumour therapy, microbubbles have a number of limitations: They are too large take advantage of the EPR effect and are also rapidly destroyed following ultrasound exposure. Recently, there has been a surge in development of nanoscopic cavitation agents such as phase change nanodroplets and nanobubbles. Here, we present novel nanoscopic particles that trap bubbles within nano-crevices, and function as nucleation agents for inertial cavitation. We demonstrate that these particles reproducibly generate inertial cavitation for extended periods of time, and facilitate enhanced drug penetration in *in vitro* and *in vivo* cancer mimicking models. Additionally, we show that the inertial cavitation threshold for these nanoparticles can be reliably predicted over a range of frequencies using a modified Raleigh-Plesset equation for crevice-stabilised bubbles.

438596 Application of New Explicit Correlation and Neural Network-Based Models for an Efficient Prediction of Natural Gas Compressibility Factor

Abstract:

This study presents two new methods for calculating the compressibility factor (z-factor) of natural gases. The first method, explicit correlation, is an efficient empirical model to calculate this important parameter. In this approach, 4158 data points obtained from the Standing and Katz z-factor chart have been used. The second method is based on Artificial Neural Networks (ANN), in which a 2:5:5:1 structure is used as an optimum network to predict Z-factor. In this approach, the proposed ANN model for predicting the Z-factor behaviour is detailed. This method employs a back-propagation algorithm, which is very effective in predicting model nonlinear variables.

In this case study, the z-factor data of a gas sample is calculated by different existing methods. Afterwards, the calculated data are graphically compared to experimental Z-factor data. In this research a new correlation for rapid estimation of the z-factor for sweet gases is studied. This correlation is developed based on 4158 points from the Standing and Katz Z-factor chart. The advantage of the developed correlation is that it is explicit in Z and thus does not require an iterative solution unlike other methods. The accuracy of the proposed Z-factor correlation has been compared to other correlations. The comparison demonstrates the superiority of the new correlation over other correlations used to calculate the Z- factor of natural gases. The results also show that the proposed ANN can be properly trained for the purpose of estimating the compressibility factor with an acceptable accuracy.

Effect of Salinity on Surfactant Enhanced Oil Recovery with special reference to Upper Assam Basin

Kalpajit Hazarika [a*] , Subrata Borgohain Gogoi [a]

[a] *Department of Petroleum Technology, Dibrugarh University, Dibrugarh,India*-786004

Abstract

The effect of salinity gives a new dimension on Surfactant Enhanced Oil Recovery (SEOR). The brine salinity varies from reservoir to reservoir and normally lies in the range of $3000 - 5000$ ppm in case of Upper Assam basin. During SEOR process brine soluble surfactants are introduced into the underground porous media. The brine should be at optimum salinity so as to generate Winsor type III microemulsion. This work aims at determining the optimum salinity of brine in which the surfactant is soluble by way of finding the solubilization ratio in between oil and surfactant and water and surfactant.

It is experimentally observed that at optimal water salinity a middle phase Winsor type III micro-emulsion exist while at salinity below optimum water salinity oil in water (O/W) Winsor type I micro-emulsion exist and at salinity above optimum water salinity water in oil (W/O) Winsor type II microemulsion exist. It is also found that in case of Winsor type I micro-emulsion the IFT between the microemulsion and insufficient water is high, while in case of Winsor type II micro-emulsion the IFT between the microemulsion and insufficient oil is high. In both these cases oil

*Corresponding Author: kalpajecm44@gmail.com

mobilization is hampered and trapping of oil results, thereby reducing the displacement efficiency in the porous media. Therefore the aim of this paper is to generate a Winsor type III micro-emulsion at optimum salinity where the two IFTs in between O/W microemulsion and oil and in between O/W microemulsion and water is less and equal which is one of the most significant parameter in SEOR.

Keywords : Surfactant Enhanced Oil Recovery, Salinity, Micro-emulsion, Interfacial tension .

1 Introduction

Since the beginning of the oil and gas industry, oil exploration and production companies have tried to maximize the oil recovery by finding new reserves or by using conventional recovery techniques. But using these conventional techniques the oil companies can recover only a fraction of initial oil in place (IOIP). A huge portion of IOIP remains in the reserves. To recover these the petroleum companies have entered an era where they have to push the bounds of technology, and think outside the box on how to produce the large quantities of remaining oil in place (ROIP), unconventional resources. To achieve this the companies think some new ideas and upgrade the existing technologies. In recent years, the field of enhanced oil recovery has grown to become more popular due to a combination of the worlds rising energy consumption, stagnant oil production, and low recoveries by conventional methods. On average, both the primary and secondary oil recovery phases account for about one-third of the (IOIP). The rest of the oil is trapped in the rock due to high capillary forces that prevent oil from flowing through the rock and into the wellbore for production. Any process that involves injection of fluid(s) to support natural reservoir energy and changing the reservoir fluid properties by interacting with the rock-oil-brine system to create favorable conditions for maximum oil recovery is known as an enhanced oil recovery (EOR) process [4]. The fluid properties may be oil swelling, lowering the interfacial tension, rock wettability modification, oil viscosity reduction, and favorable phase behavior. Chemical flooding is the one of the most effective EOR technique which uses surfactants, alkali, and/or polymers to increase oil recovery. Surfactants are used to lower the oil-water interfacial tension (IFT) and modify the wettability of the reservoir rock. Alkaline chemicals are used to react with crude oil to generate soap and increase pH it also change the wettibility of the reservoir rock. Polymer can be used to increase the viscosity of the displacing fluid. Either of these chemicals can be combined to complement each other in various forms of recovery methods. Despite the high potential of chemical EOR in increasing recovery, it only accounts for less than 1% of the US EOR production [5]. Salinity is an very important factor in SEOR process. Formation of micro-emulsion, stability of emulsions and some other factors are greatly dependent on salinity. So in this paper we discus the effect of salinity on SEOR process.

2 Material

In this paper we have used two types of Surfactant one is anionic and other is non-ionic. The anionic surfactant which is used is a was te product from Nagoan Paper mill, which is basically Sodium Ligno-Sulfonate based. Different types of Alkalis are used to established its effect on the IFT.

Sl.No	Material	Type
1	Black Liquer	Anionic Surfactant
2	TritonX−100	Non ionic Surfactant
3	Sodium Bi-Carbonate ($NaHCO_3$)	Alkali
4	Sodium Carbonate (Na_2CO_3)	Alkali
5	Sodium Hydroxide (NaOH)	Alkali
6	Sodium Chloride (NaCl)	Alkali

Table 1: List of Material

3 Results and Discussion

3.1 Effect of Brine Salinity on Interfacial Tension

Salinity is a most important parameter in EOR process. From salinity determination we establish the optimal salinity at which recovery is maximum. Optimal salinity, is the salinity required for the surfactant/brine/oil system to be at midpoint salinity (i.e., that point in the Type III phase environment where the concentration of oil equals the concentration of brine in the micro-emulsion, middle phase). Winsor describes three types of micro-emulsion among which type III micro-emulsion is most efficient for EOR. In this type middle phase micro-emulsion is formed. Examples are presented that support the proposal that chemical flooding is most efficient when conducted in a salinity gradient. For most anionic surfactants, midpoint salinity decrease as surfactant concentration decrease, particularly in the presence of multivalent cations.

The interfacial tension is nothing but the force required to separate one fluid layer from a immiscible fluid system. Lower the IFT lower will be the force required. In the hydrocarbon reservoir the pore spaces are always filled with water oil and gas. When the primary and the secondary recovery mechanisms are not efficient to recover the crude then we have to think chemical or thermal EOR processes. The main aim of chemical EOR process is to change the fluid properties. Interfacial tension is one of the fluid property which can increase the recovery of hydrocarbon. There are several factors which effect the IFT. Salinity is one of the factor. When we increase the salinity of the Surfactant solution then it reduces the IFT. But we can not increase the salinity beyond a certain limit. The solution must compatible with the reservoir fluid. Cost is an another factor which we should get importance. It has been observed that the IFT of a pure hydrocarbon versus brine system increases with salt concentration; however, when a small amount of surfactant is present, the IFT can decrease with salinity [14]; [15]. The IFT also varies with the type of hydrocarbons. Hence, the IIFT of crude oil versus reservoir brine depends on type and concentration of hydrocarbons, salts, and concentration of surfactants as well as type of surfactant present in the system. The interfacial tension also depends on temperature and pressure.

3.2 Effect of Salinity on Surfactant

The salinity effect on non ionic surfactant (TX−100) and anionic Surfactant (BL) are determined at different Salinities in Kruss Tensimeter. From the experimental values it is seen that with increases salinity IFT decreases, which is a significant effect for Chemical EOR process. It is also observed that the critical micelle concentration (CMC) have found at 0.2 and 0.4 % TX−100. Since the IFT at 0.2% TX−100 is more than at 0.4% so we can take 0.4% as CMC value. But in case of BL CMC have found at 3% concentration of BL solution

3.3 Effect of Alkali on IFT

Alkali has some effect on IFT, because it produces in-situ surfactant when it reacts with Carboxylic group of the oil. In case of NaHCO3 we have observed that interfacial tension increases with increase in $NaHCO_3$ concentration. An increase in IFT in the presence of electrolytes has been explained by negative adsorption of ions at oilwater interface, [1]. The experimentally observed increase in IFT is evidently due to the negative adsorption of sodium bicarbonate salts at the oilbrine interface. The surface tensions of the salt solutions increase with increasing salt concentration up to the saturation concentrations, which are about 1.1 mol/L for $NaHCO_3$. [2] But in case of Na_2CO_3 result is opposite, increase in concentration decreases the IFT of oil-brine interface.

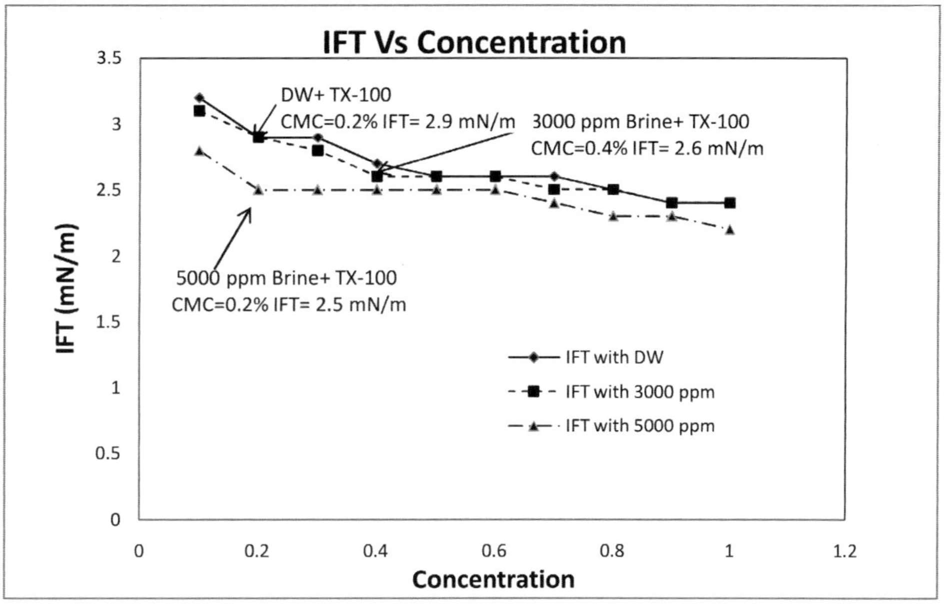

Figure 1: Effect of Salinity on TX-100

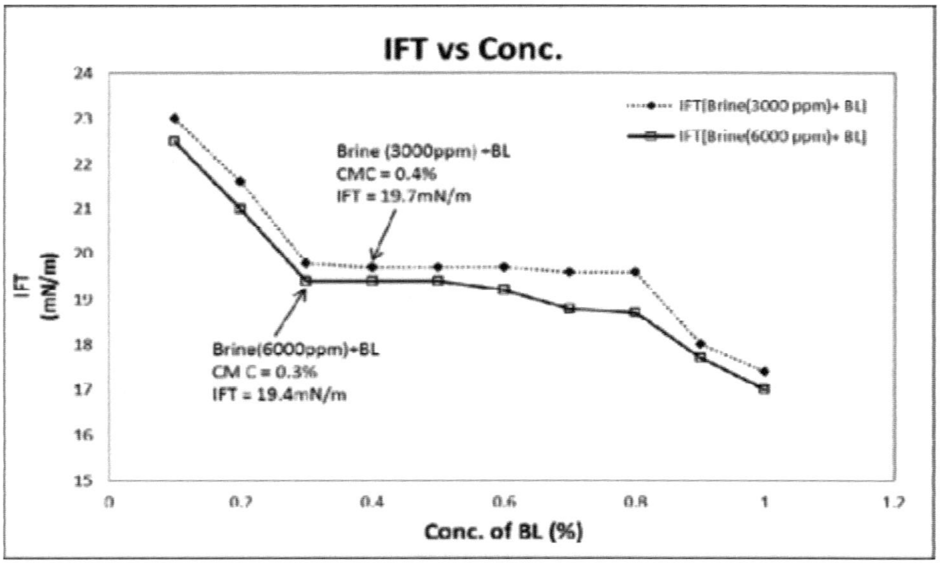

Figure 2: Effect of Salinity on BL

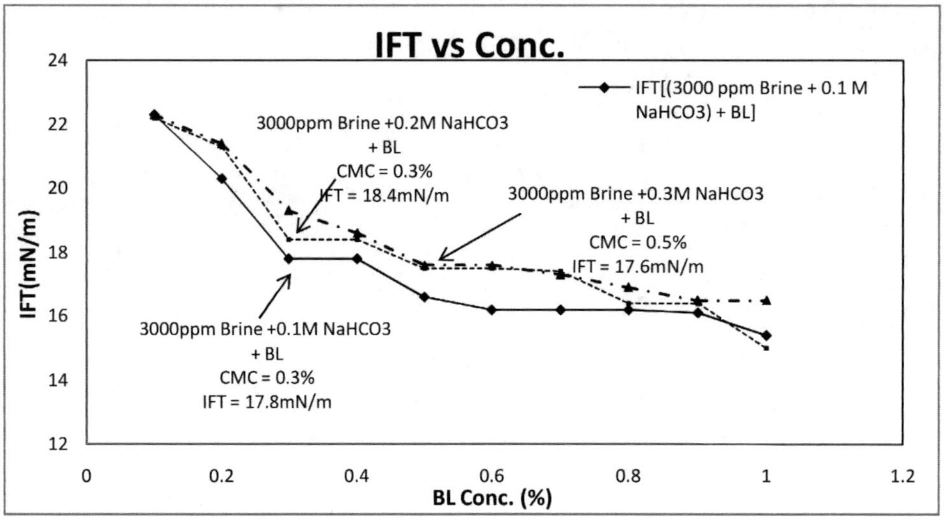

Figure 3: IFT (3000 ppm Brine+$NaHCO_3$+ BL)

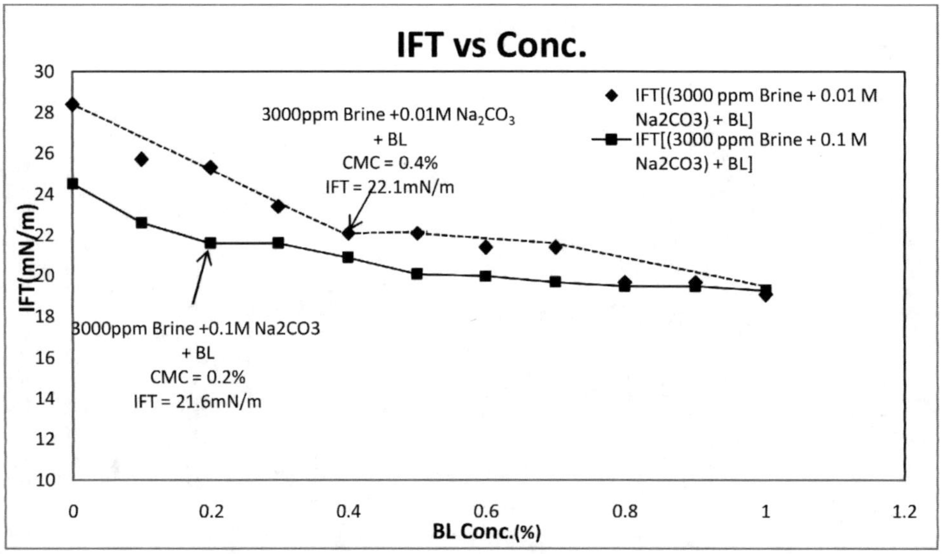

Figure 4: IFT (3000 ppm Brine +Na_2CO_3+BL)

4 Conclusion

Salinity is an important factor in EOR process. The selection of brine for EOR process is not a easy task. Efficiency of the process depends on the optimal salinity. Salinity also effects the liquid Permeability of the Reservoir rock. Before selecting the salinity we should emphasized on the reservoir salinity, because the brine should compatible with the reservoir fluid. It also effects the size of micro-emulsion. From economic point of view it is an important parameter, because it effects the recovery of crude oil.

References

[1] Weissenborn, P.K., Pugh, R.J., 1996. Surface tension of aqueous solutions of electrolytes: relationship with ion hydration, oxygen solubility, and bubble coalescence. Journal of Colloid and Interface Science 184, 550563.

[2] Orhan, Ozdemir , Stoyan, I. Karakashev , Anh V. Nguyen , Jan D. Miller, 2006. Adsorption of carbonate and bicarbonate salts at the airbrine interface, Int. J. Miner. Process. 81 149158

[3] Bansal, V.K. and D.O. Shah, 1978. The effect of ethoxylated sulfonates on salt tolerance and optimal salinity of surfactant formulations for tertiary oil recovery. SPE J., 8(3): 167-172.

[4] Jamaloei, B.Y. and M. Rafiee, 2008. Effect of monovalent and divalent ions on macroscopic behaviour of surfactant flooding. Proceeding of the Canadian International Petroleum Conference. Petroleum Society of Canada, PETSOC-2008-085, Canada.

[5] Kennedy, H.T. and E.T. Guerrero, 1953. The effect of surface and interfacial tensions on the recovery of oil by waterflooding. SPE J., 6(5): 41-48.

[6] Shabib-Asl, A., M. Abdalla Ayoub Mohammed, M. Kermanioryani and P. Paulo Jose Valentim, 2014. Effects of low salinity water ion composition on wettability alteration in sandstone reservoir rock: A laboratory investigation. J. Nat. Sci. Res., 4(13).

[7] Yousef, A.A., S.H. Al-Saleh, A. Al-Kaabi and M.S. Al-Jawfi, 2011b. Laboratory investigation of the impact of injection-water salinity and ionic content on oil recovery from carbonate reservoirs. SPE Reserv. Eval. Eng., 14(5): 578-593.

[8] Fatt, I., Pore Volume Compressibilities of Sandstone Reservoir Rocks, J. Pet. Tech., March 1958.

[9] Archie, G.E.: The Electrical Resistivity Log as an Aid in Determining Some Reservoir Characteristics, Pet. Tech., Vol. 5 (1), January 1952.

[10] Calhoun, J.C., Jr.: Fundamentals of Reservoir Engineering, University of Oklahoma Press, Norman, OK, USA (1951).

[11] Johnson, E.F., Bossler, D.P., and Naumann, V.O.: Calculation of Relative Permeability from Displacement Experiments, Trans. AIME (1959), V. 216, 370-372.

[12] Jones, S.C. and Rozelle, W.O.: Graphical Techniques for Determining Relative Permeability from Displacement Experiments, J. Pet. Tech., May 1978.

[13] Collins, R.E.: Flow of Fluids Through Porous Media, The Petroleum Publishing Co., Tulsa, OK, USA (1976).

[14] Hiemenz, P.C., and Rajagopalan, R., Principles of Colloid and Surface Chemistry, 3rd Edition, Marcel Dekker, New York, 1997.

[15] Cai, B., Yang, J. and Guo, T., Interfacial Tension of Hydrocarbon + Water/Brine Systems Under High Pressure, Journal of Chemical and Engineering Data, 1996, 41(3), 493-496.

438622 Towards a Sustainable Energy Future: The Role of Science-Driven Modeling and Systems Analysis

Abstract:

Climate change and energy security are two of the most defining challenges of our time and generations to come. Unfortunately, there is no *silver bullet* and the future energy landscape needs to encompass a suit of 'clean and green' energy technologies such as: .

- *Bioenergy (including advanced biofuels and biochemicals)*
- *Carbon capture and sequestration (technologies and infrastructure)*
- *Hydrogen driven transportation (technologies and infrastructure)*
- *Clean coal technologies (such as integrated gasification combined cycle)*
- *Renewable energy sources (including solar and wind)*
- *Energy systems of the future (e.g., urban & rural energy systems, zero-carbon homes)*
- *Water and healthcare (technologies and infrastructure)*

Together, all these technologies could potentially alleviate both the energy security and climate change concerns. My long-term research vision is to nurture these technologies and drive them from 'concept' to 'commercialization', using science-driven model-based design and optimization tools.

438599 The Applications of Spherical Polyelectrolyte Brushes

Sunday, November 8, 2015
Exhibit Hall 1 (Salt Palace Convention Center)

Yu Cang, Rui Zhang and Xuhong Guo, State Key Laboratory of Chemical Engineering, East China University of Science and Technology, Shanghai, China

The spherical polyelectrolyte brushes (SPB) are stable colloid system with polymer chains densely grafted from core. The functional polymer chains and core-shell structure provided many potential applications in many fields, such as catalyst and biology. The SPB have been reported to be used as nanoreactors for loading catalyst. The hybrid nanoparticles exhibited outstanding performances in catalytic efficiency. As nanoreactor, the SPB were firstly used as quantum dots carriers which exhibited stability and reversible photoluminescence. Both CdTe and CdS could be successfully immobilized in the shell of SPB with narrow distribution. Besides acting as nanoreactor, the SPB could be acting as crosslinkers to fabricate tough hydrogel with super water absorbency. The functional polymer chains of SPB had multi-crosslinking points and the weight content of SPB was dramatically decreased in the hydrogel. Therefore, the hydrogel was transparent and combined the characteristics of toughness and water absorbency. Furthermore, the SPB could be formed membrane to effectively adsorb dyes and filter nanoparticles which was still in the investigation. The SPB have various applications in industrial and engineering fields.

438603 Sustainable Design through Process Integration, Control and Optimization

Abstract:

T1 | MEET THE FACULTY CANDIDATE POSTER SESSION – SPONSORED BY THE EDUCATION DIVISION

Sustainable Design through Process Integration, Control and Optimization

Kazi Monzure Khoda

Post-Doctoral Researcher

Department of Chemical Engineering | Qatar University

Email: kazi.khoda@gmail.com; kazi0001@qu.edu.qa |Tel: (+974) 6640 8300

KEY CONTRIBUTION AREAS

Process Optimization	Process Design	Process Simulation
Process Integration	Process Safety	Process Control

Financial Modelling Ad hoc Reporting Market Analysis

RESEARCH INTERESTS

The focus of my research is to develop the concept of sustainable design, which involves process-design activities that lead to economic growth, environmental protection, and social progress for the current generation without compromising the potential of future generations to have an ecosystem that meets their needs. The ultimate target of my research activities is to conserve natural resources (mass and energy), recycle and reuse materials, prevent pollution, enhance yield, improve quality, advance inherent safety, and increase profitability. To achieve the desire goals a systematic framework and associated concepts and tools are necessary to guide designer methodically with the global insight of the process, to identify root causes of the problems or key areas of opportunities, to benchmark the performance of the process, and to develop a set of design recommendations that can attain the true potential of the process. My research background of process integration, process control and specially process optimization assist me for developing new concepts and tools to systematically enhance and reconcile various process objectives, such as cost effectiveness, yield enhancement, energy efficiency, and pollution prevention. I am working last few years to develop such methodical sustainable approaches for various processes starting from small scale production (producing valuable pharmaceutical drags using liquid chromatography) to large scale facility (managing industrial flares during abnormal situations). As my specialization is to work with different kinds of optimization algorithms and control concepts, I would also like to face the challenges from process uncertainty and from the black box problems which can be mitigated using sustainable process integration. My knowledge of research is not limited to the industrial processes, I have also implemented my acquire optimization knowledge to develop some adaptable financial modelling/algorithms which can be used in the most challenging and unpredictable global financial markets. The strong point of my research activity is that I can use my long experience of capital market to develop long term view for energy and commodity prices; this knowledge can be integrated with my developed framework to make the system/design more robust to face different kind of challenges including unpredictable price fluctuations (price fluctuations of oil, natural gas etc.), unpredictable incidents (industrial flaring) or unknown disturbances (sudden changes in process parameters).

QUALIFICATION SUMMARY

1. Possesses 7 years' R&D experience

2. Expert in different optimization algorithms (i.e., Genetic Algorithm, SQP, Neural Network)

3. Developed a customized separation technique and online monitoring system for chiral drug separation named as *i*-SCC

4. Working on sustainable plant design through process integration

5. Took charge of developing an integrated approach to the simultaneous design and operation of industrial facilities for abnormal situation management

6. Developed an optimization framework and some energy alternatives tools for industrial flare reduction along with a greenhouse gas calculator and the complete historical database of ethylene plant flaring with the collaboration with industrial advisory (Q-Chem)

7. Worked on developing different feed injection mechanism, initial modelling for SFC and prepared the framework for measuring adsorption isotherm in large-scale chromatography with green Solvents.

8. 6 years' experience of managing financial portfolio in-line with money management by investing in global FOREX, CFD and STOCK market

9. Contributed strategic insight in analysing experimental data for system optimization using mathematical and statistical models

10. Worked as lecturer in Green University of Bangladesh and as teaching assistant in the School of Chemical and Biomedical Engineering, NTU in four chemical engineering courses: Advanced reaction engineering, Process control, MATLAB workshop and Chemical/Bioengineering lab.

12. Worked with different software i.e., MATLAB, LINGO, LabVIEW, Aspen HYSYS, Auto CAD, Latex, Origin, EndNote, MetaQuotes Language (MQL4/MQL5), JFOREX

13. Worked with various instruments i.e., High pressure liquid chromatography (HPLC), Supercritical fluid chromatography (SFC), Gas chromatography (GC), Magnetic suspension balance

EDUCATION

Doctor of Philosophy	Nanyang Technological University (NTU), Singapore	*July 2013*
Bachelor of Science in Chemical Engineering (CGPA: 3.75/4.00)	Bangladesh University of Engineering and Technology (BUET), Bangladesh	*January 2008*

April, 2014

AWARDS AND HONORS

Microsoft Faculty Fellow – Microsoft

Certificate of Completion of Sustainable Design through Process Integration Training Course - Process Technology, Qatar — *March, 2014*

Graphical system design achievement awards - National Instruments ASEAN — *2011*

Research scholarship - Nanyang Technological University, Singapore — *January,2009 – January, 2012*

Dean's list awards - Bangladesh University of Engineering and Technology — *2003 - 2007*

Technical scholarship - Bangladesh University of Engineering and Technology — *2003 - 2007*

-

SELECTED PUBLICATIONS

Patent

[P1] Bijan Medi, **Monzure-Khoda Kazi**, and M. Amanullah. Design of an improved single-column chromatographic separation process and its online monitoring system. (Patent pending).

Journal Papers

[J1] **Monzure-Khoda Kazi**, Bijan Medi, and M. Amanullah. Optimization of an Improved Single-Column Chromatographic Process for the Separation of Enantiomers. *Journal of Chromatography A*, 2012. 1231: p. 22-30. (doi:10.1016/j.chroma.2012.01.057).

[J2] **Monzure-Khoda Kazi**, Fahd Mohammed, Ahmed Mhd Nabil AlNouss and Fadwa Eljack. Multi-objective optimization methodology to size cogeneration systems for managing flares from uncertain sources during abnormal process operation. *Computers and Chemical Engineering, 2015* vol 76, p. 76-86

Invited Talk

[T1] **Monzure-Khoda Kazi**, and Fadwa T. Eljack. Sustainable Approaches for Abnormal Situation Management. *Water and Energy Workshop, Doha, Qatar, 2015.*

438615 Engineering the Plant Microbiome to Complement Host Phenotype

Sunday, November 8, 2015
Exhibit Hall 1 (Salt Palace Convention Center)
Collin M. Timm, Biosciences Division, Oak Ridge National Laboratory, Oak Ridge, TN

Engineering the plant microbiome to complement host phenotype

Plant roots are a rich source of carbon in an otherwise nutrient starved soil environment, and as a result, a diverse community of bacteria is attracted to the plant. The complex network of beneficial and antagonistic interactions of community members, with each other and with the host, forms a microbiome that is generally a beneficial extension of the host plant. The plant microbiome can increase nutrient acquisition rates, increase resistance to pathogens, and alleviate environmental stresses such as drought. Understanding how the microbiome functions and interfaces with the host has applications in food and bioenergy agriculture, production of secondary metabolites from plants for commodity chemicals and pharmaceuticals, and will improve our general understanding of the host-microbiome relationship.

My current research is focused on the bacterial microbiome of *Populus,* a genera of fast-growing trees cultivated as a second-generation biofuel feedstock. Using metabolic models, comparative genomics, and phenotype screening we showed that the extreme diversity within 19 *Pseudomonas fluorescens* strains isolated from *Populus* was driven by the metabolite profile of the host and correlated with spatial organization of the microbiome. To gain further insight into spatial aspects of cooperation and inhibition, we designed a novel patterning approach to study distance dependent interactions between bacterial strains. We are also studying the molecular mechanisms that lead to phenotypic changes in the host when colonized by individual bacteria or mixtures using phenotype measures, metabolomics, and transcriptomics. The results show that bacteria can behave synergistically to help optimize and modify plant growth phenotypes. To study community behavior, we measured microbiome response to host stresses using 16S sequencing and identified candidate bacterial taxa that may help alleviate host response to drought, light limitation, and toxin stresses.

My research group will build on the results described by studying how constructed communities of bacteria respond to the host and environmental stresses. Specifically, we will design communities of genome-sequenced bacterial isolates using individual phenotypes, phylogenetic relationships, and predicted genomic content and test the ability of the community to mimic the natural microbiome response to stress. Our approach will include quantitative analysis of bacterial strains and their interactions, and characterization of community response to host stress. Models of bacterial kinetics, metabolism, and interactions in culture conditions will be used to predict behavior on the host plant. Integrating quantitative experimental data with predictive models will help us understand how the microbiome assembles in nature, ultimately leading to the ability to engineer constructed communities of bacteria with specific and predictable functions.

Postdoctoral Project: Interactions between *Populus* and cultured representatives of its bacterial microbiome

Mentors: Dale A Pelletier and David J Weston, Oak Ridge National Laboratory, Biosciences Division

PhD Dissertation: Kinetics of vesicular stomatitis virus mRNA and genomes during infection

Mentor: John Yin, University of Wisconsin-Madison, Department of Chemical and Biological Engineering

Research Career

My research career began as an undergraduate where I studied gas hydrate formation at the Colorado School of Mines. I helped develop software for predicting hydrate formation in gas pipelines and tested predictions using a novel reactor in the laboratory. This valuable experience familiarized me with quantitative experiments and laboratory techniques, but more importantly convinced me to go to graduate school to continue my academic career. In graduate school I changed my focus to biological engineering, where I learned how to design experiments and control the biological system to make quantitative measurements of viral replication. I used the data I collected to build kinetic models of viral infection, and showed that viral mRNA production is independent from host environment and only depends on viral genome number. This suggests that transcription is robust and insensitive and helps to identify targets for anti-viral strategies for RNA viruses. While studying interactions between two organisms (viruses and cells), I became interested in complex ecological systems with multi-member interactions. I was fortunate to find a post-doctoral position as a part of the Plant-Microbe Interfaces project at Oak Ridge National Laboratory studying the microbiome of *Populus* trees. I have focused specifically on understanding bacterial genetic and phenotypic diversity in the system to help determine how interactions between microbiome members and the host lead to overall system function. In my next step I will study how multi-member systems can be engineered to perform specific functions.

Teaching and Mentorship

As I have moved through my academic career I have become more active in my role as a teacher and mentor. During my undergraduate studies I worked closely with many of my peers, and began to appreciate the different learning and teaching styles present in the group. I had the opportunity to informally tutor younger students in heat and mass transfer classes; in both cases their grades improved after we began working together. In graduate school I got my first opportunity to formally teach classes to students. As a teaching assistant for the introductory transport course, I enjoyed spending time developing lectures and example problems for students, and focused on how I could engage the entire class in a single topic. I also worked as a TA for the process control laboratory where I helped students run experiments and taught them to present their results in concise reports. As a post-doc I have worked with undergraduate summer students and new graduate students from chemistry, biochemistry, biological systems engineering and chemical engineering. Their unique backgrounds have helped me generalize my teaching methods. I have helped them learn how to work in the laboratory, analyze data, and present results. The students helped identify metabolic pathways, isolated bacteria that grow on *Populus* metabolites, identified genes that affect plant growth, and developed methods for studying root growth in microfluidic devices. Watching students succeed with my input has been extremely important to my development and has helped me appreciate the commitment of my teachers and mentors throughout my academic career.

438708 Computational Models for Growth and Defects of Melt-Grown Crystals

Sunday, November 8, 2015
Exhibit Hall 1 (Salt Palace Convention Center)

Gaurab Samanta, QuantitativeSilicon Research, SunEdison Semiconductor, St. Louis, MO

Computational models for growth and defects of melt-grown crystals

Crystals form the basis of many technologies. While semiconductor devices and solar cell industries rely on high quality crystalline silicon (Si) wafers, laser and radiation detector technologies use high quality fiber crystals of materials like sapphire. In order to make high quality crystals, it is crucial to develop an understanding of the crystal growth processes. Computational models can provide invaluable insights into the crystal growth process. For example, a thermal-capillary model of a fiber growth process known as "Micro-pull-down" (μ-PD) can reveal operating limits and the underlying mechanisms. In an investigation of sapphire fiber growth through μ-PD, we have used a thermal-capillary model to find various failure modes of the process and identified underlying mechanisms.

On the other hand, it is a formidable challenge to control the amount of defects in crystals grown from commercial growth processes like "Czochralski" (Cz). Controlling the quantity of defects in crystals is necessary for successful development of any device technology based on them. Computational modeling of defects can provide quantitative predictions which can be used to successfully engineer them. For example, using classical nucleation theory (CNT) based model of oxygen precipitation in Cz-Si, the number density of oxygen precipitates can be quantitatively estimated as a function of time and temperature. We have developed a new nucleation rate model which is more suitable for quantitative estimation of oxygen precipitate density nucleated at temperatures lower than 750 C.

As we continue to develop new computational models and improve current ones to better predict crystal growth and defects, it is also important to teach these approaches to future generations. One effective way to teach computational methods is by engaging the students in the development of a full simulation. The full simulation problem may be disintegrated into several sub-problems which covers the entire coursework. Complexities can be reduced by using an established solver platform to develop the simulation.

438710 High Resolution Carbon Nanotube Enantiomer Separation By Specific DNA Sequences

Abstract:

Carbon nanotubes (CNTs) have remarkable electrical, thermal, mechanical, and optical properties making them uniquely suited as fundamental building blocks for electronics and biomedical applications. However, CNT polydispersity has been one of the major challenges limiting their fundamental research and technological development. Polymer aqueous two-phase (ATP) extraction has been recently demonstrated as an effective technique to sort polydisperse nanotube material. The spontaneous partition of single-stranded DNA-wrapped single-wall carbon nanotubes (ssDNA-SWCNTs) in a given ATP system is strongly sequence-dependent and can be further modulated by salt and polymer additives to efficiently isolate single nanotube structures. Additionally, circular dichroism (CD) spectroscopy shows that purified semiconducting and metallic nanotubes are also enantiomer enriched. Separating CNT enantiomers is of great scientific interest since it provides important information into the mode of DNA binding and molecular recognition towards specific nanotube structures. In this work, we have conducted an exhaustive screening of a 12-mer DNA library using a specific DNA sequence pattern. To our surprise, this new pattern enabled the identification of a super-sequence which can recognize and separate both enantiomers of a single chirality (6,5) species in record high purity and yield in an ATP system.

438712 Polymer/Nanomaterial Structural Control Using Flow and Confinement: Modeling, Experiment, and Applications

Abstract:

With the rise of innovative nanoscale material science over the past decade, efforts to combine the versatility of polymer with novel functionalities of nanofillers have substantially increased. Recent studies have found that a strategic structural placement of these materials within a polymer can dictate the composite material properties, whether it is electrical, mechanical, or thermal. Gaining control over such structures is crucial in enhancing material properties significantly to tailor to wide variety of modern applications such as battery, sensory, and protective gear. Evidently, the nanoscale structure of these nanocomposites can greatly influence their macroscopic properties. The challenge lies in controlling the structure of the so-called nanocomposite material to tailor to the application of one's choice at will.

To this end, I have sought answers to the following questions: i) how does the polymer/nanofiller composite structure change under various processing conditions? And ii) how does its functionality change with the corresponding structure and subsequently used as in a real-life application? To effectively answer these questions, I have done a comprehensive study of modeling, experiment, and application on polymer/nanofiller composite structure under various processing conditions.

Using coarse-grained molecular dynamics, multiple combinations of polymer(homopolymer or block copolymers) and nanofillers(nanoparticles, nanorods, and nanotubes) composite structures under A) shear and extensional flow[2,5] and B) cylindrical nanoscale confinement[1,3,4] were investigated. Our study reveals that depending on the control variables of the aforementioned dynamics, we can induce novel functional structures such as highly oriented polymer and nanomaterials, evenly placed nanomaterials within polymer matrix, and interconnected nanorods.

As a comparative experiment, electrospinning was used to fabricate polymer/nanofiller nanofibers which undergoes both extensional deformation and nanoscale cylindrical confinement[1,3,5]. The structural agreement between the simulation and the experiment result confirms the validity of the model. As such, some of the novel structures observed in simulation were also observed in experiment, such as well-dispersed and well-aligned nanorods within the nano-scale polymer fibers. The novel functional properties that arise from nanostructures of electrospun nanofiber are then readily utilized in various applications such as molecular sensors[5] and battery electrodes[6].

Lastly, preliminary work has been done to significantly increase the mechanical strength of electrospun polymeric fibers by improving the molecular orientation and degree of crystallinity from gel-electrospinning ultra high molecular weight polyethylene. By defining a fundamental connection between the predicted nanostructure and its functionality, we can expand much further on the capabilities of electrospun nanofibers to significantly impact the polymer processing and its application as a whole.

Selected Publications

[1] **J. H. Park**, V. Kalra, and Y. L. Joo, "Cylindrically Confined Assembly of Asymmetrical Block Copolymers with and without Nanoparticles: Simulation and Experiment", *Soft Matter*, **2012**, *8*, 1845.

[2] **J. H. Park**, V. Kalra, and Y. L. Joo, "Controlling the Dispersion and Orientation of Nanorods in Polymer Melt under Shear: Coarse-Grained Molecular Dynamics Simulation Study", *Journal of Chemical Physics*, **2014**, *140*, 124903.

[3] **J. H. Park**, J. Yin, V. Kalra, and Y. L. Joo, "Role of Nanoparticle Selectivity in the Symmetry Breaking of Cylindrically Confined Block Copolymers", *Journal of Physical Chemistry C*, **2014**, *118*, 7653–7668.

[4] **J. H. Park** and Y. L. Joo, "Formation of Interconnected Morpholgies via Nanorod Inclusion in Confined Assembly of Symmetric Block Copolymers", *Physical Chemistry Chemical Physics,* **2014**, *16*, 8865 - 8871.

[5] **J. H. Park** and Y. L. Joo, "Tailoring Nanorod Alignment in a Polymer Matrix by Elongational Flow under Confinement: Simulation, Experiments, and Surface Enhanced Raman Scattering Application", *Soft Matter,* **2014**, *10*, 3494-3505.

[6] J. Yin, J. Carlin, J. J. Kim, Z. Li, **J. H. Park**, B. Patel, S. Chakrapani, S. H. Lee, and Y. L. Joo, "Synergy between Metal Oxide Nanofibers and Carbon Substrates for Rechargeable Lithium-Oxygen Batteries", *Advanced Energy Materials* **2015,** *5,* 1401412

438713 Preparation of Multimetallic Catalytic Systems By Controlled Surface Reactions for Biomass Upgrading

Abstract:

Supported multimetallic catalysts are an important class of materials owing to their enhanced activity, selectivity and stability that surpasses their monometallic counterparts. The enhanced performance of these multimetallic catalysts can be due to synergistic effects between the constituent metals, achieved when these metals are in close proximity. Moreover, achieving a narrow distribution of structures and compositions during the preparation of these multimetallic catalysts is critical in understanding structure-property relationship of these materials and is important for improving their design and function. Preparing such a multimetallic catalyst has proven to be a challenging task. My experience and skill in preparing such a muticomponent system puts me in a unique position to synthesize novel materials for a variety of contemporary applications, including design of heterogeneous catalysts for upgrading biomass-derived compounds to value added chemicals.

I started working on bimetallic catalysts during my PhD research in the Department of Chemical Engineering, Middle East Technical University (Ankara-Turkey), under the supervision of Prof. Timur Dogu and co-supervision of Prof. Gulsen Dogu of Gazi University (Ankara-Turkey). My thesis was based on the development of active and stable bimetallic catalysts for steam reforming of ethanol for hydrogen production.

As a visiting scholar at the Iowa State University, I worked on a collaborative project between Prof. Brent Shank's Group and Prof. James A. Dumesic's Group. In this collaborative project, I synthesized bifunctional SBA-15 supported catalysts comprising of both acidic and basic sites. In parallel, I also worked on synthesizing a hydrophobic, SBA-15 supported Pt catalyst for the production of hydroxymethylfurfural (HMF) from glucose in a biphasic system.

During my postdoctoral research with Prof. James A. Dumesic, I worked on diverse projects including both homogeneous and heterogeneous catalysis. I worked on understanding the kinetics of dehydration of the hemicellulose fraction of biomass, primarily xylose, using a biomass-derived solvent, gamma-valerolactone (GVL), and I investigated the hydrogenation of furfural to form furfuryl alcohol with copper based catalysts which where stabilized by atomic layer deposition of MgO_x and AlO_x.

During my tenure as a post-doctoral researcher, I developed new methods for synthesizing supported bimetallic catalysts using controlled surface reactions (CSR). I prepared and characterized bimetallic catalysts by a new synthesis protocol developed in our group, and I studied these catalysts for hydrogenolysis of 2-(hydroxymethyl)tetrahydropyran [1]. I also worked on using these well-defined and characterized catalysts to study their improved activity for industrially relevant reactions, including the water gas shift reaction and higher alcohol synthesis from syngas. Currently, I am using the CSR technique for the synthesis of ternary systems.

During my graduate studies in the Department of Chemical Engineering at the Middle East Technical University, I worked extensively as a teaching assistant. I was a teaching assistant for all of the core chemical engineering courses both at the undergraduate and graduate levels (Chemical Process Calculations, Fluid Mechanics, Chemical Reaction Engineering, Thermodynamics, Mass Transfer and Separation Processes, Heat Transfer, Reactor Design, etc.) and chemical engineering laboratory courses (Chemical Engineering Laboratory I-II). I also mentored undergraduate students working in our group for their research projects. During these teaching assistantship experiences, I always enjoyed teaching and simultaneously learning from these enriching experiences.

For my future career, the ultimate goal of my research is to use my experience in synthesis and characterization of heterogeneous catalysts. Developing the CSR technique was important because conventional synthesis methods for the preparation of bimetallic catalysts can complicate the analysis of experimental data to elucidate the nature of the active sites. The procedure is well-suited for extension to ternary systems that can potentially enable us to bring together less compatible components for catalytic property enhancement. These materials are also ideal for comparison with theoretical studies targeted for discovery, development and understanding of supported bimetallic catalysts. Relevant reactions for producing value added chemicals from biomass derived compounds are hydrogenolysis of polyols, hydrodeoxygenation of aromatic alcohols, and reduction of 3-hydroxypropionic acid to 1,3 propanediol.

Reference:

1. Hakim S.H., Sener C., Alba-Rubio A.C., Gostanian T.M., O'Neill B.J., Ribeiro F.H., Miller J.T., Dumesic J.A., "Synthesis of Supported Bimetallic Nanoparticles with Controlled Size and Composition Distributions for Active Site Elucidation", Journal of Catalysis,328, 75-90 (2015).

438719 A Systems Biology Definition of the Core Proteome of Metabolism and Expression

Abstract:

Finding the minimal set of gene functions needed to sustain life is of both fundamental and practical importance. Minimal gene lists have been proposed using comparative genomics-based core proteome definitions. A definition of a core proteome that is supported by empirical data, is understood at the systems-level, and that provides a basis for computing essential cell functions is lacking. Here, we use a genome-scale model of metabolism and expression to define a functional core proteome consisting of 356 gene products, accounting for 44% of the *Escherichia coli* proteome by mass based on proteomics data. This systems biology core proteome includes 212 genes not found in previous comparative genomics-based core proteome definitions, accounts for 65% of known essential genes in *E. coli*, and has 78% gene function overlap with minimal genomes (*Buchnera aphidicola* and *Mycoplasma genitalium*). Based on transcriptomics data across environmental and genetic backgrounds, the core proteome is significantly enriched in non-differentially expressed genes, and depleted in differentially expressed genes. Compared to the non-core, core gene expression levels are also similar across genetic backgrounds (two times higher Spearman rank correlation), and exhibit significantly more complex transcriptional and post-transcriptional regulatory features (40% more transcription start sites per gene, 22% longer 5'UTR). Thus, genome-scale systems biology approaches rigorously identified a functional core proteome needed to support growth. This framework, validated using high-throughput datasets, facilitates a mechanistic understanding of systems-level core proteome function through *in silico* models.

438720 Thermodynamics and Kinetics for Energy, Environment and Materials: A Comprehensive Research Initiative Based on Experimental and Theoretical Investigation

Sunday, November 8, 2015
Exhibit Hall 1 (Salt Palace Convention Center)
Sanjoy Bhattacharia, Department of Chemical Engineering, Texas Tech University, Lubbock, TX

Thermodynamics and kinetics are the most fundamental aspects that need to be understood for designing new materials, fluid mixtures and chemical processes. A balanced investigation based on experiments and modeling are required to solve many fundamental and engineering problems in the area of chemical engineering science. Typical examples are the designing chemical process plant for the treatment of produced water from oil and gas production, designing hydraulic fracturing fluids, carbon-dioxide captures, hydrate formation in oil and gas pipeline, designing explosives and propellant for mining and military application, nuclear and chemical waste treatment, crystal growth, designing polymer composites etc. There are many scientific and engineering problems remains in these aforementioned fields that can be solved with the understanding of thermodynamics and kinetics of the materials and processes by experimental and theoretical investigations.

I am a seasoned experimentalist with a substantial modeling and simulation experience. I contributed 24 published article with 224 citations which include both experimental and modeling/simulation works. My post-doctoral research is the development of thermodynamic models for chemical processes associated with petrochemical, oil and gas, pharmaceutical, energetic materials and other relevant industries. My PhD research was to understand kinetics and thermodynamics of phase changes and reactions of materials. Before starting my PhD, I was employed in Bangladesh Atomic Energy Commission, which is one of the premiere National Laboratories in Bangladesh, for 3.5 years as a research engineer. Here my research was to develop radiation processed hydrogel, polymer composites, biodegradable thin films, surface coating for natural fiber and wood.

My future research project includes, but not limited to, (i) understanding the thermodynamic and kinetic properties of electrolytes, chemical and nuclear waste, petrochemicals, CO_2-captures, fluids in oil and gas reservoir condition, pharmaceutical compounds (ii) understanding the mechanism of kinetics processes and fragmentation of energetic materials for safer storage and applications (iii) Structure–kinetic–thermodynamic property relationship of materials. To fund my research, I will apply for research grant to National Science Foundation (NSF), DARPA, Department of Defense, Office of Naval Research, Department of Energy, National Labs and relevant industries. Impact of the work– (i) the new and improved fundamental understanding of materials, fluid mixtures and chemical process, (ii) designing safe and efficient chemical process and units, (iii) safe design explosives for processing, storage and applications (iv) advanced materials.

438624 Microfluidic Studies of Emulsions and Suspensions in Wall-Bounded Shear Flow

Abstract:

Microfluidics provide a promising platform to study a wide range of phenomena from fluid flow analysis to individual and collective cell behaviors. In my ongoing research, I am pursuing a fundamental study on the dynamics of individual double emulsion droplets in a wall-bounded shear flow. Initially, monodisperse double emulsion droplets of water in oil in water are generated with a diameter in the range of ~ 100 um and polydispersity index (*PDI*) of 1.002 in a flow focusing device by controlling the flow rate ratios. Using a capillary micromechanics approach, the droplets are shown to be very deformable. My preliminary experiments reveal that double emulsions are focused to a certain distance from the wall while traveling down the microcapillary channels at low *Re* numbers ($Re < 1$). The equilibrium position in the channel and the entry length required to achieve that position is being determined. In parallel, I am working on an industrially funded project to study viscoelastic entry flow phenomena in microfluidic abrupt contractions. It was observed that vortices of different lengths form at various *Re* and elasticity numbers. Micro-Particle Image Velocimetry (μ-PIV) is used to analyze flow at the upstream and in the constriction. The experimental results are compared with 3D simulations in Comsol Multiphysics.

438626 Computational and Experimental Studies for Advancement of Sustainable Energy Systems

Abstract:

A safe and secure future for humanity depends to a large extent on the development of sustainable and economic energy resources, and will require fundamental science and engineering contributions. In this talk I will present some of my work on the advancement of energy systems in electronics, oil and gas, and solar energy sectors.

Cooling of electronic systems, particularly high performance data centers, is a highly energy intensive process. In 2013, data centers consumed about 100 billion kWh of electricity in the U.S. alone and it is expected that by 2020, the carbon footprint of data centers will exceed that of the airline industry. One of the solutions to reduce energy consumption is to move from conventional air cooling to liquid cooling, and in that respect I will present a computer-aided molecular design approach for development of novel heat transfer fluids for cooling of such systems. The fluids designed in this work not only possess superior heat transfer properties, but also significantly lower global warming potentials. I will also present a critical evaluation of nanoparticle dispersions (or nanofluids) for heat transfer applications and show, contrary to widely held belief, the effective thermal conductivity of nanofluids decreases with decreasing particle size. In my future research I plan to expand on this approach and design new thermal storage materials for concentrated solar thermal and hybrid photovoltaic/thermal systems.

In conventional oil and gas industries, clathrate hydrates of natural gases are associated with many energy transportation ('flow assurance') related problems. However, gas hydrates also present a potential future energy resource, as it is estimated that more than 10^{15} m^3 of natural gas is present in oceanic and permafrost gas hydrate deposits. I will present some of my work on hydrate phase equilibria modeling, including the application of a quantum chemical approach, COSMO. In my future research I also plan to explore other applications of gas hydrates including seawater desalination and gas separation.

438721 Hydrodynamics and Phase Separation in Complex Fluids

Abstract:

A fundamental understanding of soft matter and complex fluids is crucial for the development of emerging technologies such as 3D printing, artificial blood, and engineered biological tissues, as well as solving existing industrial challenges in enhanced oil recovery, water recycling, food processing, and others. Here I provide a brief overview of my research on fluid mechanics and soft matter physics in complex fluid systems. In particular, this overview includes:

(i) The first dynamic force measurements of the adhesion between two vesicles in suspension, including the impact of membrane tension and separation velocity [1, 2].

(ii) A scaling analysis to account for dimple formation in the drainage of thin films prior to phase separation in emulsions [3].

(iii) Measurements of the impact of interfacial rheology on the amount of liquid entrained during the formation of foam films.

References:

[1] J.M. Frostad, M.C. Collins, L.G. Leal. "Cantilevered-capillary force apparatus for measuring multi-phase fluid interactions" Langmuir 29, 47154725 (2013).

[2] J.M. Frostad, M. Seth, S.M. Bernasek, L.G. Leal. "Direct measurement of interaction forces between charged, multilamellar vesicles". Soft Matter 10, 7769-80 (2014).

[3] J.M. Frostad, J. Walter, L.G. Leal. "A scaling relation for the capillary-pressure driven drainage of thin films" Physics of Fluids 25, 052108 (2013).

438723 Porous Materials: A Unique Platform for Separations and Catalysis

Sunday, November 8, 2015
Exhibit Hall 1 (Salt Palace Convention Center)

Gokhan Barin, Chemistry, University of California, Berkeley, Berkeley, CA; Chemistry, Northwestern University, Evanston, IL

Metal-organic frameworks (MOFs) and porous organic polymers as solid-state adsorbents have advanced significantly during the last two decades on account of their high surface areas and pore volumes, chemical and thermal stabilities, and the flexibility in their design and composition. Such features render these materials attractive and promising alternatives over traditional adsorbents in applications including gas storage, gas separations and capture, and catalysis. For instance, in order to facilitate the widespread use of alternative energy sources, new materials that can store large quantities of energy needs to be developed. Highly efficient and selective adsorption of contaminants at vanishingly low concentrations is also a significant environmental and industrial challenge that requires the discovery of solid-state adsorbents with high affinities. In the field of catalysis for energy-related conversions, there is a necessity to develop platforms that can achieve selective transformations in a level that biological systems accomplish. In order to address these challenges, MOFs and porous polymers have the potential to provide the desired combination of structural and chemical composition; therefore the fundamental understanding of structure-property relationship in these materials is highly important in identifying the next-generation materials in energy research.

In this poster, I will present an overview of my efforts during my graduate and postdoctoral studies towards developing and understanding functional materials at the molecular and macromolecular level. My graduate research focused on the chemistry of mechanically interlocked molecules, which provided a platform to investigate the properties of molecular species, i.e., radical dimers, in confined environments, and also their incorporation into a range of integrated systems to develop functional materials. Moving from the self-assembled systems at the molecular level to the ones at the macromolecular level, my postdoctoral studies have so far involved the development of porous materials for applications such as natural gas storage, hydrocarbon separations, and the removal of toxic contaminants from air and in solution. Developing new organic/inorganic synthetic strategies to make these materials and tools to investigate their performance have been the central theme of my postdoctoral research.

My research interests lie primarily at the intersection of organic/inorganic synthesis, physical chemistry, and materials science. In the future, my goal is to establish a research program focusing on (i) the development of soft porous materials for the capture/separation of industrially, environmentally, and biologically relevant chemicals and (ii) the exploration of novel catalysis concepts by taking advantage of the unique structural features (e.g., tunable pore environment and chemical composition) of porous materials in order to produce fuels and chemical feedstocks in an efficient and selective manner. These initial endeavors will encompass areas of chemistry, materials science, and chemical engineering to synthesize, characterize, and evaluate the performance of such materials.

438627 Mesoscale Modeling of 2D Materials for Energy and Biomedical Applications

Sunday, November 8, 2015
Exhibit Hall 1 (Salt Palace Convention Center)

Sanket A. Deshmukh, Center for Nanoscale Materials, Argonne National Laboratory, Argonne, IL

Mesoscale Modeling of 2-D Materials for Energy and Biomedical Application

Sanket A. Deshmukh

Center for Nanoscale Materials, Argonne National Laboratory, Argonne, IL

We have shown recently that 2-D materials ranging from one atom thick graphene layer to one nanoparticle film of a closely packed ligated metal nanoparticle can have dramatically different functionality than bulk counterpart. The potential applications of these 2-D materials include, but not limited to, lubricants, chemical and pressure sensors, and catalysis. I will specifically highlight two case studies where we employ mesoscale modeling to probe the existing 2-D materials and design a new class of hybrid 2-D materials with specific emphasis on their mechanical properties. In the first case study, we discovered the role played by graphene to achieve superlubricity - almost zero friction - at macroscale. To probe the mechanism of superlubricity we have performed meso-scale all-atom simulations of a system consisting of graphene, nanodiamond, and diamond like carbon (DLC).[1] Our simulations suggest that the macroscopic superlubricity originates from an intriguing nanomechanical phenomenon: graphene patches at a sliding interface, wrap around the tiny nanodiamond particles and form nanoscrolls with reduced contact area that slide easily against the amorphous DLC surface, achieving an incommensurate contact (a necessary condition to achieve superlubricity) and near zero coefficient of friction (0.004). In the second case study, we probe the ligand dynamics during the self-assembly process of 2-D nano-thin membrane of ligated metal nanoparticles.[2] We have conducted meso-scale simulations of self-assembly of ligand-modified gold nanoparticles at the air-water interface to elucidate the exact role of ligands that contributes to their asymmetric distribution during the membrane formation process. Our simulations suggest that the experimentally observed anisotropic distribution of ligands originates from ligand re-distribution that is greatly facilitated by the mobility of ligands at gold surface. The mobility of the ligands in turn is a strong function of their surface coverage. Fully ligated nanoparticles have much lower ligand mobility and do not display an asymmetric distribution in the self-assembled membranes. On the other hand, for lower ligand coverages, we find the ligands to be highly mobile which reorganize to form Janus-like membranes. This asymmetry is shown to have interesting consequences on the mechanical and bending properties of these 2-D membranes.

Reference:

1. Macroscale superlubricity enabled by graphene nanoscroll formation

Diana Berman*, Sanket A. Deshmukh*, Subramanian KRS Sankaranarayanan, Ali Erdemir, Anirudha V Sumant (* Shared First Author)

Science, *348, 6239, 1118-1122, 2015*

2. Subnanometre ligand-shell asymmetry leads to Janus-like nanoparticle membranes

Zhang Jiang, Jinbo He, Sanket A. Deshmukh, Pongsakorn Kanjanaboos, Ganesh Kamath, Yifan Wang, Subramanian K. R. S. Sankaranarayanan, Jin Wang, Heinrich M. Jaeger□and Xiao-Min Lin

Nature Materials, *doi:10.1038/nmat4321, 2015.*

438628 Microstructure and Collective Dynamics of Cytoskeletal Assemblies

Abstract:

Complex fluids are materials that are driven out of their equilibrium structure by means of internally generated, or externally imposed forces. The coupling of microstructure and the macroscopic behavior is the hallmark of complex fluids, and the origin of many intriguing behaviors of these materials. I am interested in studying complex fluids and complex assemblies from a microscopic perspective using theories and simulations. My goal is to bridge the gap between micro-scale dynamics and the collective behavior in these systems; such findings potentially provide the blueprint for design and control of novel materials in microscopic scales. One fascinating example of complex fluids is the cell cytoskeleton which is the primary machinery for performing many vital cellular processes such as motility, intracellular transport, and cell division. My postdoctoral research concerns the positioning and dynamics of the mitotic spindle apparatus during cell division which is crucial to the accurate segregation of chromosomes and the subsequent development of the organism. I use biophysical models to describe the interactions of microscopic tubular polymers of spindle structure, i.e. microtubules (MTs), with the molecular motors, the other organelles, and the cell boundaries; these models are then used as inputs in the dynamic simulations of this highly active cellular assembly.

The dynamics of mitotic spindle is regulated by several factors including the growth and shrinking of MTs, their interactions with molecular motors, and their flexibility and strong hydrodynamic coupling. In part, I will talk about the numerical toolbox I have developed that incorporates all these physical variables. Our method is unparalleled in a number of aspects; most notably it is the first technique to include many-body hydrodynamic interactions and the resulting cytoplasmic flows in cellular assemblies. I use boundary integral methods in Stokes flow for describing the hydrodynamic interactions between the MTs and other bodies, coupled with the state-of-the-art technology on fast summation techniques for Stokes flow. The deformation of the fibers is described by Euler-Bernoulli theory for flexible rods and their polymerization is modeled by reparametrization of the dynamic equations in the appropriate Lagrangian frame. The use of pseudo-spectral spatial discretization and implicit time-stepping makes this method highly accurate and efficient. The entire method is parallelized and scalable which allows simulating assemblies composed on thousands of fibers.

We use simulations to study the migration dynamics of pronuclear complex (PNC) in early stages of cell division of *C-elegans* embryo. We find that although the proposed force transduction mechanisms properly position and align the PNC, each mechanism leaves its unique *fingerprint* on the generated cytoplasmic flows. This feature can be utilized to differentiate between, and identify the mechanisms more relevant to the migration process. Considering that these flow signatures are generic features of each model, they can equally be applied to identifying the active mechanisms involved in other stages of cell division, as well as other organisms.

I will also talk about an ongoing collaborative work on studying the mechanics of mitotic spindle in *C-elegans* using a 3D reconstruction of spindle structure with electron tomography. In short, the results of the tomography show that the distribution of length and curvature of the MTs attached to chromosomes, i.e. kinetochore MTs, are different from the non-kinetochore MTs. Based on these observations we propose models that delineate the dynamics of the kinetochore MTs; we corroborate these models with detailed simulations.

Finally, I will showcase a few applications of the developed numerical framework in studying other problems in complex fluids and soft materials which I will be pursuing as a faculty member.

438631 Conversion of Space Crew's Wastes into Biofuel Using Thermophiles

Abstract:

The manned space exploration is a highly complex and challenging mission, and one of the difficulties is the logistic load which can be costly and limit for travelling time and distance. Therefore for long term space flight, the reduction of logistical requirement is necessary and the related technologies are under investigation. The Trash to Gas Fuel (TGF) is one of the promising processes in which crew's waste generated during space mission can be converted into gaseous biofuels such as hydrogen. Consolidated bioprocessing (CBP) has been applied for biohydrogen production from different kinds of wastes, and it offers promising alternatives over traditional multistep processes. In the present research, CBP was used to convert untreated crew's waste simulants into hydrogen through thermophilic microbial consortia and anaerobic fermentation processes. Four consortia were developed using water/sediment/soil samples collected from hot springs, water treatment plant, and compost facility, and used for biohydrogen production from lignocellulosic substrate or food waste as a carbon source. The results demonstrated that all of the consortia could produce biohydrogen (1.137 mmol) after 7 days culturing at 70. The biohydrogen production from different kinds of human waste simulants as sole nutrient source by the same consortia using thermophilic CBP is currently being tested. Optimization of various parameters including composition and concentration of simulants for better yields of biohydrogen will also be carried out. The biohydrogen obtained could be used as propellants or energy source for power system during space missions. Additionally, Trash to Gas Fuel process can diminish the volume requirement to store the waste generated by the crew members during long-term space missions.

438632 Computational Modeling and Experimental Investigation for Membranes at the Water – Energy Nexus

Abstract:

This poster will provide an overview of my computational and experimental research about the design and fabrications of different types of membrane (microfiltration, ultrafiltration, reverse osmosis and forward osmosis) for water treatment. My Ph.D. dissertation focused on the application of polymeric nano-composite membranes to water purification. During the first part of my research, I worked on understanding the mechanism of natural organic matter (NOM) fouling on an ultrafiltration membrane by relating the operational viewpoint for fouling (macroscopic) to the microscopic viewpoint. Then, in the second part of my research, I used the power of nanoparticles to increase the antifouling properties of the polysulfone membrane. The mixture of titanium dioxide nanoparticles and multiwall carbon nanotube with the optimum mass ratio was embedded with the novel and simple method into the polymeric membrane and interestingly increased both the membrane permeability and rejection two times compared to the pure polysulfide membrane. Also, the nanocomposite membranes showed the higher antifouling behavior compared to the pure membrane at different operational conditions and flaunt physicochemical properties.In my post-doctoral research, I have continued to work on the membrane application for water treatment and energy production. Forward osmosis (FO) (also, pressure retarded osmosis) is a prime membrane technology of interest due to its potential for low energy use in water – energy nexus. The main goal of my post-doctoral project is investigation of transport phenomena in forward osmosis to improve FO membrane design and process efficiency. This project combines computational fluid dynamics (CFD) modeling with experimental fabrication of FO membranes to examine the assumptions imbedded in current transport equations and identify sources of inaccuracies in water flux predictions. This effort can lead us to the development of transport equations that can be used by all to support FO research, enabling the deployment of FO for real-world problem solving with economically competitive performance.Based on my knowledge of membranes, instrumentation design and componential fluid dynamic (CFD) modeling, my future research will focus on both experimental and modeling aspects of membrane technology issues for the water – energy –food nexus. In details, I plan to mainly focus my academic career on parallel research area of the computational modeling and experimental synthesis, modification and fabrication of different types of novel mixed matrix membranes for using at different connected interdisciplinary areas at the water – energy – food nexuses.

438724 Phase Transitions and Self-Assembly of Block Copolymers, Colloids and Proteins

Sunday, November 8, 2015
Exhibit Hall 1 (Salt Palace Convention Center)
Jens Glaser, Chemical Engineering, University of Michigan, Ann Arbor, MI

Entropy maximization and the emergence of order from disorder are highly universal principles for the assembly of materials. Self-assembly allows the creation of novel, functional materials from nanoscale building blocks. We discuss how entropy controls the formation of structure in polymeric and colloidal materials. In block copolymers, the order-disorder transition isdriven by collective composition fluctuations, and we show how they stabiliz the disordered phase in a universal way. Computer simulations of various chemically distinct coarse-grained models of diblock copolymers show how their phase behavior can be mapped onto each other using a minimal set of parameters. In colloidal self-assembly, the depletion interaction mediates an emergent, attractive force between large colloids in a sea of smaller particle. We show how these forces introduce strong entropic bonds between faceted particles that favor the creation of open structures. Our results point towards novel ways of understanding the role of particle shape and entropy for structure formation also in biological systems. One intriguing yet poorly understood phenomenon is the crystallization and aggregation of folded proteins.

438725 Polymerization within Porous Media: Transformative Coatings and Interfaces Lab (TCIL)

Abstract:

Polymers are versatile materials with attractive properties ranging from optoelectronic to electrochromic and electrocatalytic. The development of effective approaches for polymer integration into current device architecture is of paramount importance to properly leverage the attractive properties of polymeric materials in diverse contexts. The current emerging energy conversion and storage technologies rely on physical and chemical design of component at nanometer length scale. This translates into a need for the development of an effective method to integrate polymeric materials with accuracy at nanometer length scales. In my research, I rely on Initiated and Oxidative Chemical Vapor Deposition (iCVD/oCVD) as two unique, solvent free methods that enable polymerization and coating in one step. I aim to fundamentally understand polymerization reactions within nanoscale and confined structures, and to study material properties confined within these spaces in a precise and systematic fashion. I envision that the outcome of my research will lead to the discovery and development of new metamaterials and interfaces with attractive properties for applications in clean energy and water resources managements.

During my Ph.D. at the Thin Film and Devices Lab, I contributed to the development of a new branch of oCVD, where the dehydrocoupling reaction of heterocyclic monomer was initiated using readily vaporizable liquid oxidants of weak Lewis acid nature. My experimental observations have led me to conclude that not only can surface conditions significantly change nucleation mechanisms, but that the physical properties of the underlying substrate can define the rate determining process and change the overall polymerization growth rate and final material properties. Additionally I realized that the tight and complex combination of the processes with dynamic time constants spanning a few orders of magnitude, down to the rate of charge transfer at the interface, plays a key role in synthesis and final device properties. Therefore by adjusting the dynamic rate constant and delicately manipulating the interfaces using polymeric materials, I was able to show how one can benefit from the tight integration of the polymeric materials within the nanoscale network of the electrodes typical to many electrochemical devices. The significant improvement in the device performance, the observation of anomalies in charge storage capacity of polymers within nanostructures, and the change in the chemistry of the oxides surface that enables controlling the heterojunction energetic were the significant outcomes of my research. Currently I am doing my postdoctoral training under the supervision of Chinedum Osuji and Menachem Elimelech at Yale University. My research has been focused on transport phenomena within porous media and design and development of processes targeting challenges in the Water-Energy Nexus. I have been actively working on fabrication of ordered and porous polymeric media at micro- and nanoscale range. This exciting experience has broadened my scientific scope and also equipped me with skills for fabricating polymeric based templates.

In my lab, TCIL, I plan to pursue three principal directions and aim at (i) polymer synthesis through vacuum based pathway and within porous matrix (ii) understanding the scale dependence of polymer properties (iii) generating new polymeric interfaces and domains with unprecedented properties.

Selected publications:

Nejati, S. and K. K. S. Lau "Pore Filling of Nanostructured Electrodes in Dye Sensitized Solar Cells by Initiated Chemical Vapor Deposition", Nano letters 2011 11 (2), 419–423.

Nejati, S. and K. K. S. Lau "Chemical Vapor Deposition Synthesis of Tunable Unsubstituted Polythiophene", Langmuir 2011 27(24), 15223-15229.

Nejati, S.; Minford, T.; Smolin, Y. Y. and Lau, K. K. S., "Enhanced Charge Storage of Ultrathin Polythiophene Films within Porous Nanostructures". ACS Nano 2014 8 (6), 5413–5422.

Nejati, S. and K. K. S. Lau "Integration of polymer electrolytes in dye sensitized solar cells by initiated chemical vapor deposition", Thin Solid Film 2011 519(14), 4151-4155.

Nejati, S.; Patel, A; Wallowitch G. R. and Lau, K. K. S., Electrical Conductivity and Stability of oCVD Copolymer Thin Films of Thiophene and Pyrrole. Nanoscience and Nanotechnology letters 2015 7 (1) 50-55.

Nejati, S.; Boo, C.; Osuji, C. O. and Elimelech, M. "Engineered Poly(Vinylidene Fluoride) Membrane for Membrane Distillation" Journal of Membrane Science 2015.

Bavarian, M.; Nejati, S.; Lau, K. K.S.; Lee, D; and Soroush M., "Effects of Critical Dye Sensitized Solar Cell Parameters on Cell Performance: a Theoretical Study" Industrial Engineering and chemistry research 2013 53 (13), 5234-5247.

Smolin, Y. Y.; Nejati, S.; Bavarian, M.; Lau, K. K.S.; Lee, D; and Soroush M., "Effects of Polymer Chemistry on Polymer-Electrolyte Dye Sensitized Solar Cell Performance" Journal of Power Sources 2015 274, 156-164

Lin, S.; Nejati, S.; Boo, C.; Hu, Y.; Osuji, C. O. and Elimelech, M. "Omniphobic Membrane for Robust Membrane Distillation" Environmental science and Technology 2014 1 (11), 443-447.

Bose, R.; Nejati, S.; Stufflet, D. and K. K. S. Lau "Grafted Poly(ethylene oxide) (PEO) Anti-fouling Surfaces using Initiated Chemical Vapor Deposition (iCVD)", Macromolecules 2012 45 (17), 6915–6922.

438726 Design of Functional Polymeric Materials: From Ion Transport to Bio-Inspired Assembly

Sunday, November 8, 2015
Exhibit Hall 1 (Salt Palace Convention Center)

Katherine P. Barteau, Materials Science and Engineering, Cornell University, Ithaca, NY

Lithium polymer batteries offer a number of advantages to standard lithium ion batteries, including an all-solid state structure, increased safety, and the potential to be combined with lithium metal anodes for increased energy density over lithium intercalation anodes. However, the low-temperature (< 80 °C) ionic conductivity of polymer electrolytes has remained a major limitation, and progress in understanding strategies for systematic improvement in ionic conductivity has been dominated a single polymer for elucidating structure-property relationships in ionic conducting polymeric solids, i.e., poly(ethylene oxide) (PEO).

While earning my Ph.D. in Chemical Engineering at UC Santa Barbara under the guidance of Profs. Glenn Fredrickson, Edward Kramer, and Craig Hawker, I focused on the development of poly(allyl glycidyl ether) (PAGE) as an alternative platform for lithium battery polymer electrolytes. PAGE was found to have peak conductivities at [O]/[Li] = 16, with $\sigma > 3\times10^{-5}$ S/cm at 25 °C and $\sigma > 5\times10^{-4}$ S/cm at 80 °C. Below 60 °C, PAGE has a conductivity that is 10–100 times higher than that of PEO at equivalent salt concentrations with this disparity in conductivities between PAGE and PEO increasing with decreasing temperature. In addition, the synthetic versatility of allyl glycidyl ether as a building block was demonstrated by the preparation and evaluation of various AGE-EO macromolecular architectures that show superior performance to both PAGE and PEO by utilizing copolymerization and thiolene coupling chemistry. In order to develop design rules for polymer electrolytes, a library of structurally similar polyethers, poly(glycidyl ether)s (PGEs) were synthesized by anionic polymerization with controlled molecular weight and polydispersity index. Polymer electrolytes were prepared with lithium bis(trifluoromethanesulfonyl)imide (LiTFSI) at a molar ratio of [O]/[Li] = 16. Significantly, no correlation between glass transition temperature or room temperature viscosity and ionic conductivity over the range 20–80 °C could be established. Similar to conventional small molecule organic electrolytes, the ionic conductivity increases with the dielectric constant (ε) of the parent polymers over the range $\varepsilon = 4.8–6.0$, indicating a strong dependence of ion conductivity on the ion dissociation equilibrium in the polymer matrix. This hypothesis was further supported by pulse field gradient (PFG) NMR studies of the lithium and fluorine diffusion behavior. Our results suggest that synthetic efforts to create highly conductive polymer electrolytes should focus on increasing the dielectric constant of the parent polymeric material.

In my current research as a postdoctoral associate with Prof. Uli Wiesner and Prof. Lara Estroff at Cornell University, I am developing block copolymer based peptide brushes to investigate the self-assembly of these polymers and their ability to locally control crystallization of small molecules. Proteins, and their smaller peptide components, offer a diversity of self-assembly pathways: not only of the amino acid-based polymer itself, but in biomineralization, a protein matrix directs crystal phase and orientation of inorganic materials. The rich sequence diversity of proteins, however, has not been demonstrated for synthetic three-dimensional (3-D), continuous frameworks. Block copolymers offer long range periodic order of 0-D to 3-D structures through self-assembly and have been utilized to template high-surface area inorganic materials. Nevertheless, control has primarily been spatial while direction of crystal type and crystallographic orientation has remained challenging. By combining the long-range periodic structure direction of block copolymers with the chemical and sequence specificity of peptides, new highly-controlled composite materials might be created. Here I will discuss the synthesis of peptide and peptoid functionalized block copolymers. By altering the amino acid content and order, we investigate the role of sequence control on polymer self-assembly and on directed and templated inorganic materials growth. Ultimately, the development of composite three-dimensional structures with controlled single-crystal facets and large surface area could be expanded beyond biological and mechanical applications to new energy materials with high selectivity towards catalytically active surfaces and to pharmaceutical design.

438730 Structure, Deformation, and Flow of Soft Materials

Sunday, November 8, 2015
Exhibit Hall 1 (Salt Palace Convention Center)
Vikram Jadhao, Physics and Astronomy, Johns Hopkins University, Baltimore, MD

Soft materials are the basic components of biological systems and form an integral part of our modern lifestyle. Advances in the design of supramolecular systems that can mimic the structure of innate biological matter and recent progress in the synthesis of polymers for display and lighting technologies (light-emitting polymers), together with the advancements in the ability to design and assemble materials such as colloids, hydrogels, molecular electrolytes, and polymeric lubricants demonstrate that soft materials will rapidly become the foundation of future devices and applications in medicine, energy, and electronics. Simulations and theory have an important role in guiding this emerging field by assisting the design of soft materials and accurately evaluating their properties as well as providing predictions of their behavior in applications. I have explored the fundamental physics associated with the structure, deformation and flow of soft materials by examining a number of model soft-matter systems such as flexible membranes, colloidal suspensions, polymeric lubricants, and electrolytic solutions. At the same time, I have evaluated the prospects of these systems as applied materials.

I have demonstrated how long-range repulsive interactions can be used as a tool to manipulate shapes of soft membranes formed in emulsions and synthetic nanocontainers by tuning environmental conditions such as the electrolyte concentration. The proposed shell-design mechanism can be used as a template to synthesize nanoparticle-based drug-delivery carriers, and it can help understand shape changes in micelles and emulsions that form during metal-extraction processes. I have developed a variational formulation and an associated simulation procedure that can accurately and efficiently predict the structural and dynamical behavior of ions in biological and synthetic materials that often exhibit varying dielectric response. Applications of this method accurately described the non-monotonic structural features in the ionic density near dielectric surfaces and showed how these features can significantly alter the inter-surface interactions. I have also advanced our understanding of flow in disordered systems which many soft materials can be identified as. I investigated the shear-thinning response of simple binary Lennard-Jones fluids and lubricant squalane over a wide range of strain rates and uncovered the underlying mechanism responsible for shear thinning as the systems are driven through the glass transition. In addition, I showed the shortcomings of methods such as time-temperature superposition that are frequently employed to understand such phenomena. These results are important to models of elastohydrodynamic lubrication and will be useful in superior polymeric lubricant design.

My poster will present the key aspects and results of my research summarized above along with an outline of future projects which will use the aforementioned investigations as a platform. A few questions that I anticipate will influence my future project designs include: Given a particular set of tunable environmental parameters and constraints, what are the typical morphologies that can be tailored in a soft material? How does a soft material adapt to changes in its shape and composition; are there ways to make the material more robust and/or functional by controlling its environmental properties such as pH or salt concentration? How can we accurately extract and then efficiently exploit the dynamical behavior of soft materials that are often amorphous in nature and are characterized with rheological properties that lie in between liquids and solids? In the investigations of these problems I envision the development and application of both computational and theoretical methods that will encompass a wide range of concepts from statistical mechanics, supramolecular chemistry, electrostatics, polymer science, and rheology.

438732 Molybdenum Dioxide-Based Catalysts for the Generation of Electrical Power from Biofuels

Abstract:

Molybdenum dioxide (MoO_2) is a mixed conductivity oxide that has been found to exhibit high activity as well as coking resistance and sulfur tolerance for the partial oxidation of liquid fuels such as jet fuel and gasoline. However, stability issues prevent the use of MoO_2 for long periods of time. This problem seems to be related to the phase transition resulting from the interaction between the oxide and the carbon-containing fuel, which results in the formation of a carbide phase. Different approaches are discussed in this work to increase the stability of the oxide phase such as doping and the use of supporting materials. Thus, titanium doping showed to be able to broaden the oxygen-to-carbon ratio operating range of MoO_2 for the partial oxidation of long-chain hydrocarbons by increasing the redox stability. The modification in the structure causes the hydrogen reduction mechanism to change from a three-dimensional nuclei growth to a three-dimensional hydrogen diffusion limited model.

Unsupported MoO_2 nanoparticles synthesized at 180°C using a hydrothermal approach exhibit loss of surface area when exposed for long periods of time to typical reforming temperatures (~850C). However, the particle size of MoO_2 nanoparticles was found to become stable when supported on silica through a synthesis process involving a sol gel approach. Compared to large particle size MoO_2, the selectivity to syngas of silica-supported MoO_2 seems to be significantly enhanced, as deduced from the results obtained using dodecane as probe molecule. Long-term activity tests using methyl oleate as biodiesel surrogate showed a stable performance for at least 200 h, using an $O_2/C=0.75$, producing a H_2 yield of 80% and 100% conversion.

Mixed conductivity and high reforming activity, as well as significant coking resistance and sulfur tolerance displayed by this metal oxide, suggest MoO_2 as a potential candidate as anode material for solid oxide fuel cells. Conventional materials such as nickel–yttrium-stabilized zirconia (Ni–YSZ) cermet anodes operating with complex liquid hydrocarbon fuels display severe carbon formation on the Ni surface and sulfur poisoning. Molybdenum dioxide (MoO_2)-based anodes with an optimum microporous structure were able to produce a maximum power density over 3 W cm^{-2} at 750 °C under the direct feeding of *n*-dodecane fuel mixture. The cell resisted the carbon (coke) formation for over 24 h of continuous cell testing and was also resistant to deactivation with 500 ppm of sulfur in the fuel stream. Using this anode, a fuel cell running with commercial gasoline demonstrated a power density >3.0 W cm^{-2} at 0.6 V. Over a 24 h period of operation, the open cell voltage remained stable at 0.9 V. At the cell voltage of 0.6 V, its current density dropped over the first 7 h to a value of 3.0 A cm^{-2}, where it stayed for the remaining 17 h of the test with a minor fluctuation. Power density of 2.0 W cm^{-2} at 0.6 V was still measured after 24 h on stream with a continuous feed of gasoline. Scanning electron microscopy (SEM) examination of the anode surface pre- and post-testing showed no evidence of coking, which hints at the reason for the observed stability under such harsh cell operating conditions. Thus, a SOFC using a MoO_2-based anode has potential for generating electrical power from liquid fuels for future hybrid electric vehicles.

(6ca) Polymeric Mechanical Amplifiers of Tumor Cell Death

Mitchell, M. J., MIT
Langer, R., Massachusetts Institute of Technology

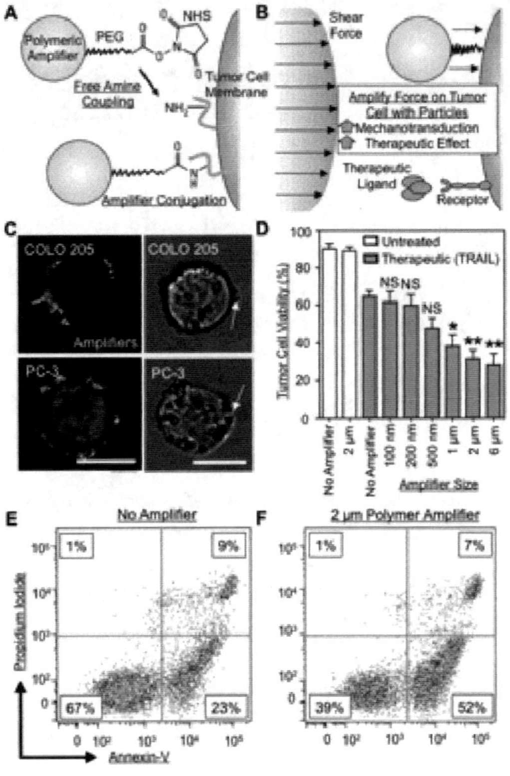

Polymeric Mechanical Amplifiers of Tumor Cell Mechanotransduction and Cell Death

Michael J. Mitchell[1], Robert Langer[1]

[1]Chemical Engineering and David H. Koch Institute for Integrative Cancer Research, MIT, Cambridge MA 02139

Introduction: It has become evident that tumor cells are responsive to mechanical forces *in vivo*, and prove critical to tumor cell proliferation and death. Recent work has shown that tumor cells exposed to fluid shear forces increase receptor-mediated signaling. We hypothesized that biocompatible, polymeric micro- and nanoparticles conjugated to the tumor cell surface via free amine coupling act as mechanical amplifiers in presence of fluid shear forces to increase mechanotransduction, and can be exploited to increase the efficacy of therapeutic ligands.

Materials and Methods: Polymeric particles ranging from 100 nm–6 µm in size were conjugated to free amines on tumor cells via NHS crosslinker chemistry **(Fig. 1A)**. Nondegradable polystyrene and degradable (PLGA, PCL) particles bound to tumor cells were assessed using flow cytometry, brightfield, and confocal fluorescence microscopy. A cone-and-plate viscometer was used to apply a fluid shear force (2.0 dyn/cm^2) to tumor cell suspensions and to amplify the force exerted by polymeric particles on tumor cells **(Fig.1B)**. Tumor cells (COLO 205, PC-3) were treated with 1 µg/mL of a TNF-related apoptosis-inducing ligand (TRAIL) to assess amplified mechanotransduction and receptor-mediated apoptosis in the presence of polymeric particles. An annexin-V apoptosis assay was used to characterize the mode of cell death. Caspase colorimetric assays and inhibitors (Z-VAD-FMK) were utilized to assess caspase-dependence in the mechanotransduction response.

<u>Figure 1</u>: (**A**) Conjugation of polymer amplifiers to tumor cell surface via NHS crosslinker chemistry. (**B**) Polymeric particles amplify force exerted on the tumor cell surface under fluid forces to increase mechanotransduction and therapeutic efficacy. (**C**) Confocal and brightfield images of 1 μm polymeric particles conjugated to the surface of tumor cells. Scale bars: 10 μm. (**D**) Particle-functionalized tumor cell viability after TRAIL treatment in the presence of shear force.$^{*}P<0.05^{**}P<0.01$. NS: not significant. **(E,F)** Annexin-V apoptosis assays of non-functionalized and functionalized tumor cells treated with TRAIL in the presence of shear forces. Lower left/right and upper left/right quadrants denote viable, early apoptotic, late apoptotic, and necrotic cells, respectively.

Results and Discussion: NHS crosslinker chemistry was successfully used to conjugate polymeric particles to the tumor cell surface **(Fig. 1C)**. In the presence of fluid shear forces, it was found that polymeric particles act to mechanical amplify tumor cell mechanotransduction, as evidenced by increased receptor-mediated apoptosis and decreased tumor cell viability in the presence of the therapeutic ligand TRAIL (**Fig. 1D**). Additionally, amplification of TRAIL-mediated apoptosis was increased with particles of increasing size (**Fig. 1D**), demonstrating that increasing the force exerted on the cell surface with larger particles amplified the therapeutic response. Annexin-V apoptosis assays showed the addition of conjugated polymeric particles to the cell surfaces nearly doubled tumor cell apoptosis in the presence of TRAIL under shear forces **(Fig.1E,F)**, and inhibition assays revealed the response to be caspase-dependent apoptosis.

Conclusions: These data demonstrate that polymeric particles, both degradable and non-degradable, act as mechanical amplifiers of tumor mechanotransduction in the presence of shear forces, and are exploited to increase therapeutic efficacy. Clinically, this approach shows that increased mechanical force applied to target tumor cells can increase sensitivity to therapeutic ligands.

438735 Bottom-up Design of Nanostructured Thermoelectric Materials from Solution Phase Synthesized Nanowires, Nanocrystals and Heterostructures

Abstract:

Thermoelectric (TE) devices convert heat flow into electrical current or vice versa, which have many applications including waste heat recovery and solid state cooling. Recently, significant progress has been made in the development of nanostructured TE materials with high figure of merit (ZT), which is a measure of the performance of a thermoelectric material. By reducing the grain size of a polycrystalline thermoelectric material, or by incorporating nanoinclusions into a host thermoelectric material, the thermal conductivity can be greatly reduced due to the phonon scattering effect, which is the main reason for most of the experimentally observed ZT enhancements. To achieve the desired nanostructures, a top-down approach involving high temperature (above 1000 ºC) melt synthesis followed by ball milling and thermal sintering is commonly used, which is a tedious and energy intensive process. Meanwhile, a bottom-up approach based on solution phase synthesized nano building blocks could also be utilized to obtain similar nanostructures, which is, however, advantageous to the top-down approach due to the low-energy, low-cost and scalable features of the solution synthesis. Moreover, with the synthetic tools developed in the past decades, nano-sized building blocks with tunable size, composition and morphology are possible, which provide a versatile platform for the design of novel TE materials. Here, we show the application of bottom-up approach towards nanostructure TE materials in different systems including nanocrystalline Ag_2Te from Ag_2Te nanowires, PbTe-Ag_2Te nanocomposites from PbTe-Ag_2Te dumbbell heterostructures, and AuAg-PbSe superlattice from AuAg alloy nanoparticles and PbSe nanocrystals, etc. We demonstrate thermoelectric properties of the bottom-up fabricated nanomaterials are tunable by controlling the synthesis condition and post-synthesis treatment of the nano building blocks. We also show that the binary heterogeneous structures allow us to modulate the charge carrier transport through electronic band engineering, which may provide another mean to enhance ZT in addition to the phonon scattering effect.

438736 Bioengineered Personalized Disease Models for Precision Medicine

Abstract:

The 215 million dollar precision medicine initiative by President Obama intends to "*pioneer a new model of patient-powered research that promises to accelerate biomedical discoveries and provide clinicians with new tools, knowledge, and therapies to select which treatments will work best for which patients*". It is very exciting to see the recognition for personalized medicine at a level which can enable significant innovations to solve practical problems. In the scientific world, problems are solved by generating an understanding through basic research, creating solutions through innovations, and bringing concept to product through applied research. As an educator and scientist, I see myself utilizing all three aspects of science to teach the next generation of engineers and to develop solutions for outstanding problems.

Based on my postdoctoral and graduate research experience, I believe personalized medicine is one of the research area that I can contribute significantly. I have experience in developing one of most complicated *in vitro* lung models recapitulating both the bronchiole and the alveoli, miniaturizing PCR platforms, forming complex microvasculature network in polymer substrates and performing rare cell separation for cancer diagnostics. To perform my research which resulted into 8 number of high impact publications and 4 numbers of invention disclosures/patent application, I have utilized anywhere from conventional microfabrication techniques to laser based microfabrication and rapid prototyping methods. To support these platforms and to perform proof of principles, I have developed diverse skills including numerous analysis/development tools (Scanning Electron Microscopy, Finite Element Modeling, Computer Aided designs, etc.) and cell/molecular bio-techniques (cell/tissue culturing, staining, genetic modification and expression, etc.) While I prefer utilizing existing capabilities to support my research, I am also used to developing my own tools for development and analysis.

Looking ahead, I wish to combine the knowledge obtained from my graduate and postdoctoral research experience to develop a unique program to further biological and mechanistic understanding of the complex nature of disease pathology and apply it to personalized medicine. A significant challenge would be bridging the chasm between engineering and biology which is necessary to address the needs in human health and biomedicine. Ultimately, my objective is to acquire the necessary capabilities and expertise to develop technologies and test new strategies capable of overcoming the physiological barriers for improved detection and treatment. What is more interesting is that the nature of multidisciplinary approach in my research will also equip me to develop unique academic/training environment for my students in the classroom and in the laboratory. I strongly believe that the type of academic/research capability that I can bring to an institution will also be greatly appreciated by several programs with in different sponsors such as the NSF and the NIH. I would like to pursue programs like the NSF Career award, NIH young investigator award, NIH R01 grants to support the development of my research interests. I understand that in these days of multidisciplinary research, it is not possible for a single investigator to provide a comprehensive solution to a specific problem. Therefore, I will actively initiate collaborations with other investigators to fulfil any inadequacy with my research efforts. 'Balance' is what I will be seeking between my teaching effort, research activities, and personal interests with my new career as a faculty member in an academic institution.

438759 Responsible Environment and Energy Engineering; Systems, Complexity, Sustainability

Sunday, November 8, 2015
Exhibit Hall 1 (Salt Palace Convention Center)

Cory Jensen, Colorado School of Mines, Golden, CO

Catastrophic challenges, such as water scarcity and energy availability, compounded with growing environmental pollution concerns requires continued innovation among science, technology, engineering, arts, mathematics, and leadership fields. Novel complex sensors incorporating meso-, nano-, and bio-materials have the potential to reduce energy used to produce fuels-chemicals, can be applied to address water quality issues that compromise environment and human health, and can be extended to fields such as health care. For example, by identifying critical components of a complex biosensing system, emergent properties have been identified that were used to direct improvements that result in overall more efficient design (Jensen and Müeller 2015) and commercialization. Further, by working with model environmental systems, biometabolic complexity can be unraveled when developing specific analytical chemistry techniques that also make use of natural methods to improve water quality (Jensen and Gujarathi 2015). In both situations, theory and experiment are required to validate outputs, require complex signal interpretation, and provide for fundamental skills that are applicable to energy-environment problems, and fields (e.g. biomedical) that can make use of related transferable skills to meet the challenges of the future.

This poster will demonstrate:

1) The vision for my innovative energy and environmental research group,

2) provide insight into publications that demonstrate biosensor applications (Jensen and Müeller 2015) and low energy, natural, water treatment systems (Jensen and Gujarathi 2015),

3) and highlight professional service with examples of outreach combined with teaching in multi-year, collaborative, and hands-on laboratory approaches (Jensen 2015).

Jensen, CD. "Graduate Student Action Research to Help Fill Gaps for Formal Engineering Education for Sustainable Development Program Formation: Vignettes As Action." Paper #112. Nesbit, S. & Froese, T. M. (Eds.) (2015). Proceedings of EESD15: The 7th Conference on Engineering Education for Sustainable Development, University of British Columbia, Vancouver, Canada. June 9-12.

Jensen CD, Gujarathi N. "Characterization of a Macrophyte Microcosm as a Surface Water Treatment System for Antibiotics." Environmental Progress and Sustainable Energy, 2015. DOI 10.1002/ep.12158.

Jensen CD, Müeller C. "Benchmark Characteristics of pH based Optoelectronic Enzymatic Biosensors with Non-modified E. coli BL21 pGELAF+." Applied Biochemistry and Biotechnology, 2015, DOI 10.1007/s12010-015-1623-1.

(6am) Multi-Scale Process Systems Engineering

Calfa, B. A., University of Wisconsin-Madison

Multi-Scale Process Systems Engineering

In this poster session, I will discuss my Ph.D. work in the area of Process Systems Engineering (PSE) at Carnegie Mellon University (advisor: Dr. Ignacio E. Grossmann), the research topics and projects I am proposing to pursue, and my teaching interests and experience. For more information, please visit my website: http://bacalfa.com/.

Research

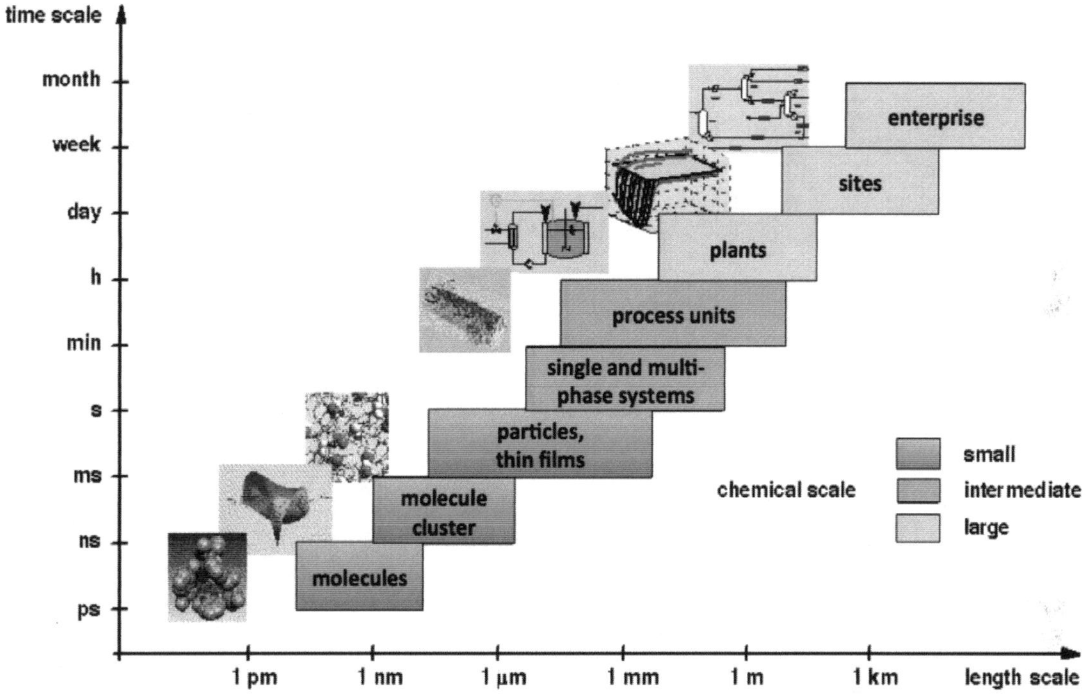

Figure 1. Multiple scales in Process Systems Engineering (PSE) research.

My Ph.D. research focused on the "large" scale top blocks in Figure 1. I **integrated** and **efficiently solved** planning and scheduling models for a network of batch plants, and developed **data-driven** approaches for modeling uncertainty in Enterprise-wide Optimization (EWO) problems [1-8]. I am interested in developing PSE methods (modeling, simulation, optimization, and control) to solve multi-scale problems of practical importance.

á **Large Scale:** novel *data-driven* models for uncertainty in sales and operations planning; multilevel optimization with contracts and pricing.

á **Intermediate Scale:** reduced-order modeling and optimization; analysis of *sustainable* technologies (e.g., solar fuels); material and energy integration through water and wastewater network optimization (e.g, water desalination, renewables).

á **Small Scale:** *property prediction* and computer-aided *material design* (e.g., crystals) via optimal inverse problems; microkinetics and optimal catalyst design.

I received the 2015 Ken Meyer Award for Excellence in Graduate Research, which is given by the Department of Chemical Engineering at CMU in which the faculty base their selection of the student on research quality, productivity, recognition, and impact.

Teaching

I had extensive teaching experience as a TA at CMU. I prepared several teaching materials, and gave guest lectures and tutorials. I am particularly interested in enhancing the students' exposure to computing, as well as helping them develop teamwork and communication skills. In addition to process design, I am also very interested in teaching introduction to chemical engineering, numerical methods, thermodynamics, and unit operations.

I received the 2012 Mark Dennis Karl Outstanding Graduate Teaching Award, which is given by the Department of Chemical Engineering at CMU to a student judged by the faculty to have done an outstanding job as a teaching assistant.

References

[1] B. A. Calfa, A. Agarwal, S. J. Bury, J. M. Wassick, and I. E. Grossmann. "Data-Driven Simulation and Optimization Approaches to Incorporate Production Variability in Sales and Operations Planning". In: Industrial & Engineering Chemistry Research. (2015). Just Accepted. DOI: 10.1021/acs.iecr.5b01273.

[2] B. A. Calfa and I. E. Grossmann. "Optimal Procurement Contract Selection with Price Optimization under Uncertainty for Process Networks". In: Computers & Chemical Engineering. (2015). Submitted.

[3] B. A. Calfa, I. E. Grossmann, A. Agarwal, S. J. Bury, and J. M. Wassick. "Data-Driven Individual and Joint Chance-Constrained Optimization via Kernel Smoothing". In: Computers & Chemical Engineering. 78.1 (2015), pp. 51–69.

[4] I. E. Grossmann, R. M. Apap, B. A. Calfa, P. García-Herreros, and Q. Zhang. "Recent Advances in Mathematical Programming Techniques for the Optimization of Process Systems under Uncertainty". In: 12th International Symposium on Process Systems Engineering and 25th European Symposium on Computer Aided Process Engineering. Ed. by Jakob K. Huusom Krist V. Gernaey and Rafiqul Gani. To appear in Proceedings. . 2015.

[5] B. A. Calfa. A Memory-Efficient Implementation of Multi-Period Two- and Multi-Stage Stochastic Programming Models. Carnegie Mellon University. Technical Report,

2014. URL: http://repository.cmu.edu/cheme/246/.

[6] B. A. Calfa, A. Agarwal, I. E. Grossmann, and J. M. Wassick. "Data-Driven Multi-Stage Scenario Tree Generation via Statistical Property and Distribution Matching". In: Computers & Chemical Engineering. 68.1 (2014), pp. 7–23.

[7] I. E. Grossmann, B. A. Calfa, and P. García-Herreros. "Evolution of Concepts and Models for Quantifying Resiliency and Flexibility of Chemical Processes". In: Computers & Chemical Engineering. 70 (2014), pp. 22–34.

[8] B. A. Calfa, A. Agarwal, I. E. Grossmann, and J. M. Wassick. "Hybrid Bilevel-Lagrangean Decomposition Scheme for the Integration of Planning and Scheduling of a Network of Batch Plants". In: Industrial & Engineering Chemistry Research. 52.5 (2013), pp. 2152–2167.

(6an) Realistic and Affordable Ab Initio Calculations for Electrochemistry

Schwarz, K.,

Electrocatalysis currently plays a significant role in the world economy, and it holds tremendous promise to help meet future needs in areas including energy materials development and energy storage/conversion. However, even the most basic electrochemical processes are incompletely understood. To realize the full potential of electrocatalytic reactions, the fundamental reaction processes must be better characterized. My previous work uniquely positions me to develop and utilize new modeling techniques that describe the nonidealities of surfaces and fluids, to re-evaluate the reaction mechanisms occurring in real electrochemical experiments. In the poster session I will present my previous work, along with my recent findings in these areas.

My Ph.D. thesis explored different approaches to partitioning a molecular system from its environment, using both established and newly developed computational techniques. Computational methods can easily become prohibitively expensive, especially when they are utilized to describe large and complex systems. Choosing or developing the appropriate level of theory to describe the phenomenon of interest allows for better comparison against experiment. I developed a new method for efficiently computing the dielectric properties of a molecular system from its component molecules, and tested it for ice and solid benzene. The method successfully approximated the dielectric properties of ice at considerably reduced computational cost. I also developed a framework for integrating new solvation models into highly accurate electronic structure methods, and I am a contributor to the open-source electronic structure code, JDFTx. I have applied these methods and others to explore the optical properties of pentacene defects and to evaluate the properties of surfaces, ions, and molecules important for lithium sulfur batteries.

My postdoctoral research focuses on applying solvation methods to electrochemical systems to develop a deeper understanding of fundamental reaction processes. I proposed a new mechanism for formic acid oxidation on Pt(111) under electrocatalytic conditions, in which I provide evidence that formate is the reactant, rather than the previously accepted formic acid. This mechanism provides a new explanation for both the experimentally observed surface species and the lack of experimental evidence for other surface-bound species. I also am currently developing a better understanding of the relationship between underpotential deposition of hydrogen and the hydrogen oxidation/evolution reactions on noble metals.

Leveraging my understanding of the theoretical underpinning of electrochemical modeling, my future research will target challenges in electrochemical metal deposition and electrocatalysis for fuel production and consumption.

438643 Control and Manipulation of Molecular Interactions for Nanobiotechnology, Energy, and Biopharmaceutical Applications: Control of Self-Assembly in Micro- and Nano-Scale Systems

Abstract:

Molecular Interactions are the basis of the nature, from the early evolution of life. Control of self-assembling systems, and control and manipulation of molecular interactions provide new opportunities for the engineering of novel materials for the vast variety of applications. Amino acids, nucleotides, sugars, and lipids as the building blocks of life formed through chemical reaction and interaction of molecules. Investigation of molecular interactions in a complex matrix yields the definition of control parameters for manipulating the structure, and design of desired organizing processes. The underlying design principles could also be applied to engineer materials and systems with a robustness and adaptability approaching that of biological systems. Understanding the mechanism of materials formation at a molecular level, with the bottom-up engineering of self-assembly systems enables manufacturing materials and devices with novel optical, mechanical, and electronic properties. Control of self-assembly processes is key to the manufacture of materials with unique properties. However, controlling and manipulation of self-assembly of the systems are highly challenging due to high-dimensional and stochastic nonlinear dynamics of the systems and supramolecular interactions, limited sensors for real-time measurements, and implementing the control mechanism in atomic and molecular length scale.

Postdoctoral Project: Molecular Interactions, Process Modeling, and Continuous Manufacturing of Pharmaceutical Agents

I am trying to apply Next Generation Algorithms to investigate nucleation and crystallization of pharmaceutical molecules from solution and for continuous crystallization.

Postdoctoral Supervisors: Prof. Bernhardt Trout, Prof. Allan Myerson

Department of Chemical Engineering, Massachusetts Institute of Technology

PhD Dissertation: Solid-phase Crystallization, Molecular Interactions in a Complex System of Proteins and Sugars and Manipulating Particles' Microstructure

PhD Supervisor: Prof. Tim Langrish, School of Chemical and Biomolecular Engineering, The University of Sydney

Grants, Publications, Awards: 2 successful Research Grants; 1 Book Chapter; 14 first author peer-reviewed Journal Papers; 1 US Patent; Numerous peer-reviewed Conference Presentations; 11 Prestigious Awards.

Professional Experience: 5 years Industrial Experience, Process Design Engineer.

My professional background will help me in designing the courses full of practical examples from chemical industries to foster better understanding of the topics by students.

Teaching Experience:

- *Lecturer (3 semesters):* Chemical Engineering Foundation for Master of Professional Engineering.

Prepared lecture notes and teaching materials, taught the course, manage the class of 20-25, graded homework and assignments.

- *Teaching Assistant (7 semesters):* Reaction Design and Kinetics, Separation Processes, Thermodynamics, Process Engineering Economics and Design.

Helped student with course materials, grade homework and final exams, helped with term projects, and conducting private tutoring sessions. Average class size of 40.

- *Co-supervisor and Mentor (4 years):*

- Co-supervised and mentored 4 undergraduate students for their fourth year research thesis.

- Research mentor and co-supervisor for 3 PhD and master students.

Future Direction:

My research core is the study of Molecular Interactions of large and small molecules, Assembly, and Self-assembly in Micro- and Nano-scale Systems. I would like to use this bottom-up approach in the field of Nanbiotechnology, Energy, Biopharmaceutical, and Novel Material application. The fundamental investigation in the molecular and biomolecular domains will be employed for designing novel technique to address multi scale issues. I would like to incorporate the fundamental study's results in developing predictive models to selectively tailor novel materials and properties, and processes with high impact for industry and science. This interdisciplinary research of supramolecular self-assembly and supramolecular interactions will aim for rational synthesis of materials through molecular design, wherein I am interested in fundamental interactions at material interfaces. Few application examples would be: developing of Nanostructured Materials for Energy Storage and Conversion; Application of Conducting Polymer and inorganic devices in Biomedical Application; Controlling Nucleation, Crystallization, and Polymorphism in Organic Semiconductor Thin Films; Materials for Bioelectronics; Virus/Bacteria-Based Method for Directed Synthesis of Nanowires and Nanostructures.

438646 Engineering Biomimetic Membranes

Abstract:

Cells coordinate a variety of physical and chemical processes to yield complex and dynamic behaviors. My research examines the most basic physical and chemical processes that underlie cellular behaviors. I build synthetic models of cellular membranes from phospholipids, fatty acids, or diblock-copolymers and use molecular biology and synthetic chemistry techniques to generate chemically activated nucleotides and RNA polynucleotides. With these building blocks, I construct bilayer membranes containing RNA and study how biophysical events, like membrane deformation, growth, and division can be replicated and even triggered in synthetic systems, and study how genetic processes like non-enzymatic RNA replication can be conducted without the aide of complicated cellular machinery. By studying and recreating how simple biophysical and genetic processes can be conducted with a minimal number of known components, I aim to advance our understanding of the basic thermodynamic processes that guide higher-order cellular behavior as well as create a tool kit of cellular behaviors that we can assemble to create intelligent devices for future applications in medicine and biotechnology. My research, therefore, focuses on two complementary experimental approaches 1) Designing responsive synthetic membranes and 2) Studying how RNA processes can be coupled to synthetic membranes.

The bilayer membrane, typically formed from phospholipids, provides a powerful platform to assemble biological and synthetic molecules alike. Beyond phospholipids, diblock-copolymers have emerged as a powerful class of amphiphiles to generate membranes with a wide variety of physical and chemical properties. As a doctoral student with Professor Daniel Hammer, I have studied to how to design diblock-copolymer membranes. Specifically, I designed membranes responsive to both light and mechanical stimuli. In addition to vesicle-membrane design, I have used soft lithography and microfluidic techniques to create micro-patterned surfaces and controlled vesicle architectures. These techniques enable the construction of larger scale, organized patterns of vesicles to better study vesicle-vesicle and vesicle-cell behaviors in spatially and chemically controlled environments.

As a postdoctoral researcher with Professor Jack Szostak, I have studied how short RNA polynucleotides can be complexed with fatty acid and phospholipid membranes. Using synthetic peptides, I have shown that RNA can be localized to the surface of both fatty acid and phospholipid membranes via electrostatic interactions with short peptides. Our work with synthetic membranes enables a platform to study how membrane composition and peptide sequence influence RNA-membrane binding and opens the door for studying how membranes can enhance reactions with surface-localized polynucleotides. I have also engineered vesicle systems that display temperature and divalent cation-gated RNA permeability, enabling the transport of short RNA polynucleotides into vesicle membranes to participate in non-enzymatic RNA replication chemistry. The construction of a responsive membrane that enables polynucleotide passage could advance the design of vesicle bioreactors or sensors that utilize short oligonucleotides as substrates.

Bringing together emerging engineering methods in material science and synthetic biology, I aim to construct macromolecular assemblies that can coordinate both membrane biophysical processes and genetic transcription and translation events to yield particles capable of complex signaling and responsive behaviors.

439002 Metal-ion-based materials chemistry for hydrocarbon separations, energy, and electronics applications

Sunday, November 8, 2015
Exhibit Hall 1 (Salt Palace Convention Center)

Matthew G. Cowan, Dept. of Chemical and Biological Engineering, University of Colorado at Boulder, Boulder, CO; Dept. of Chemistry & Biochemistry, University of Colorado at Boulder, Boulder, CO

3rd year postdoctoral fellow

Research Interests:

My primary research interest is to use coordination chemistry, the science of metal-ions, to develop materials for olefin/paraffin separations. I choose this as the focus of my research because olefins, and the plastics and polymers made from these molecules, are vital to society. Furthermore, olefins are currently purified by cryogenic distillation—in the United States alone this process consumes the equivalent energy to that produced by over 7 'average' nuclear power plants. Obviously, the electrical energy, economic, and environmental benefits that could be achieved by developing a 'low-energy' method of achieving olefin/paraffin separations make this a worthy goal.

My secondary research interest is to use soft-materials chemistry to develop technologies for the organization of magnetically and electrochemically active molecules. I choose this as a focus of my research because these organized materials will play a key role in developing advanced electronics devices and battery technologies. In particular, many research groups worldwide focus on developing molecules suitable for molecular electronics. However, there is little effort at developing materials from these molecules or developing engineering practices to see the successful inclusion of these molecules into functional devices. For these reasons, I consider that the development of such technologies and techniques will be of particular scientific and industrial value in the near future.

Successful Proposals and Awards:

Elman Poole Travelling Scholarship (2011; $25,000 award for international PhD research; 3 awardees)

Claude McCarthy Fellowship (2010, $5,000 award for collaborative international research, 23 awardees)

North American Membrane Society Travel Award (2013, conference attendance, 12 awardees)

Wiley Outstanding Poster Award at ISMSC-7 (2012, 1 awardee)

Finalist for the Don Stranks Award (2011, conference attendance, 6 finalists)

Postdoctoral Projects: "Olefin/paraffin separation materials" ($135,000 funded by the Saudi Basic Industries Corporation); "A high-performance membrane for CO_2/N_2 separations" ($3.4 million funded by the US Department of Energy); "An electrochemical pump for CO_2 capture" (ca. $1 million funded by the US Department of Energy).

Under supervision of Richard D. Noble and Doug L. Gin, Department of Chemical and Biological Engineering, University of Colorado at Boulder

PhD Dissertation: "The Development of Imide Coordination Chemistry"

Under supervision of Sally Brooker, Chemistry Department, University of Otago, New Zealand

Research Experience:

My research experience combines chemical engineering in materials chemistry and applications with pure chemistry studying small-molecule design and molecular electronics. My transition from PhD studies in metal-ion coordination chemistry into polymeric and ionic liquid materials design, has allowed me to excel in a series of interdisciplinary projects—and will continue to do so in the future.

In particular, I have also been part of a successful academic/industrial research team funded by the US Department of Energy that developed high-performance membrane materials for CO_2 separations (in conjunction with the 3M Company and Los Alamos National Lab, 5 publications).

As the lead researcher on smaller projects, I have achieved breakthroughs in developing solid sorbents for olefin/paraffin separations (*Angew. Chem.*) and developing switchable molecules for molecular electronics (*JACS*).

Teaching Experience:

I have taught 8 undergraduate chemistry laboratory courses, supervised the accredited research projects of 2 undergraduate chemical engineering students, and mentored 4 chemical engineering PhD students—2 of whom have graduated. I have also completed courses in mentoring research students and teaching laboratory courses.

Future Direction:

As a faculty member I will continue to direct my research to identifying and solving problems where materials and coordination chemistry can be combined into effective solutions. In particular, I intend to seek out industrial and governmental partners to identify and accurately define targets for this research, as well as collaborate with them to develop the resulting technologies into commercially viable products.

Selected Publications:

M. G. Cowan, W. M. McDanel, H. H. Funke, Y. Kohno, D. L. Gin, R. D. Noble, "High Ethene/Ethane Selectivity in 2,2-Bipyridine-based Silver(I) Complexes via Removal of Coordinated Solvent", *Angew. Chem. Int. Ed.* **2015**, 54, 5740-5743. *'VIP' (Assessed as top 5% of papers by two referees).*

M. G. Cowan, J. Olguín, S. Narayanaswamy, J. L. Tallon, S. Brooker, "Reversible Switching of a Cobalt Complex by Thermal, Pressure and Electrochemical Stimuli: Abrupt, Complete, Hysteretic Spin Crossover", *J. Am. Chem. Soc.* **2012**, *134*, 2892-2894. *Front Cover Feature.*

X. Feng, M. E. Tousley, **M. G. Cowan**, B. R. Wiesenauer, S. Nejati, Y. Choo, R. D. Noble, M. Elimelech, D. L. Gin, C. O. Osuji, "Scalable Fabrication of Polymer Membranes with Vertically Aligned 1-nm Pores by Magnetic Field Directed Self-Assembly", *ACS Nano,* **2014**, *8*, 11977-11986. *ACS Editors Choice.*

J. Zhou, M. M. Mok, **M. G. Cowan**, W. M. McDanel, T. K. Carlisle, D. L. Gin, R. D. Noble, "High-permeance Room-temperature Ionic Liquid-based Membranes for CO_2/N_2 Separation", *Ind. Eng. Chem. Res.* **2014**, *53*, 20064-20067. *Highlighted on the US DOE website.*

Review Articles:

M. G. Cowan, "The Application of Membrane Technology to Post-Combustion Separation of Carbon Dioxide from Coal-Fired Power Plant Flue Gas", *Chem. N.Z.*, **2015**, 79: 79-83.

M. G. Cowan, S. Brooker, "Syntheses, Structures and Properties of Structurally Characterised Complexes of Imide-based Ligands", *Coord. Chem. Rev.* **2012**, *256*, 2944-2971.

439666 Optimizing Metabolic Pathways for the Improved Production of Natural Products

Abstract:

The fields of metabolic engineering and synthetic biology have enabled scientists and engineers to harness the power of biology to produce a wide variety of difficult to synthesize natural products. The high-titer microbial synthesis of these compounds is possible through the careful optimization of both internal genetic elements and external fermentation conditions. Although, recent advances have shown proof-of-concept microbial production of many classes of compounds, many systems still result in non-industrially feasible titers. This less than desirable production is, in part, due to the high metabolic burden of expressing extensive extrinsic pathways in a single microbial strain. My research is focused on the development of novel strategies for reducing metabolic burden while optimizing the level of pathway overexpression to maximize the production of natural products *in vivo*.

My graduate work has concentrated on the development of strategies to improve and optimize metabolic pathways from a genetic and fermentation perspective. I have improved the bioconversion of flavanones to catechins through gene copy number balancing, elevated the production of violacein through balancing the transcriptional profile of the pathway genes, and developed a novel co-culture strategy for the production of catechins from their phenylpropanoic acid precursors. Each of the aforementioned projects has resulted in the highest reported product titer at the time of publication. Furthermore, to illustrate my passion for undergraduate education, I have developed a series of laboratory experiments to probe gene expression, pathway balancing, and fermentation optimization using the colorful violacein pathway. These laboratory exercises were designed to be cheaply implemented in a variety of biotechnology and bioengineering lab courses. Future applications of these technologies towards improving the microbial production of biofuels, pharmaceuticals, and commodity chemicals could expedite the transition from the laboratory to the marketplace.

As a graduate student at RPI I have had the opportunity to work with and mentor many talented undergraduate students. This rewarding and beneficial experience has pushed me to pursue a career teaching and mentoring undergraduate students at an undergraduate-focused university. While continuing to enhance my teaching effectiveness in the classroom, I will also work with undergraduates in the laboratory to help develop a competence in and a love for biotechnology research. Through my mentorship, I hope to develop young minds into life-long learners with a passion for improving the world around them through service, scholarship, and scientific research.

438649 Micro/Nanoarchitectured Materials of Novel Surface Properties

Abstract:

My research interests focus on understanding the basic principles of – natural and synthetic – micro/nanostructured materials of intriguing surface properties. Working at an interdisciplinary intersection of materials, surface, and nano-science, I aim to incorporate propositional and experiential knowledge for design and fabrication of micro/nanostructured materials and devices of novel surface properties, such as adhesive/non-adhesive and wetting/non-wetting surfaces.

My PhD research at University of Waterloo (ON, Canada) was focused on design, fabrication, and characterization of synthetic bio-inspired dry adhesives. By fabricating the first fibrillar dry adhesive from flouropolymers and studying the interfacial interactions of both natural and synthetic dry adhesives, I showed that a large factor behind the phenomenon of dry adhesion derives from electrostatic interactions generated via surface charging. This discovery represented a major digression from long-accepted theories, leading to a significant shift in the fundamental understanding of the phenomenon of dry adhesion and also in the design concepts and directions for fabrication of new synthetic dry adhesives.

During my postdoctoral training at Yale University (CT, USA), I designed and developed a state-of-the-art particulate contamination cleaning method using triboelectrically-chargeable micro-scale fibrillar structures made from an elastic and non-sticky polymer. With various implications in microelectronics, aerospace, optics, xerography, and adhesive bonding, polymeric fibrillar materials can offer a new paradigm for thorough yet nondestructive cleaning of micrometric and sub-micrometric contaminant particles from solid surfaces.

Building on these pursuits, I intend to develop a research program based on integration of objectively new and innovative designs in fabrication and modification of micro/nanoarchitectured materials of novel surface properties.

Simulation of Concentrated Suspensions in Thin Film Processing

Mahyar Javidi, Western University

Last year Ph.D. Student

Research Interest:

Particle-laden flows are important in a wide range of industrial fields, such as oil and gas refinement, paper manufacturing, waste water treatment, biological and polymer processes where transport and manipulation of suspensions occur. In addition, the capability of developing a thin uniform suspension layer with evenly distributed particles is essential in many applications. In coating processes of suspensions, the particle distribution pattern can enhance the performance of the final product by changing the bulk and surface characteristics. Applying the high performance computer modeling in conjunction with advanced experimental techniques gives insights into the processes of thin film by suspensions, and provides practical solutions to industrial challenges and the fundamental understanding of the process. The dynamics of suspension flow can be modeled based on the density of solid particles in the system, where macroscopic methods are used for tracking the volume fraction of particles in the flow or using the particle tracking methods following the individual particles and their interactions. In addition, coupling the particle distribution in suspensions with the surface tracking methods such as volume of fluid technique enables us to study the multiphase flows, composite materials, thin film and coating systems. The models can be incorporated into a finite volume or finite element method formulation to simulate homogenous and non-homogenous systems for exploring the dispersed particles behavior under various processing conditions.

Ph.D. Dissertation:

Modeling and simulation of dip coating process for dispersions

Under the supervision of *Dr. Andrew Hrymak*, Dean of Engineering at Western University, Associate Editor, Computers and Chemical Engineering and Editor-in-Chief, International Polymer Processing

Research Experience:

My research experience is on the combination of many industrial and scientific fields. In my Ph.D., I have learned and applied many computational fluid dynamic tools such as: Finite volume methods, Volume of Fluid techniques, Linux, C++ programming to develop and implement the new multiphase systems in OpenFOAM and CFDEM (combination of CFD and DEM numerical calculations) open source software packages, discrete element methods, LIGGGHTS (Open source discrete element method particle simulation software) based on the classical molecular dynamics code of LAMMPS. As a result of intensive research using all these computational systems, I became proficient in high performance computer modeling. Moreover, experimental investigation was always an inseparable part of the simulation procedure and I worked with many polymer processing equipment, performing the rheological tests and close collaboration with researchers at the surface science center of Western university. Investigation in both simulation and experimental fields, provided me a versatile experience in many research areas such as: coating, material science, biomedical, engineering design, polymer processing and image processing.

Teaching Experience:

I have extensive teaching experience alongside my research background. I was the teaching assistant (TA) for the fluid flow course (two semesters) at Western university and also the TA of the plant design and safety course for 5 semesters which involved teaching the Hysys software in the computer lab and supervising/mentoring several groups of students in the design of assigned chemical production plant, feasibility study for the plant, coat estimation, calculation of mass and energy balance and directing the students in preparing the relating reports and presentations.

Future Direction:

As a faculty I would like to continue applying the computational fluid dynamic methods to study the multiphase, thin film, free surface and particle/fiber filled systems in diverse processes. In many of the chemical engineering procedures it is essential to understand the behavior of multiphase processes, particles distribution and fibers orientation to manipulate and control the properties of the final products. I would like to combine the capability of computational fluid dynamics with the discrete element methods which I learned during my Ph.D. to investigate and optimize numerous industrial chemical processes, porous systems, enhanced oil recovery procedures, reactors, biomedical sensors, coating systems, composite materials and polymer processing techniques. Beside the high performance computer modeling, image processing and rheological study of the dense and dilute suspensions which consists of rigid and flexible fibers and particle would be a significant part of my research.

My goal for future investigation is to pursue research in both fields of simulation methods via a computational center and experimental study using the high speed, high resolution cameras and specific lab equipment. I believe combining these two provides the valuable details on desired processes. In addition study a system closely through the actual physical data and experimental results will be a great help to develop the practical and realistic simulations.

Publications and oral presentations:

Mahyar Javidi, Michael A. pope, Andrew N. Hrymak, Investigation on dip coating process by mathematical modeling of non-Newtonian fluid coating on cylindrical substrate, submitted to the journal of Industrial and Engineering Chemical Research

Mahyar Javidi, Andrew N. Hrymak, Numerical simulation of the dip coating process with wall effects on the coating film thickness, Journal of Coatings Technology and Research, Vol. 12 (5), pp 843-853, 2015

Mahyar Javidi, Andrew N. Hrymak, Simulation of concentrated suspensions in thin film processing, ready for the submission in an international journal

Mahyar Javidi, Michael A. pope, Andrew N. Hrymak, Modeling of free-withdrawal coating with UV curing TiO_2 filled resins, Polymer Processing Society (PPS), Morocco, 2011

Mahyar Javidi, Andrew N. Hrymak, Withdrawal of a cylinder from an Ellis fluid, International Coating Science and Technology Symposium (ISCST), US, 2012

Mahyar Javidi, Andrew N. Hrymak, Simulation of dip coating process with OpenFOAM software, European Coating Symposium (ECS), Belgiium, 2013

Mahyar Javidi, Andrew N. Hrymak, Numerical simulation of three dimensional free coating process with wall effect study, International Coating Science and Technology Symposium (ISCST), US, 2014

438651 Nature-Inspired Approaches to Catalytic Materials Design

Abstract:

The precise location of the active site in a catalytic material, specifically the environment that surrounds the active site, very often directly influences the activity and selectivity of a catalyst. The best examples of this environmental control around catalytic active sites are found in enzymes. In enzymes, the active sites are frequently located in a nano-confined environment in which there is control of electrostatic interactions, precise placement of chemical functional groups, and control of hydrophobicity/hydrophilicity, to name some common examples. These nano-confinement and ligand effects in different catalytic systems have been explored through various studies.

The first studies used a shell of organic ligands, namely calixarene ligands, to allow for stable (sinter-resistant), accessible (able to bind small probe molecules and reactants), and electronically tunable sites on gold clusters. Further studies used organic ligands on gold nanoparticles to determine site requirements for two different reactions: reduction of 4-nitrophenol and reduction of resazurin. In both systems, only a small fraction of gold catalyst was responsible for the majority of the catalytic activity.

Additional studies investigated metal clusters/complexes as well as enzymes that were immobilized inside of the pores of mesoporous materials, such as SBA-15. These immobilized systems both exhibited enhanced catalytic activities over their measured activities when free in solution under specific reaction conditions.

These examples from both my research as PhD student at UC Berkeley (supervised by Alexander Katz) and as a post-doctoral research associate at University College London (supervised by Marc-Olivier Coppens) will be presented and will form the foundation of future research. These approaches will be used to synthesize novel materials for applications in catalysis, energy, and medicine.

439758 First-Principles Approaches to Fuel Cell Catalyst Design

Abstract:

Low-temperature proton-exchange membrane (PEM) fuel cells are a promising candidate to replace combustion engines in portable applications with renewable technologies. Their commercialization is limited primarily by the high cost of electrode catalysts, for which high loadings of Pt-based nanoparticles are presently required to yield suitable activity. My doctoral research has focused on obtaining a fundamental understanding of key aspects of low-temperature fuel cell catalysis through the use of highly accurate density functional theory (DFT) methods, providing atomic-scale insight into the nature of key components of PEM fuel cell catalysis.

One crucial aspect of improving fuel cell viability is the improvement of the oxygen reduction reaction (ORR) at the cathode. Although Pt is the best monometallic ORR catalyst, its expense and the requirement of a high overpotential to obtain significant catalytic activity make it unsuitable for large scale production. One strategy to address these points is to deposit Pt as conformal layers over a substrate such as Pd, which introduces compressive strain into the Pt layers and destabilizes adsorbed O and OH to improve catalytic activity. The Pt-mass activity and durability of such catalysts are simultaneously improved relative to pure Pt particles due to the core-shell templating. Novel DFT models are used to connect closely with experiments, rationalizing and predicting observed activity trends on a range of shape-selected nanocrystals toward optimal catalyst design.

Another significant aspect of fuel cell catalyst design is the identification of a viable chemical fuel to be oxidized at the fuel cell anode. Hydrogen oxidation is a clean process, but practical considerations (most notably, storage) limit its potential in portable applications. One promising candidate to replace hydrogen is dimethyl ether (DME), perhaps most notably due to its compatibility with existing fuel infrastructures for ease of transportation and storage. A detailed investigation of DME electro-oxidation, including determination of the thermodynamics and kinetics of a complete network of elementary reaction steps, on three Pt facets demonstrates the structure sensitivity of DME electro-oxidation and informs catalyst design for improving DME electro-oxidation.

In addition to engineering surface reactivity, DFT also provides critical insights into practical considerations governing the formation of catalysts. In one example, calculations show that Pt and Pd intermix during the layer-by-layer formation of core-shell alloy structures due to favorable substitution of Pt into the Pd surface. As a result, a small amount of Pd is dispersed into Pt layers as they are added. Calculations further show that this dispersion of Pd into the Pt layers is a necessary and sufficient condition for the formation of a new class of hollow Pt nanostructures with well-defined facets and excellent catalytic activity for reactions such as the ORR. A broader DFT study has applied these same principles to a number of transition metal surfaces, allowing the prediction of substitution and intermixing for dozens of permutations of adlayer and substrate metals.

My future work will extend these catalyst design principles to a wider range of chemical processes, including biomass conversion (both electrochemical and non-electrochemical) and other strategies for sustainable energy storage and production.

439783 Low temperature carbon dioxide reduction to carbon monoxide on perovskite-type oxides

Abstract:

In the US, around 6000 million mega tonnes of CO_2 are released yearly to the atmosphere, yet less than 300 mega tonnes are re-utilized in industrial processes such as urea and salicylic acid synthesis. Here, an alternative to CO_2 utilization is proposed, where CO_2 can be transformed to high-value chemicals that could be used as a source of C1 chemistry for synthetic fuels synthesis. Being a very stable molecule, CO_2 requires extreme conditions (very high temperatures or reducing environments) for its activation. The reverse water gas shift (RWGS) reaction can convert CO_2 to CO at ~ 600 °C but requires excess hydrogen, is equilibrium limited and commonly produces undesired CH_4. In spite of its disadvantages, it is cost competitive and the incorporation of renewable energy has been proven promising [1].

In this work, the Reverse Water Gas Shift Chemical Looping (RWGS-CL) process is demonstrated [2,3,4]. Here, the RWGS is intensified to avoid its most common drawbacks (use of excess hydrogen, equilibrium limitations and side reactions) using mixed oxides with high oxygen mobility. The process consists on a hydrogen pulse, which creates oxygen vacancies in the materials that are later re-oxidized by a flow of carbon dioxide, while yielding CO [2,3,4].

Perovskites (ABO_3) are ideal for RWGS-CL because they have the ability to form oxygen vacancies while maintaining their crystalline structure and their catalytic properties can be tailored by partial substitution of the A and B site metals. Various concentrations of lanthanum and strontium [2], and cobalt and iron [3] were tested in the A and B site of the perovskite respectively to design a process that allows maximum CO formation, at the lowest temperatures whit high CO yields. The tested materials proved stable during the five cycles studied.

The RWGS-CL process transforms CO_2 to CO starting at 500 °C (~100 °C lower than conventional RWGS) [3], which could allow for a temperature integration of the process for CO_2 conversion and fuels synthesis through Fischer-Tropsch, lowering overall energy use and environmental impact if heat is provided by renewable sources.

References

[1] Mallapragada, D. S.; Singh, N. R.; Curteanu, V.; Agrawal, R. Sun-to-Fuel Assessment of Routes for Fixing CO2 as Liquid Fuel. Ind. Eng. Chem. Res. 2013, 52, 5136−5144.

[2] Daza, Y.A., Kent, R.A., Yung, M.M., Kuhn, J.N., Carbon dioxide conversion by reverse water gas shift chemical looping on perovskite-type oxides. Ind. Eng. Chem. Res. 2014, 53, 5828−5837.

[3] Daza, Y.A., Maiti, D., Kent, R.A., Bhethanabotla, V.R., Kuhn, J.N., Isothermal reverse water gas shift chemical looping on $La_{0.75}Sr_{0.25}Co_{(1-Y)}Fe_YO_3$ perovskite-type oxides. In press Catalysis Today, 2015.

[4] Daza, Y.A., Maiti, D., Hare, B.J., Bhethanabotla, V.R., Kuhn, J.N., More Cu, more problems: Decreased CO_2 conversion ability by Cu-doped $La_{0.75}Sr_{0.25}FeO_3$ reduced perovskite oxides. Submitted to Surface Science for a special issue honoring Gabor Somorjai.

438653 Simple, Novel and Applicable Strategies for Innovative Medical Solutions

Sunday, November 8, 2015
Exhibit Hall 1 (Salt Palace Convention Center)

Shiyi Zhang, MIT, Cambridge, MA, Robert Langer, Koch Institute for Cancer Research, Massachusetts Institute of Technology, Cambridge, MA and Karen L. Wooley, Department of Chemistry, Texas A&M University, College Station, TX

During my PhD research in Prof. Karen Wooley's laboratory, my interests ranged from degradable polyphosphoester-based nanomaterials for nanomedicine to the self-assembly of block copolymers for complex functional nanostructures. The development of synthetic methodologies for the preparation of complex, functional polymer materials by simple strategies has enormous potential for biological applications. My past research on the polyphosphoester-based nanomaterials has built a platform that enables a rapid and versatile construction of two families of degradable nanoparticles in treating diseases, such as pulmonary cancers, infections and injuries. Two degradable nanoparticle systems based on one alkyne-functionalized polyphosphoester and their broad applicability, will be discussed. The highly water soluble backbone of the polyphosphoester endowed the first system with ultrahigh paclitaxel loading capacity (55 wt%) and a maximum paclitaxel concentration of 6.2 mg/mL in water, for the treatment of lung cancers. In the second system, the easily functionalizable side-chain of the polyphosphoester allowed us to prepare a family of nanoparticles with diverse surface charge types, non-ionic, anionic, cationic and zwitterionic, in a rapid and versatile manner. The anionic nanoparticles are designed to load silver ions to treat pulmonary infections, while the cationic nanoparticles are being applied to regulate lung injuries by their degradation products.

My postdoc research in Prof. Bob Langer's laboratory is focusing on the development of novel technologies for prolonged gastric residence time and in turn extended release of loaded drugs, potentially to facilitate the oral administration on an infrequent - potentially even a one-time - basis in order to maximize medication adherence. Devices resident in the stomach - which are used for a variety of clinical applications including nutritional modulation for bariatrics, ingestible electronics for diagnosis and monitoring, and gastric-retentive dosage forms for prolonged drug delivery - typically incorporate elastic polymers to compress the devices during delivery through the oesophagus and other narrow orifices in the digestive system. However, in the event of accidental device fracture or migration, the non-degradable nature of these materials risks intestinal obstruction. Here, we show that an elastic, pH-responsive supramolecular gel remains stable and elastic in the acidic environment of the stomach but can be dissolved in the neutral-pH environment of the small and large intestines. In a large animal model, prototype devices with these materials as the key component demonstrated prolonged gastric retention and safe passage. These enteric elastomers should increase the safety profile for a wide range of gastric-retentive devices. We have been working on loading various drugs into such gastric device for the extended release to solve the poor medication adherence problem.

My PhD and postdoc research utilized different polymeric materials for distinct applications, but shared the same design principle that is simple, novel and applicable.

438654 Merging Electrochemical Devices, Protein Engineering and Tissue Engineering: A Multi-Disciplinary Approach to Electrode Structures and Stem Cell Culture

Abstract:

My research interests are derived from my diverse experiences working in multiple fields: protein engineering, tissue engineering, stem cell differentiation and membrane-based electrochemical technology. These fields are growing, for example, the market for stem cells was estimated at $26.3 billion in 2013, and is expected to grow annually by 24% (Transparency Market Research). The market for membrane-based fuel cells is expected to grow at a similar rate. I want to uniquely combine these fields to create new electrochemical processes and devices, as well as provide fresh approaches to stem cell culture. Not only will my research create new technology, but it will support the demand for educated professionals to work and innovate in these fast growing markets.

Initially, my group will work in three areas: 1) protein-engineered electrode structures, 2) electrochemically-modulated protein-engineered materials and 3) electrochemical devices for more controlled and automated stem cell culture. These projects will result in increased understanding and structuring of electrodes used in energy applications such as fuel cells or electrolyzers, and will also improve the performance of enzyme-based electrodes enabling implantable fuel cells, efficient chemical generation, or even wastewater treatment technologies. In addition, new insights into stem cell culture and differentiation will be gained by continuous monitoring and control of culture medium components, and unique tools for more effective 3D tissue engineering will be developed.

My past experiences provide a strong foundation for my future research. For my doctoral work at Purdue University, I designed proteins to mimic cartilage tissue and direct human bone marrow-derived stem cells to produce cartilage extracellular matrix. Notably, my work resulted in materials that had a compressive modulus comparable to natural cartilage. As an industrial postdoctoral researcher at Proton OnSite, a world leader in proton exchange membrane electrolysis, I specialized in conceptualizing, developing and evaluating new and emerging electrochemical technologies. To date, I've obtained SBIR funding to develop an electrochemical ammonia generation device and a microbial fuel cell for the remediation of nitrate-contaminated water. I possess a truly unique toolkit of electrochemical device design and evaluation, electrode characterization, protein design, protein manufacturing and characterization, as well as expertise in stem cell differentiation analysis.

In addition to supporting my own independent research program, I will flex the experience gained in my postdoctoral position to encourage strong government-industry-academic collaborations. Also, as a committed mentor supervising over 17 student projects to date, I will continue to encourage creative design, multidisciplinary education, and a collaborative team-oriented environment in my future group and classroom.

438655 Advanced Materials for the Water-Energy Nexus

Abstract:

The U.S. Department of Energy (DOE) released a report in June 2014 that frames an integrated challenge and opportunity space around the so called "water-energy nexus." Water is used in all phases of energy production and electricity generation. Conversely, energy is required to extract, convey, and deliver water of appropriate quality for diverse human uses, and then again to treat wastewaters prior to their return to the environment. High cost, together with limited supply and functionality of traditional materials, are bottlenecks for the advancement in energy and water systems. My research interests are motivated by the need for advanced intelligent materials for applications in energy and water systems - from enhancing efficiency in energy generation, recovery, and storage to developing efficient and reliable technologies for water purification. I will work on developing inexpensive, widely available, and high performance smart materials, compatible with high throughput manufacturing, and on innovative design to increase durability and functionality to save energy and water.

438656 Development of Functional Materials for siRNA Delivery and Neural Tissue Engineering

Sunday, November 8, 2015
Exhibit Hall 1 (Salt Palace Convention Center)

Metin Uz[1], Sacide Alsoy Altinkaya[2] and Surya K. Mallapragada[1], (1)Department of Chemical and Biological Engineering, Iowa State University, Ames, IA, (2)Chemical Engineering, Izmir Institute of Technology, Urla-Izmir, Turkey

The current nonviral siRNA delivery systems in the literature face many problems such as, cellular entry, endosomal escape and efficient siRNA release. Considering this motive, we developed gold nanoparticles (AuNPs) and temperature/pH responsive pentablock copolymer based siRNA delivery systems to address these problems. The temperature and pH responsive cationic and amphiphilic pentablock copolymers, which were consisted of the temperature responsive Pluronic F127 middle block constructed by PEO-PPO-PEO ((poly(ethyleneoxide)-block-poly(propyleneoxide)-block-poly(ethyleneoxide))) blocks contributing cellular entry through temperature responsive micellization and pH responsive cationic PDEAEM (poly(2-diethylaminoethyl methacrylate)) end blocks facilitating nucleic acid condensation and endosomal escape, were used for the first time in the development of polyplex and AuNP based multicomponent siRNA delivery systems (MCSs). The results indicated that systems managed to protect siRNA from external effects, maintain the system stability, facilitate cellular entry and enhance endosomal escape. It was noted that the transfection efficiency of the MCSs, which were boosted by the presence of cleavable disulfide bond, was ~15% higher than the commercial product RNAiMax while the efficacy of polyplexes alone were similar to the RNAiMax. In addition, a hydrogel system based on these responsive copolymers were also used as depot for MCSs to provide sustained siRNA release, which is another current issue to be overcome. As an alternative to responsive pentablock copolymers, TAT-HA2 fusion peptide consisting of influenza A virus hemagglutinin protein (HA2) sequence providing endosomal escape and cell permeable HIV Trans-Activator of Transcription (TAT) protein transduction domain (PTD) enabling cellular entry, was used to develop AuNP based MCSs, peptideplex and conjugate siRNA delivery systems. The results indicated that each system provided good siRNA protection and stability, cellular uptake and enhanced endosomal escape properties. The siRNA activity of siRNA-peptide conjugates bearing cleavable disulfide bonds with 70% siRNA efficiency were found to be 5% more efficient than that MSCs and peptideplex systems. As a parallel research, the lack of detailed information about the potential effects of surface modified or bare AuNPs on cell division and replication steps, programmed cell death and DNA damage motivated us to pursue a further detailed study. We have investigated the influence of the PEG coated gold nanoparticles (AuNPs) not only on cellular uptake and toxicity but also on cell cycle phase arrest, apoptosis and DNA damage against different cells. The overall results illustrated that it is possible to minimize the alteration in cell cycle upon exposure to nanoparticles by changing PEG grafting density and hydrodynamic volume. It seems that an optimum PEG grafting density (~0.65 g/nm^2) exists at which particles do not severely alter the cell cycle phases.

Another aspect of our ongoing research involves the development of functional biodegradable particles and films/conduits for controlled release of nerve growth factors targeting the problems in peripheral nerve regeneration. In this study, the nerve growth factor (β-NGF) encapsulated amphiphilic and biodegradable polyanhydride microparticles with different compositions (50:50 CPTEG-CPH and 20:80 CPTEG-CPH microparticles (1,6-bis-(p-carboxyphenoxy)hexane (CPH) and 1,8-bis-(p-carboxyphenoxy)-3,6-dioxaocatane (CPTEG))) were prepared and distributed in porous poly-L-lactic acid (PLLA) films with longitudinal micro-patterns on the surface to manipulate and provide sustained β-NGF release. The results indicated that the initial burst NGF release from the film surface was further followed by a sustained NGF release from the polyanhydride particles distributed in the film matrix. The release characteristics changed with respect to particle type and encapsulated NGF amount. The cumulative released NGF amount showed enhanced neurite extension activity on PC12 cells. As an alternative, transdifferentiated mesenchymal stem cells (tMSCs) capable of continuously secreting β-NGF were used along with the lumen of different conduits (gelatin, PLLA, PLGA etc.) with different morphologies to provide sustained NGF release and activity *in vitro* and *in vivo*. Our preliminary experiments showed very promising results indicating the importance of conduit morphology.

438658 Rational Catalyst Design for Renewable Energy Technologies

Abstract:

Today two of the most prominent issues, associated with a growing world population and increased standards of living, are climate change caused by CO_2 emissions and the availability of clean drinking water. There is general interest in replacing fossil fuels to supply mankind energy needs sustainably on a long-term basis. Ideally this is done by 1) producing energy using alternatives such as fuel cells, electrolyzers or batteries, 2) synthesizing fuels using biomass, methane oxidation or electrochemical conversion of $CO_2$to hydrocarbons. Heterogeneous catalysis is at the heart of all these renewable energy technologies.

There are, however, severe shortcomings of the present technologies, which need to be overcome to make these processes more economically attractive. One of the most important problems is the lack of active, selective and inexpensive catalysts for these processes. Any progress in this field is connected to the concept of scaling relations[1], which is the correlation between surface bond energies of different adsorbed species for specific reaction. The scaling relations pose limitations on many of the known catalysts for reactions such as oxygen reduction (ORR), oxygen evolution (OER), CO_2 reduction and N_2 reduction reactions.[2] Therefore, catalyst design requires development of new, effective and non-precious catalysts with active site motifs, which obey different scaling relations than the so far recognized ones. *My work utilizes theoretical tools in particular density functional theory (DFT) to model graphene, boron-nitride and oxide materials with different modifications strategies including doping and defect design to describe the surface reactions. These strategies have not been explored so far. I seek to find a breakthrough for scaling relations and my goal is to discover novel catalysts for the following applications: 1) improving the efficiency of the devices that are related to sustainable energy applications, i.e. fuel cells[3], electrolyzers[4], and metal-air batteries[5], and 2) to discover/design new, effective and non-precious catalysts for ORR, OER, CO_2 and N_2 electrochemical reduction reactions, fuel synthesis from biomass, and methane oxidation.*

Another important issue facing humanity is related to the potable water. Millions of people in the developing world are lacking access to drinking water due to its contamination with urban, industrial and agricultural waste. These pollutants can in principle be removed by their total oxidation by hydrogen peroxide (H_2O_2) as a strong oxidizing agent without leaving any residues behind. Production and transportation of H_2O_2 to the developing world, however, is challenging due to high costs and safety issues, making this potentially game-changing chemical largely inaccessible to the massive number of people who need it the most. *My research combines the expertise from fundamental understanding of the reaction and DFT with versatile synthetic methods, electrochemical evaluation and catalyst characterization to develop an electrochemical route by designing an efficient and cost effective electrocatalyst for selectively reducing O_2 to H_2O_2 to enable continuous, small-scale and decentralized production of H_2O_2 with minimal danger of explosion.*[6,7]

---

[1] Abild-Pedersen, F., Greeley, J.P. and Studt, F. *et al. Phys. Rev. Lett.*, 99, 2007, 016105.

[2] Vojvodic, A., Nørskov, J.K., *National Sci. Rev.*, 2, 2015, 140.

[3] Siahrostami, S., Bjorketun, M.E. *et al. Phys. Chem. Chem. Phys.* 15, 2013, 9326.

[4] Siahrostami, S., Vojvodic, A., *J. Phys. Chem. C*, , 2015, 119, 1032.

[5] Siahrostami, S., Tripkovic, V., *et al. Phys. Chem. Chem. Phys.* 15, 2013, 6416.

[6] Siahrostami, S., Verdaguer-Casadevall, A., *et al. Nat. Mater.* 12, 2013, 1137.

[7] Verdaguer-Casadevall, A., Deiana, D., Karamad, M., Siahrostami, S., *et al. Nano Lett.* 14, 2014, 1603.

438659 Computational Catalysis Design for Fuel Synthesis

Abstract:

Heterogeneous catalysis is currently attracting much attention mainly for energy-conversion purposes. In particular, many future energy transformation processes rely on electro-catalysis. Few important examples are oxygen reduction reaction (ORR), CO electrooxidation and the electrochemical reactions of CO_2. With advances in density functional theory (DFT) mean, it is now possible to gain some insight into chemical and electrochemical reactions. Moreover, the extraordinary progress in density functional theory (DFT) calculations provides the possibility of computer-based catalyst design. Different approaches have been developed in understanding catalytic activity and selectivity trends as well as designing new catalysts design criteria[1,2]. Probably scaling relations is one of the most important one that could be used in understanding the trends as well as limitations that are exposed by the scaling relations in catalysts performance[1]. *My research is using new strategies to design catalysts that do not follow the existing scaling relations. This approach will be extended to other chemical and electrochemical reactions.*

The electrochemical reactions of CO_2 are of interest for the synthesis of chemicals and for approaches to decrease global warming[3]. However, there are two main problems associated to this reaction that need to be addressed[4]:

1) Lack of the catalysts with high activity toward desired reaction products. Cu is the only metal on which CO is reduced to hydrocarbons in significant amounts, albeit accompanied by a very high overpotential making copper an inefficient catalyst for this reaction. To move ahead, new ways to circumvent the scaling relations is needed. *My work will utilize existing theoretical and experimental results to discover new catalysts with improved activity using high-throughput computational screening of different compounds. My focus in the screening will be discovering new compounds that could break the existing scaling relations. In another approach, I seek to find new compounds on which CO_2 reduction proceeds through a different mechanism. Therefore, scaling relation limitations will not be imposed[5,6]. This approach will be extended to other reactions.*

2) Lack of catalysts with high selectivity towards desired product: CO_2 reduction will be in competition with the hydrogen evolution reaction (HER) at all negative potentials where CO_2 reduction occurs, and a key criterion for selective catalysts in CO_2 reduction will be comparatively poor HER activity in the presence of CO_2. However, no strategy to suppress HER has been proposed. *My research will focus on designing new catalysts through changing chemical composition of active sites to suppress HER and thereby promote CO_2 reduction. In this approach, I will use new strategies that have not been explored so far. One promising approach is functionalizing and tethering the catalysts to suppress hydrogen evolution reaction and improve the catalytic activity at the same time.*

The electrochemical reduction of oxygen to hydrogen peroxide is a particularly promising means for continuous on-site production of hydrogen peroxide[7]. However, it would require active, selective and stable materials to catalyse the reaction. Although progress has been made in this respect, further improvements through the development of new electrocatalysts are needed. In particular, designing non-precious electrocatalysts for direct synthesis of hydrogen peroxide is of paramount importance. Using DFT, *my research will seek to design new non-precious catalysts for the electrochemical reduction of oxygen to hydrogen peroxide using my atomic scale understanding of ORR.*

---

[1] Abild-Pedersen, F; Greeley, J; Studt, F., *et al.* Phys Rev Lett 2007, 99, 016105.

[2] Nørskov, JK.; Abild-Pedersen, F; Studt, F., *et al.* Proc Natl Acad Sci USA 2011; 108 :937–43.

[3] Gattrell, M.; Gupta, N.; Co, A., J. Electroanal. Chem. 2006, 594, 1– 19.

[4] Peterson, A. A.; Nørskov, JK., J. Phys. Chem. Lett. 2012, 3, 251– 258.

[5] Karamad, M.; Tripkovic, V.; Rossmeisl, J., ACS Catal. 2014, 4, 2268– 2273.

[6] Karamad, M.; Hansen, HA.; Rossmeisl J.; Nørskov, JK., ACS Catal. 2015, 5, 4075–4081.

[7] Siahrostami, S.; Verdaguer-Casadevall, A.; Karamad, M., *et al.* Nat. Mater. 2013, 12 (12) 1137– 1143.

438660 Engineering Cell Metabolism for Better Health, Safe Environment and Efficient Fuels

Sunday, November 8, 2015
Exhibit Hall 1 (Salt Palace Convention Center)

Peng Xu, Chemical Engineering, MIT, Cambridge, MA, Mattheos A.G. Koffas, Chemical and Biological Engineering, Rensselaer Polytechnic Institute, Troy, NY and Gregory N. Stephanopoulos, Chemical Engineering, Massachusetts Institute of Technology, Cambridge, MA

The diversity and richness of cell metabolism provides us with enormous opportunity to implement sustainability and make this world a better place. However, the complexity underlying natural biological systems restrained our ability for targeted manipulation and optimal control of cell metabolism. To bridge this gap, synthetic biology has emerged aiming at development of novel tools that support the design, construction and control of cell metabolism and maximize the performance of the engineered cell. Undoubtedly, cell metabolism engineering is going to address some of the critical challenges facing the human kind and will provide us novel solutions for the treatment of complex diseases and sustainable production of fuels and green chemicals.

Despite the development of advanced cloning and gene assembly platforms, we are still limited by the speed and scales at which we can engineer microbes before we can realize the full potenital of cell metabolism. To address this challenge, multiplexed combinatorial pathway engineering coupled with regulatory elements evolution will allow us to rapidly optimize the efficiency of a specified pathway. As one of the promising approaches, I will engineer synthetic regulatory RNAs to post-transcriptionally and translationally modulate gene expression and improve pathway efficiency. Other challenge in cellular engineering is to engineer smart cells that integrate metabolite-sensing genetic cirucits with the physiological states of cell and achieve self-adaption. Cellular resources (NADPH or ATP) will be autonomoulsy partitioned to either cell growth pathway or metabolite production pathway depending on the state of the cell. To achieve these goals, I will engineer metabolite bio-sensors, valves, transporters and controllers that integrate both negative feedback control and positive feed-forward control to improve the production of therapeutical or fuel molecules.

I am fascinated and excited by the possibilities that metabolic engineering and synthetic biology have afforded to transform the world and improve our life. I am particularly interested in applying my knowledge and expertise to engineer human gut microbiota for self-tuning therapeutics and diagnostics; engineer microorganisms for cost-efficient production of pharmaceuticals, green chemicals and fuel molecules; develop synthetic biology tools that facilitate cell metabolism evolution and optimization; integrate synthetic feedback control and metabolic circuits to improve therapeutic and pathway efficiency. Taken together, these strategies will allow us to rapidly interrogate and control cell metabolism and develop innovative sustainable solutions to upgrade low-value carbons to high-value therapeutics or green chemicals.

438661 Interfacial Processes in Energy Storage and Conversion Devices

Abstract:

My research has been focused on characterization and control of the interfacial processes in advanced electrochemical systems presented by solid acid fuel cells (SAFC) and Li ion batteries. The performance of these electrochemical systems is closely related to the processes occurring at the interface of electrodes and electrolytes.

The electrocatalysis at the interface of the super-protonic (at 245 °C) CsH_2PO_4 solid acid and electrodes occurs at active sites which are also accessible to the humidified gases. Pt remains the best catalyst for both H
O_2 reduction reactions in the anodes and cathodes of the solid acid fuel cells. Although the electrocatalysis of the $H_2 \rightarrow H^+$ is an efficient process at 245 °C, the structure of the interface plays a major role in overall performance of the anodes. 3D nanostructures of carbon nanotubes decorated carbon fibers coated with Pt provide an efficient contact between electrolyte, Pt coatings, and H_2 gas. Interestingly, the activity of such structures correlates with the film thickness, where thinner films show higher activity. On the cathode side, because of the strong O=O bond, oxygen reduction is a slow reaction even on a Pt surface at 245 °C, requiring high loadings of the precious Pt catalyst. Solid state TiO_2 catalysts are an attractive alternative. The activity of the TiO_2 films grown on Ti metal under slightly oxidizing environment at elevated temperature (600-900 °C) is controlled by the phase and thickness of the films. Thinner films with more rutile phase show higher activity.

In Li ion batteries, our analysis shows that high molecular weight oligomerized species are formed as a part of solid electrolyte interphase (SEI). Nature of the formed SEI and degree of oligomerization can be controlled by using electrolyte additives such as vinylene carbonate and Vinyl ethylene carbonate.

Interfacial and bulk mechanical effects also play a major role in capacity retention, and battery performance. Advanced Sn based anodes, depending on their SnO_x content, exhibit significant changes in compressive and tensile surface stresses which directly influences their performance. In graphite anodes, our analysis shows that the mechanical effects scale with the charging rate, which explains the chemo-mechanical degradation effects common during fast charging of batteries.

438663 Modeling-Inspired Membrane and Particle Devices for Solar Fuels and Environmental Remediation

Abstract:

The current energy and environmental issues associated with using fossil energy sources has one way out eventually, harnessing the free and abundant energy from the sun. While solar photovoltaics and electrochemical energy storage burgeon due to growing industries, the missing piece in this sustainable picture is to build efficient and scalable chemical devices that produce transportation fuels and commodity chemicals directly from sunlight. Development of such chemical devices for photoelectrochemical (PEC) solar-fuel production and photocatalytic environmental remediation often relies on modeling and simulation capabilities for both their design and implementation.

My research vision is that modeling and simulation has been proven, and will continue to be instrumental to guide device design and accelerate experimental development. For sunlight-driven chemical fuel production, membrane-bound wire-array devices have shown the potential to achieve efficiency, stability and low-cost all at once. Particle devices, e.g. by liberating the membrane-bound wires into free-floating particle in suspension, promise disruptive cost advantages with further theoretical understanding of catalytic selectivity. For environmental remediation, similar photocatalytic membrane- and particle-based devices can be adapted to photo-decompose urea or methane in waste water. Previous modeling analysis has shown to understand requirements of light-capture materials selection, to manage mass transport, to optimize light absorber geometry and catalyst placement, and to determine local electrolyte pH. My future work will involve a proper blend of theory, modeling and experiments for demonstration of solar-fuel and environmental devices. Coupled modeling and simulation of multiple physical phenomena in 2D and 3D will be developed as a powerful and necessary tool for such complex devices. Furthermore, a TiO_2 coating strategy which is realized by atomic-layer deposition, a surface chemistry technique, will be established for stabilization of the otherwise unstable electrochemical interfaces. This emerging technique together with a selection of technologically important semiconductor materials provide pathways forward for a sustainable future via scalable solar-driven chemical devices.

439796 Performance Advances in Electrochemical Energy Storage and GHG Recycling

Abstract:

Global energy production faces a rising challenge to drastically reduce greenhouse gas output. My research develops a spectrum of technologies to achieve this goal, including highly safe and affordable electrochemical energy storage devices for facilitating massive adoption of photovoltaic and wind energy production, and advanced methane/CO recycling bioreactors.

My work on zinc anode rechargeable batteries focusses on morphological control of zinc electrodeposits in paste and flow electrodes.

Nanoscale to microscale material characterization has been performed, including grazing-incidence in-situ x-ray work at Brookhaven's NSLS I.

Nanoscale metal oxides are seen to affect the efficiency of zinc energy storage on nickel substrates. In pasted zinc anodes, pressure and oxidation rate are seen to play a key role in the longevity of the electrode. Several successful grant applications, papers, and two international patents have resulted from my work on this subject.

My work on advanced bioreactors for methane and CO recycling focusses on gas-liquid mass transfer in fast-reaction systems. Energy efficient means of improving the gas-liquid transfer coefficient are poorly researched yet very important for the cost-effectiveness of methane recycling (or other GHG recycling). New sparging and gas mixing devices are developed and tested to most efficiently disperse gas-liquid surface area throughout the reactor without harming microbes or expending unnecessary cost. Additionally, improved understanding of gas-liquid mass transfer in turbulent flows is researched from a fundamental perspective.

In each of these research areas, my research takes a laboratory experimental approach, allowing, for example, patents on new battery designs or gas reactor designs, or analysis of leading mechanistic theories of the underlying physics or chemistry. These presented results are from my research as PhD student at UC Santa Barbara (supervised by Sanjoy Banerjee), as a post-doctoral research associate at Brookhaven National Laboratory (supervised by Vasilis Fthenakis) and as a senior scientist at the City University of New York Energy Institute.Global energy production faces a rising challenge to drastically reduce greenhouse gas output. My research develops a spectrum of technologies to achieve this goal, including highly safe and affordable electrochemical energy storage devices for facilitating massive adoption of photovoltaic and wind energy production, and advanced methane/CO recycling bioreactors.

My work on zinc anode rechargeable batteries focusses on morphological control of zinc electrodeposits in paste and flow electrodes.

Nanoscale to microscale material characterization has been performed, including grazing-incidence in-situ x-ray work at Brookhaven's NSLS I.

Nanoscale metal oxides are seen to affect the efficiency of zinc energy storage on nickel substrates. In pasted zinc anodes, pressure and oxidation rate are seen to play a key role in the longevity of the electrode. Several successful grant applications, papers, and two international patents have resulted from my work on this subject.

My work on advanced bioreactors for methane and CO recycling focusses on gas-liquid mass transfer in fast-reaction systems. Energy efficient means of improving the gas-liquid transfer coefficient are poorly researched yet very important for the cost-effectiveness of methane recycling (or other GHG recycling). New sparging and gas mixing devices are developed and tested to most efficiently disperse gas-liquid surface area throughout the reactor without harming microbes or expending unnecessary cost. Additionally, improved understanding of gas-liquid mass transfer in turbulent flows is researched from a fundamental perspective.

In each of these research areas, my research takes a laboratory experimental approach, allowing, for example, patents on new battery designs or gas reactor designs, or analysis of leading mechanistic theories of the underlying physics or chemistry. These presented results are from my research as PhD student at UC Santa Barbara (supervised by Sanjoy Banerjee), as a post-doctoral research associate at Brookhaven National Laboratory (supervised by Vasilis Fthenakis) and as a senior scientist at the City University of New York Energy Institute.

439898 Photoelectrochemical solar energy conversion system : New materials and concepts

Sunday, November 8, 2015
Exhibit Hall 1 (Salt Palace Convention Center)

Aravind Kumar Chandiran, University of California, Berkeley, Berkeley, CA

Growing energy demand and the anthropogenic climate change, resulting from the use of fossil fuels, drives research both in academia and industries towards identifying a cheap and clean energy source. Solar energy, with its enormous power resource accounting to about 120,000TW far exceeds the global energy demand (13-15TW) and an efficient conversion of solar photon energy to electricity/fuels can potentially alleviate the energy problem. In addition to the efficient conversion, their cost factor plays an important role in determining their deployment on a TW scale. Third generation molecular photovoltaics including dye-sensitized photoelectrochemical devices because of simple device fabrication techniques and the use of low cost device materials, offer a significant potential for the construction of cheap solar cells /solar fuel systems.

During my PhD, I worked on understanding the photogenerated electron loss mechanisms and finding ways to rectify the losses, in such a photoelectrochemical system. The parasitic electron recombination at the semiconductor-electrolyte interface reduces the efficiency of the solar conversion. I extensively investigated the carrier losses at this interface and developed quantum tunneling layers to block the unwanted reaction, for high efficiency dye-sensitized solar cells. One of the key research outcome is, we redefined a key theory of electron recombination. In the photoelectrochemical solar cell community, there existed a common (mis)conception that 'any insulating oxide that is sufficiently thin to allow electron tunneling, should block the electron recombination at the semiconductor-electrolyte interface'. However, we discovered that only certain insulating oxides can block the charge-carrier recombination while other oxides don't. By a systematic and thorough investigation, we developed a set of criteria to select an efficient tunneling layer for high efficiency solar cells. This work, in addition to the semiconductor-electrolyte interface, has been extended to the dye/perovskite semiconductor-solid state hole conducting devices and the criteria we developed, is also found to be valid for the latter systems. With the research background in the photoelectrochemistry, I moved to UC Berkeley for a postdoctoral research to develop photoconducting metal-organic frameworks (MOFs) that can act as an absorber and charge-transporter in solar cells/solar fuels systems. The possibility of combining wide range of metal cations and linkers make MOFs an interesting candidate for tuning the optical and electronic properties. Currently I am investigating the charge transport and photophysical properties of porphyrin and thiol based MOFs for their potential application in solar energy conversion.

In future, I want to be a part of the greatest scientific community that explores the interdisciplinary aspects of science in solving the energy crisis. As an independent faculty member, I am interested in investigating new materials and device concepts for an effective solar fuel generation using photoelectrochemical systems, that uses three abundant resources on earth, namely, water, carbon dioxide and sunlight. Two key chemical reactions that I specifically would like to explore are carbon dioxide reduction to hydrocarbons and water splitting to hydrogen fuel. Considering the pitfalls/limitations in the existing technologies for the solar fuel generation, I developed new device concepts and material synthesis pathways by blending novel concepts in ferroelectrics, semiconductor electrochemistry, photochemistry, catalysis, coordination chemistry and material science, that will bring in new dimension to energy sciences. In the AIChE poster, I will present my new ideas that can potentially revolutionize the field of solar fuels.

440343 Innovating Multiphase Contactor Design with Advanced Experimental and Simulation Tools

Abstract:

Research Interests:

Multiphase flows are encountered in most of the chemical, biological and energy generation processes. By using state-of-the-art experimental techniques in conjunction with high performance computer modeling enables process innovation to increase profit while reducing the environmental impact of the process. For example, optimization of flow pattern has resulted into reduction of the two phase contactor volume at least 10 folds for same throughput. My philosophy is to use combined CFD/experimental approach to rigorously validate the CFD model with subsequent use of CFD to achieve design objective. Main advantage of such approach is savings in terms of cost of new device development without compromising the reliability of the design. I have applied such approach to a wide range of equipment size from micro-fluidized bed to large scale methane fermentation reactor.

Successful Proposals:

- High performance computing: 2000000 SU (processor hours) granted on LSU HPC (Supermike II and SuperMIC), 2000000 SU granted on LONI clusters (QueenBee 2).
- University of Newcastle postdoctoral fellowship, Australia. 3 year award, AUD $87878 per annum.
- Faculty strategic pilot grant (AUD $14600), 'Flotation in fluidized bed with turbulence tuning particles' (Mayur Sathe).
- Startup grant (AUD $10000), 'Aeration of High Solids Concentration Wastewater Streams', (Geoffrey Evans, Elham Doroodchi, Mayur Sathe).

Postdoctoral Project: "Advanced velocimetry in turbulent two and three phase flow."

Under supervision of Geoffrey Evans (Chemical Engineering, University of Newcastle, Australia).

PhD Dissertation: "Development of Experimental Techniques and CFD Simulations of Multiphase Dispersions." Under supervision of J. B. Joshi (Institute of Chemical Technology, Mumbai, India) & B. D. Kulkarni (National chemical laboratory, Pune, India)

Research Experience:

My research work is balance of experimental and numerical study of multiphase systems. Multiphase flows are unique since the medium is discontinuous yet fluid. I probed multiphase dispersions with advanced experimental techniques like Particle Image velocimetry, Laser Doppler Anemometry, Hot film anemometry and Phase Doppler Particle analyzer. I customized these instruments using in-house electronics and software to adapt them for measurements considered beyond the range of these instruments. The experimental program was parallel to CFD simulations, providing better insight into transport phenomena in multiphase equipment. So far, I have worked on all combinations of gas, liquid and solid phases. As Research scientist, my current principle project at LSU is design of a novel bioreactor for fermentation of stray methane to liquid products (ARPAE-REMOTE program). The project budget is $4000000. I am involved with CFD modeling and coming up with innovative designs to exceed the ARPA-E targets. I am also involved with experimental studies which are primarily located at City College, New York. As IT consultant at CCT, LSU I am involved with use of supercomputing resources for applications relevant to chemical industry. I am a part of EPIC (Enabling Process Innovation through Computation) consortium. Under the aegis of EPIC, I have worked to use CFD tools like Ansys FLUENT and OpenFOAM to solve problems of industrial as well as fundamental interest.

Teaching Experience:

I am currently mentoring 7 PhD students at University of Newcastle as PhD co-supervisor. I am mentoring 4 PhD students at Louisiana state university. I have taught "Green Engineering" (UG course, with Geoffrey Evans) and "Advanced Particle Processing II" (PG course).

Future Direction:

As a faculty I would like to continue using experiments and simulation for process intensification in chemical as well as in other industries. In particular I would like to optimize the turbulence generation in multiphase contactors. The turbulent motion of fluid enhances transport processes. It also enhances viscous dissipation of energy. To maximize 'bang for the buck' the energy must be provided at right length scale, right intensity and at right location. While micro reactors offered a solution which met such criteria, they are not economical at industrial scale. I would like to design decimeter scale devices which facilitate mass, heat and momentum transport via carefully tuned turbulence. I would like to demonstrate the use of homogeneous, tunable turbulence to achieve unsurpassed energy efficiency for multiphase process. To achieve this, I would work to make a balanced experimental and computational research group. I would collaborate with people in Petroleum engineering, Bioengineering, mineral engineering and mechanical engineering to make the working prototypes of multiphase contactors for industries including chemical, biological, petroleum, energy, water processing and agriculture.

Selected Publications: (total number-29)

1. **1. Sathe, M. J.,** Thaker, I. H., Strand, T. E. and Joshi, J. B., 2010. Advanced PIV/LIF and Shadowgraphy System to Visualize Flow Structure in Two-Phase Bubbly Flows. Chemical Engineering Science, 65 (8), 2431-2442.
2. **2. Sathe, M. J.,** Deshmukh, S. S., Joshi, J. B. and Koganti, S. B., 2010. Computational fluid dynamics simulation and experimental investigation: Study of two-phase liquid-liquid flow in a vertical Taylor-couette contactor. Industrial and Engineering Chemistry Research, 49 (1), 14-28.
3. **3. Sathe, M. J.,** Mathpati, C. S., Deshpande, S.S., Khan, Z., Ekambara, K. and Joshi, J. B., 2011. Investigation of flow structures and transport phenomena in bubble columns using particle image velocimetry and miniature pressure sensors. Chemical Engineering Science, 66 (14), 3087-3107.

4. **4.** **Sathe, M.,** Joshi, J., Evans, G., 2013. Characterization of turbulence in rectangular bubble column. Chemical Engineering Science, 100, 52-68.
5. **5.** Deshmukh, S. S., **Sathe, M. J.,** Joshi, J. B. and Koganti, S. B., 2009. Residence Time Distribution and Flow Patterns in the Single Phase, Annular Region of Annular Centrifugal Extractor. Industrial and Engineering Chemistry Research, *48* (1), 37–46.
6. **6.** Tabib, M. V., **Sathe, M. J.,** Deshpande, S. S. and Joshi, J. B., 2009. A Hybridized Snapshot Proper Orthogonal Decomposition-Discrete Wavelet Transform Technique for the Analysis of Flow Structures and Their Time Evolution. Chemical Engineering Science, 64 (21), 4319-4340.
7. **7.** Mitra, S., **Sathe, M.J.,** Doroodchi, E., Utikar, R., Shah, M.K., Pareek, V., Joshi, J.B., Evans, G.M., 2013. Droplet impact dynamics on a spherical particle. Chemical Engineering Science, 100, 105-119.
8. **8.** Doroodchi, E., **Sathe, M.,** Evans, G., Moghtaderi, B. 2013. Liquid-liquid mixing using micro-fluidised beds. Chemical Engineering Research and Design, 91(11), 2235-2242.
9. **9.** Hoque, M. M., **Sathe, M. J.,** Mitra, S., Joshi, J. B., Evans, G. M., 2015. Comparison of specific energy dissipation rate calculation methodologies utilising 2D PIV velocity measurement. Chemical Engineering Science, 137, 752-767.

440413 Development of a Versatile Drop-based High-Throughput Single-Cell/Molecule Study platform

Sunday, November 8, 2015
Exhibit Hall 1 (Salt Palace Convention Center)

Huidan Zhang, Experimental Soft Condensed Matter Group, John A. Paulson School of Engineering and Applied Sciences, Harvard University, Cambridge, MA

High-throughput single cell/molecule analysis can dramatically benefit from drop-based microfluidics. This high-throughput technique addresses the need for lower cost, shorter time, higher accuracy and sensitivity using water-in-oil emulsion drops to compartmentalize reactants, such as cells and molecules, into picoliter volume. In our recent research work, we have developed a variety of in-drop reactions to study the biological processes at all the levels of single cells/molecules, including single-cell/molecule PCR to detect genomic variation, single-cell/molecule RT-PCR to analyze gene expression and single-cell protein quantification to screen target protein secreting cells. Such a highly efficient platform enables faster and more reliable exploration of cell heterogeneity which is very important for many fields including development, tumorigenesis and immunology. Furthermore, our method has been employed to study rare but vital viral evolution events and also perform early diagnosis of infectious diseases and forensic marker-short tandem repeats (STR) identification. The further improvement of the whole platform will be largely driven by many requirement form biology and medicine.

Integrated single-cell genomics: Combined epigenome and transcriptome sequencing of single cells to understand cellular differentiation

Siddharth S. Dey, Ph.D.
Alexander van Oudenaarden's group, Hubrecht Institute, The Netherlands.
Email: s.dey@hubrecht.eu. Website: http://siddharthdey.com

Education

- University of California, Berkeley, USA. Ph.D., Chemical and Biomolecular Engineering (May 2012).
- Institute of Chemical Technology, Mumbai, India. B.S., Chemical Engineering (May 2006).

Research Appointments

- Post-doctoral researcher (11/2012 - Present).
  Quantitative Biology of Development & Stem Cells, Hubrecht Institute, The Netherlands.
  Advisor: Alexander van Oudenaarden.
  Project: Developing novel single-cell genomics methods to quantify cell-to-cell heterogeneity in cancers and stem cells.

- Graduate Research Assistant (08/2006 - 10/2012).
  Department of Chemical and Biomolecular Engineering, University of California, Berkeley, USA.
  Advisor: David V. Schaffer.
  Ph.D. Dissertation: A computational and experimental approach to understanding HIV-1 evolution and latency for the design of improved antiviral therapies.

- Undergraduate Research Assistant (05/2004 - 07/2004).
  Department of Organic Chemistry, Indian Institute of Science, Bangalore, India.
  Advisor: Goverdhan Mehta.
  Project: Total synthesis of cyclitols and investigation of their crystal structures.

Research Interests

The genome within all cell-types of a multicellular organism is identical, yet different cell-types display varied functions within an organism due to differences in other factors, collectively termed as the epigenome. Similarly, in specific cases, such as cancers, viral infections and in certain cell-types during normal development, mutations and other structural variations within the genome also influences cellular functions. Thus, a central question in biology is to understand how the genome or epigenome influences cellular phenotypes. From an engineering perspective, gene expression regulation can be viewed as the output of a network of complex chemical and physical processes, and understanding how these processes interact and integrate to govern cellular phenotypes has been a major focus of my graduate and postdoctoral research. More specifically, regenerative medicine has recently emerged as a promising therapeutic strategy for treating degenerating tissues using different types of stem cells. However, the full promise of regenerative medicine has been difficult to achieve so far, partly due to our incomplete understanding of the genetic and epigenetic mechanisms regulating differentiation of stem cells to specific lineages and tissues.

To understand how the genome and epigenome regulates cellular function, development of high-throughput sequencing methods, known as next-generation sequencing, are beginning to unravel genome-wide correlations between the genome, epigenome and transcriptome within a large population of cells or tissues. However, as these measurements are made from a bulk population, they only provide an average description of the system. Because tissues are composed of several cell-types and even cells within the same cell-type have been shown to display dramatic cell-to-cell variability in gene expression, bulk measurements obscure quantification of how genetic or epigenetic features directly influence the function of individual cells. To overcome this limitation, recent advances in molecular biology have enabled genome-wide single-cell measurements of the transcriptome, genome or certain epigenetic marks that capture this cell-to-cell heterogeneity. Thus, my research interests are to develop novel single-cell genomics methods to better understand how changes in the epigenetic landscape during normal development regulates cellular differentiation, information that is critical towards realizing the full potential of regenerative medicine.

Research Experience

My graduate and postdoctoral research has focused on investigating how the genome and epigenome regulates the dynamics of gene expression in viral and mammalian systems[8]. During my graduate studies, I used a systems biology approach to demonstrate that chromatin environments at different genomic loci decouple transcription factor mediated initiation of gene expression from subsequent gene activation[6]. Further, while cell-to-cell heterogeneity in gene expression has been shown to drive dramatic phenotypic variations, the upstream epigenetic mechanisms regulating this heterogeneity remain largely unknown[8]. I used DNA accessibility assays and single-molecule mRNA FISH (smFISH) to discover that repressed chromatin is associated with increased gene expression noise[3]. This project sparked my interest in single-cell biology. However, traditional single-cell techniques, such as smFISH, allowed quantification of only a handful of genes concurrently while next-generation sequencing techniques only enabled ensemble averaged genome-wide measurements starting from a bulk population. Therefore, as a post-doctoral researcher, I was interested in developing genome-wide quantification techniques in single cells. In particular, I developed the first genome-wide technology that enables sequencing both genomic DNA and mRNA from the same cell[4]. I found that low genomic copy numbers drive increased gene expression variability between tumor cells. Next, to understand the extent of cell-to-cell heterogeneity in the epigenome, I have been involved in several projects. First, we developed a method to map the 3D organization of the genome in single cells[2]. Similarly, I developed another technique to quantify the epigenetic modification, 5-hydroxymethylcytosine in single mouse embryonic stem cells[1].

Future Directions

These studies have shown that a major unanswered question is to precisely understand how upstream epigenetic mechanisms regulate cell-to-cell heterogeneity in gene expression. This question has been difficult to study and the mechanisms of regulation are poorly understood because current technologies are limited to quantifying either the genome, epigenome or transcriptome from a single cell. To unambiguously understand how a particular gene expression program in a cell is regulated will require direct measurement of both the transcriptome together with the epigenome from the same cell. Historically, development of new technologies is closely followed by giant leaps in science and therefore in my laboratory, we will develop novel integrated technologies that enable simultaneous genome-wide measurements of the epigenome and transcriptome from the same cell to gain insights into early mammalian development and mechanisms contributing to maintenance and regeneration of adult tissues.

Specifically, we propose to quantify multiple epigenetic marks, such as DNA methylation, DNA hydroxymethylation, DNA accessibility and transcription factor occupancy together with the transcriptome of single cells, innovations that will require advances in protein engineering, molecular biology and robotics. These integrated single-cell sequencing techniques will provide a unique opportunity to explore differentiation of pluripotent stem cells into the three germ layers of the developing mouse embryo to address fundamental questions in the field and lay foundations for the design of better cell-based therapies in regenerative medicine.

Publications

1. Mooijman D, **Dey SS (co-first author)**, Boisset JC, Crosetto N, van Oudenaarden A. Single-cell 5hmC sequencing reveals extensive chromosome-wide epigenetic heterogeneity. ***Nature Biotechnology*** (Manuscript under review).

2. Kind J, Pagie L, de Vries S, Azar LN, **Dey SS**, Bienko M, Zhan Y, Lajoie B, de Graaf CA, Amendola M, Imakaev M, Fudenberg G, Mirny L, Jalink K, Dekker J, van Oudenaarden A, van Steensel B (2015). Genome-wide maps of nuclear lamina interactions in single human cells. ***Cell*** (Accepted, in press).

3. **Dey SS**, Foley JE, Limsirichai P, Schaffer DV, Arkin AP (2015). Orthogonal control of expression mean and variance by epigenetic features at different genomic loci. ***Molecular Systems Biology*** 11:806.
   - Study highlighted in: Tyagi S (2015). Tuning noise in gene expression. *Molecular Systems Biology* 11:805.

4. **Dey SS**, Kester L, Spanjaard B, Bienko M, van Oudenaarden A (2015). Integrated genome and transcriptome sequencing from the same cell. ***Nature Biotechnology*** 33:285-289.
   - Study highlighted in: The genome and transcriptome of a single cell (2015). *Nature Methods* 12:173.
   - Study highlighted in: One cell at a time (2015). *Cell* 161: 1479.

5. Wong VC, Fong LE (co-first author), Adams NM, Xue Q, **Dey SS**, Miller-Jensen K (2014). Quantitative evaluation and optimization of co-drugging to improve anti-HIV latency therapy. ***Cellular and Molecular Bioengineering*** 7:320-333.

6. Miller-Jensen K, **Dey SS (co-first author)**, Pham N, Foley JE, Arkin AP, Schaffer DV (2012). Chromatin accessibility at the HIV-1 LTR promoter sets a threshold for NF-κB mediated viral gene expression. ***Integrative Biology*** 4:661-671.

7. **Dey SS**, Xue Y, Joachimiak MP, Friedland GD, Burnett JC, Zhou Q, Arkin AP, Schaffer DV (2012). Mutual information analysis reveals coevolving residues in Tat that compensate for two distinct functions in HIV-1 gene expression. ***Journal of Biological Chemistry*** 287:7945-7955.

8. Miller-Jensen K, **Dey SS (co-first author)**, Schaffer DV, Arkin AP (2011). Varying Virulence: Epigenetic control of expression noise and disease processes. ***Trends in Biotechnology*** 29:517-525.

9. **Dey SS**, Prausnitz JM (2011). Opportunities for chemical engineering thermodynamics in biotechnology: Some examples. ***Industrial & Engineering Chemistry Research*** 50:3-15.

10. Mehta G, Sen S, **Dey SS** (2005). (1S*,2S*,4S*,5S*)-Cyclohexane-1,2,4,5-tetrol monohydrate. ***Acta Crystallographica Section C*** 61 (Pt 6):o358-360.

11. Mehta G, Sen S, **Dey SS** (2005). (1*R**,2*S**,4*S**,5*S**)-Cyclohexane-1,2,4,5-tetrol. ***Acta Crystallographica Section E*** 61 (Pt 4):o920-922.

Teaching and Mentoring Experience

- Mentored a Masters student, Department of Molecular and Cellular Life Sciences, Utrecht University (01/2014 - 08/2014).
- Mentored two Graduate students, Department of Plant and Microbial Biology, University of California, Berkeley (11/2011 - 05/2012).
- Mentored two Graduate students, Department of Bioengineering, University of California, Berkeley (01/2009 - 05/2009 & 08/2011 - 12/2011).
- Teaching assistant, Chemical engineering thermodynamics, University of California, Berkeley (01/2009 - 05/2009).
- Teaching assistant, Chemical process design, University of California, Berkeley (01/2008 - 05/2008).

Awards and Honors

- Keystone Symposia Scholarship: Conference Travel Scholarship (2011).
- JRD Tata Trust Scholarship: All round academic excellence (2004 - 2005).
- Sir Ratan Tata Trust Scholarship: All round academic excellence (2003 - 2004).
- Gujarat Ambuja Cement Award: First rank in Semester I of Chemical Engineering (2002 - 2003).
- Certificate of Honor from Maharashtra State Board: Ninth rank out of 200,000 students in High School (2001 - 2002).

Water-Energy Nexus Focusing on the Application and Modification of Membrane-Based Desalination Processes

Leila Karimi, New Mexico State University

Research Interests

Fresh water resources are inherently limited by their quantity, and this has drawn the world's attention. As an independent scientist, I will focus on improving membrane-based desalination technologies, such as ED/EDR, EDI, and FO, as well as other desalination technologies, in order to reduce the energy and cost requirements of these processes without impacting their performance. Additionally, I will work on using renewable energies, especially solar energy, to provide off-grid desalination systems for rural, remote off-grid areas and provide water for people who are suffering from fresh water scarcity.

Participated in Submission of Proposals

Center of Sustainable Energy for Sustainable Water, National Science Foundation
Sustainable Aggie Approach for Water Resources, submitted to United States Department of Agriculture (USDA)
Global Innovation Initiative Grant

Ph.D. Dissertation

"Theoretical, Experimental, and Predictive Models for Ion Removal in Electrodialysis and Electrodialysis Reversal"

Research Experience

During my Ph.D. studies, I conducted experiments on pilot-scale EDR at the Brackish Groundwater National Desalination Research Facility (BGNDRF) in Alamogordo, NM. These experiments provided me with a general idea of how operating parameters affect the removal of different ions in the EDR process. Utilizing this data, I designed and built a complete laboratory-scale ED system equipped with a complete data acquisition system. The reason that the data acquisition system was designed and used in this system was to continuously and accurately monitor and record the operating conditions during the entire set of experiments. The collected data was used in developing an empirical model to predict the removal of monovalent and divalent ions in the ED/EDR process at various operating conditions. The developed dimensionless model can be widely used for full-scale electrodialysis operations for which the removal of specific ions is desired.

Teaching Experience

Prior to attending NMSU, I independently instructed a number of undergraduate courses, such as Mass and Energy Balance, Organic Laboratory, and Chemistry Laboratory. During my Ph.D. studies, in the role of Future Faculty Fellow, I independently taught 5 sessions on heat exchangers, held review sessions, and assisted in quiz and exam question development. Additionally, I individually taught a number of chapters and sessions and desiged the quizzes of an Engineering Data Analysis course. I have been assigned as the instructor of this course for spring semester of 2016. Additionally, I have served as a TA and guest lecturer for several undergraduate courses.

Future Direction

As a faculty member, I would like to continue my research in the water-energy nexus. Desalination process optimization, which is one of my interests, can be viewed from several perspectives. The optimization process can be conducted either by decreasing the cost of the process or by increasing the sustainability of the process. Moreover, the desalination process as a whole can be improved by the hybridization of different processes. The utilization of renewable energy resources in desalination processes, which is my second research area, can reduce the operating cost of the process.

My short- and long-term research objectives give me confidence that I will be able to obtain funding and add significant improvements in areas related to the water-energy nexus; this wide research area will provide enough opportunity for me to improve the efficiency and sustainability of desalination systems to help provide fresh, safe water for people who currently do not have access to it. Based on my experience in the area of the water-energy nexus, I will be able to establish a strong collaboration with industry and academia to make my research objectives achievable.

Selected Publications and Presentations

L. Karimi, L. Abkar, M. Aghajani, A. Ghassemi, "Technical Feasibility Comparison of Off-Grid PV-EDR and PV-RO Desalination Systems via Their Energy Consumption," Separation and Purification Technology, 2015, 151, pp 82-94.

L. Karimi, A. Ghassemi, "Effects of Operating Conditions on Ion Removal from Brackish Water Using a Pilot-Scale Electrodialysis Reversal System," Desalination and Water Treatment, 2015, (In Press).

C. Hanrahan, **L. Karimi**, A. Ghassemi, A. Sharbat, "High-Recovery Electrodialysis Reversal for the Desalination of Inland Brackish Waters," Desalination and Water Treatment," 2015, (In Press).

L. Karimi, A. Ghassemi, "Investigating and Understanding the Selectivity of the Conventional Ion-exchange Membranes used in Electrodialysis Process," Desalination & Water Purification Research, Prepared for Reclamation under Cooperative Agreement No. R10AC80283, 2014.

L. Karimi, A. Ghassemi, "Investigating the Effect of Velocity on Ion Removal at Different Conditions in Electrodialysis," IDA World Congress, Aug.-Sep. 2015, San Diego, CA (Accepted).

L. Karimi, A. Ghassemi, "The Preferential Ion Removal in Electrodialysis Process," World Environmental and Water Resources Congress, May 2015, Austin, TX.

L. Karimi, A. Ghorbani, A. Ghassemi, J. Loya, "Selectivity Comparison for Two Cation Exchange Membranes in the Electrodialysis Process," AICHE Annual Meeting, Nov. 2013, San Francisco, CA.

Honors and Awards

Preparing Future Faculty Graduate Assistantship Award, 2014
International Desalination Association's 2013 Channabasappa Memorial Scholarship
IEE/WERC Graduate Research Education Fellowship, 2013
Helen Zimmerman Memorial Scholarship, 2013
Reenergize the Americas Scholarship, 2012
NSF CMMI Engineering Research and Innovation Conference Scholarship, 2012
Outstanding Graduate of Chemical Engineering Department at Sahand University of Technology, 2006-2009

440715 Biomaterials-based Charge Storage Devices for Edible Electronics

Abstract:

Organic compounds represent promising alternatives to inorganic materials as renewable and sustainable energy storage devices. Organic molecules can be processed into non-conventional form factors, are cost-effective, and exhibit reduced toxicity compared to other exotic inorganic materials. Naturally-occurring biomaterials would be more advantageous due to exceptional biodegradability/biocompatibility and known cytotoxicity. Many researches have shown the potential use of biologically-derived materials into electronics devices. Such devices include organic transistors, biological sensors, and thermal-electric materials.

I am especially interested in using eumelanin pigments as charge storage materials.[1] Eumelanins are extended heterogeneous biopolymers composed of molecular subunits of dihydroxyindole (DHI) and dihydroxyindole-carboxylic acid (DHICA) with ambiguous macromolecular topology. Eumelanins exhibit unique properties such as hydration-dependent hybrid ionic electronic conductivity as well as electrochemical redox activity, which lead to promising biomaterial-based electrodes for aqueous Na^+ or Mg^{2+} charge storage devices.[2] Although the molecular composition of natural eumelanin pigments is understood, little is known about the macromolecular topology that links DHI and DHICA subunits. Understanding the macromolecular structure of eumelanins could provide insight into the self-assembly mechanisms employed by protein templates during natural melanogenesis and the origin of bulk optoelectronic properties.

Biomaterials-based charge storage devices will further be applied to power "edible electronics" devices for biomedical applications.[3] Edible electronics is the emerging strategy of delivering electronics devices through GI-tract. A variety of potential advantages include safer diagnosis, quicker therapies, and programmable drug delivery. Eumelanin pigments would be beneficial for powering these classes of transient electronics devices.

Interconnection between biological systems and electrical devices require interdisciplinary approach that incorporates the aspect of synthesis of polymeric network, microfabrication techniques, device design, and biotic/abiotic interface. My research program will lead an interdisciplinary group of understanding the functionality of biomaterials and applying them into a synthetic approach to build the charge storage devices and further develop biomedical and electronics devices.

References

(1) Kim, YJ; Wu, W; Chun, S-E; Whitacre, JF; Bettinger, CJ *Proc. Natl. Acad. Sci.* **2013**, 20912.

(2) Kim, YJ; Wu, W; Chun, S-E; Whitacre, JF; Bettinger, CJ *Adv. Mater.* **2014**, *26*, 6572.

(3) Kim, YJ; Chun, S-E; Whitacre, J; Bettinger, CJ *J. Mater. Chem. B* **2013**, *1*, 3781

441165 Porous Nano-Structured Doped Materials for Energy-Related Applications

Sunday, November 8, 2015
Exhibit Hall 1 (Salt Palace Convention Center)
Maryam Peer, Chemical Engineering, Massachusetts Institute of Technology, Cambridge, MA

The global increasing demand for energy calls for cost-effective and sustainable approaches towards energy storage and conversion. Porous catalytic materials are the heart of the process in a broad range of applications including energy conversion, gas adsorption and fine chemicals' synthesis. Porous nano-structured materials (carbon, zeolite, metal oxides, etc.) have been extensively studied and used either as high surface area supports for dispersing nanoparticles or as metal-free catalysts. Chemical composition and textural properties of these materials significantly affect their performance and stability.

Heteroatom doped carbon-based nanostructures are among the most promising metal free catalysts studied recently. It has been shown that doping different functionalities (nitrogen, sulfur, etc.) into the carbon structure can improve its activity when used in catalytic and electrocatalytic applications. On the other hand, doping increases the dispersion of nanoparticles when these materials are used as catalyst support. In addition to chemical composition, textural properties (BET surface area and pore size) are equally important in dictating the catalyst effectiveness in different processes. Post synthesis approaches along with templating are typically utilized to tailor the chemical composition and textural characteristics of porous materials. However, it is challenging to maintain the dopant level while increasing the BET surface area of porous carbons. Usually materials with high level of doping suffer from low surface area and undeveloped porosity; while those with higher surface area and desired hierarchical porous network are usually low in dopant level. This challenge, however, maybe addressed by applying bottom-up approaches starting from carefully chosen building blocks, containing desired functionalities and composition.

As a graduate student, I worked on synthesis of polymer-derived hierarchical carbon materials as porous supports for embedding catalytic nanoparticles. I extended my knowledge by working on nitrogen-doped carbon materials for energy conversion applications and gained valuable experiences on reactions in microreactors and flow chemistry. I would like to use the experience and knowledge I obtained during my PhD and postdoc and combine them in order to develop simple and cost-effective synthesis routes for chemically tuned, doped porous carbon structures possessing desired physical and chemical properties. Uniform distribution of heteroatoms along with desired pore architecture would provide interesting and unique features such as preferred adsorption/affinity with different substrates, size exclusion and enhanced interaction with secondary nanoparticles. I would like to specifically tailor nanostructured materials for energy-related applications including oxygen reduction (fuel cells), sodium ion batteries, fine chemical's synthesis and CO_2capture; Where surface area, pore size, chemical composition and morphology play crucial roles.

Postdoctoral Projects:

- Soft-templated synthesis of high surface area nitrogen-doped carbon

- Green, continuous biphasic oxidation of alcohols using polyoxometalates as the catalyst

- Under supervision of Professor Klavs. F. Jensen, Chemical Engineering, Massachusetts Institute of Technology

PhD Dissertation:"The design of a shape selective platinum-carbon catalyst with high effectiveness"

- Under supervision of Professor Henry C. Foley, Chemical Engineering, The Pennsylvania State University

Selected Publications:

M. Peer, A. Qajar, R. Rajagopalan, H.C. Foley, "On the effects of confinement within a catalyst consisting of platinum embedded within nanoporous carbon for the hydrogenation of alkenes", **Carbon**, 66, 459-466 (2014).

M. Peer, A. Qajar, R. Rajagopalan, H.C. Foley, "Synthesis of carbon with bimodal porosity by simultaneous polymerization of furfuryl alcohol and phloroglucinol", **Microporous and Mesoporous Materials**, 196, 235-242 (2014).

M. Peer, A. Qajar, B.P.M. Holbrook, R. Rajagopalan, H.C. Foley, "Platinum embedded within carbon nanospheres for shape selective liquid phase hydrogenation", **Carbon**, 57, 485-497 (2013).

M. Peer, A. Qajar, R. Rajagopalan, H.C. Foley, "On the effects of emulsion polymerization of furfuryl alcohol on the formation of carbon spheres and other structures derived by pyrolysis of polyfurfuryl alcohol ", **Carbon**, 51, 85-93 (2013).

A. Qajar, **M. Peer**, R. Rajagopalan, H.C. Foley, **"**Characterization of micro- and mesoporous materials using accelerated dynamics adsorotion", **Langmuir,** 29, 12400-12409 (2013).

A. Qajar, **M. Peer**, R. Rajagopalan and H.C. Foley, "High-pressure hydrogen adsorption apparatus: Design and error analysis", **International Journal of Hydrogen Energy**, 37, 9123-9136 (2012).

441282 Supramolecular Mesochemistry: Engineering Materials from the Bottom Up

Abstract:

One of the most inspiring aspects about the chemistry of life is the hierarchical organization by which biomolecules self-organize into complex structures across vast length scales. The grand vision of 'supramolecular mesochemistry' is to emulate Nature's capacity to build intricate structures of virtually any size using bottom-up self-assembly. Toward this ultimate goal I propose three projects aimed at ***engineering well-defined mesoscale aggregates and functional materials from colloidal particles and reversible supramolecular crosslinks.***

Project I. Retrosynthetic Self-Assembly of Functionalized Colloids. Retrosynthetic thinking will be applied to the rational self-assembly of mesoscale objects, guided by noncovalent interactions between complementary molecular recognition motifs mounted on the surfaces of shape-defined colloidal particles. Initial target mesostructures include sequence-defined colloidal polymers and 3D colloidal networks. These assemblies may be observed directly by microscopy and reconfigured by manipulating the reversible, stimulus-responsive crosslinking interactions.

Project II. Competition Driven Evolution of Dissipative Self-Organizing Aggregates. Surface-functionalized, shape-defined microparticles displaying various supramolecular elements will be passed through iterations of an evolutionary algorithm (manual or automated) involving (i) ***competition*** among particles for noncovalent crosslinks, which induce aggregation via interfacial supramolecular interactions, (ii) ***selection*** of aggregates that dissipate energy according to imposed fitness criteria, (iii) ***mutation*** involving the introduction of new microparticle geometries, materials, stoichiometries, surface motifs, etc., and (iv) ***reproduction***, i.e., repeated cycles of competition, selection, and mutation. The evolution of aggregates exhibiting 'life-like' behaviors such as stimulus response, adaptation, templation, and autocatalysis will be prioritized.

Project III. Host-Guest Organic Semiconductors. Supramolecular design motifs are numerous and diverse, yet few of them are suited for applications involving organic semiconductors because of unfavorable optoelectronic properties. Novel semiconducting host-guest complexes based on fully π-conjugated hosts (e.g., cycloparaphenylenes) and guests (e.g., polycyclic aromatic hydrocarbons) could be of use for mediating charge transfer in self-assembling organic electronics and organic photovoltaics. If substituted appropriately, these design motifs could also help establish electronic communication between functionalized colloidal particles when implemented in Projects I–II.

441283 Energy Storage Devices and Advanced Electrochemical Separation Processes

Abstract:

Research Interests

Renewable energy and access to clean water are among the grand challenges of our century. As a future faculty member, I plan to develop a research program that will focus on pushing the boundaries on these challenges through the strategic design of: (1) next generation electrode materials and safe electrolytes towards their utilization in batteries and supercapacitors, (2) nanoscale structural networks towards their integration for advanced separation processes for water treatment (e.g., metal removal, desalination).

Research Proposals Submitted

L'Oréal USA for Women in Science Fellowship, 2015, "*Durable and Highly Absorbent Dressings for Chronic Wounds*"

NASA Postdoctoral Research Program, Glenn Research Center, Electrochemical Energy Storage Program, 2014, "*Ionic Liquid based non-flammable electrolytes for energy devices*"

Post-doctoral Research Experiences

University of Akron (Supervisor: Bryan D. Vogt)

Projects: (i) Design of high capacity anode material for Na^+ - battery and its integration with high-temperature electrolyte, (ii) Improving mechanical integrity of electrospun thermoplastic polyurethanes

Massachusetts Institute of Technology (Supervisor: T. Alan Hatton)

Projects: (i) Electrochemically mediated CO_2 separation, (ii) Non-volatile liquid sorbents with enhanced gravimetric CO_2 capacity

PhD Thesis

University of Notre Dame (Supervisor: Joan F. Brennecke)

Dissertation: Kinetics of CO_2 absorption by chemically reactive ionic liquids

Side project: Vapor-liquid phase behavior of hydrazine and its derivatives by Monte Carlo simulations (Supervisor: Edward Maginn)

Teaching Experience

Supervised 4 undergraduate researchers, 1 first year graduate student and 2 high school students. Assisted in courses of Thermodynamics, Reaction Engineering, Physical Chemistry, Materials Science and Unit Operation Laboratory: held office hours, prepared homework solutions, helped in grading.

Selected Publications

1. **1. Gurkan, B.**, Simeon, F., Hatton, T. A., "Quinone Reduction in Ionic Liquids for Electrochemical CO_2 separation", *ACS Sustainable Chemistry & Engineering* 7 (2015) 1394-1405
2. **2. Gurkan, B. E.,** Gohndrone, T. R., McCready, M. J., Brennecke, J. F., "Reaction kinetics of CO_2 absorption into phosphonium based anion-functionalized ionic liquids", *Physical Chemistry Chemical Physics* 15 (2013) 7796-7811
3. **3. Gurkan, B.**, Goodrich, B. F., Mindrup, E., Ficke, L., Massel, M., Seo, S., Senftle, T., Wu, H., Shah, J. K., Maginn, E. J., Brennecke J. F., Schneider, W.F., "Molecular Design of High Capacity, Low Viscosity, Chemically Tunable Ionic Liquids for CO_2 Capture", *Journal of Physical Chemistry Letters* 24 (2010) 3494-3499. Editor's choice in Science Magazine - December 17, 2010 (DOI: 10.1126/science.330.6011.1588-b) 110+ citations
4. **4. Gurkan, B.**, de la Fuente, J. C., Mindrup, E., Ficke, L., Goodrich, B. F., Price, E., Schneider, W. F., Brennecke, J. F., "Equimolar CO_2 Absorption by Anion-functionalized Ionic Liquids", *Journal of American Chemical Society Communications* 132 (2010) 2116-+. 280+ citations
5. **5. Eksioglu, B.,** Nadarajah, A.,"Structural Analysis of Carbon Nanofibers", *Carbon* 44 (2006) 360-373.

Awards/Honors

Bayer Pre-doctoral Research Fellowship, Notre Dame Indiana 2010-11

RECS (Research Experience on Carbon Sequestration) Fellow, Birmingham Alabama 2011

Travel Grant for Gordon Research Conference on Green Chemistry, Davidson North Carolina 2010

Honorable Mention, The Link Energy Foundation Energy Fellowship, 2009

Poster Presentation Award, Cairns Australia 2009

ACS Sustainability and Green Chemistry Summer School Fellow, Golden Colorado 2008

Best Master Thesis, Toledo Ohio 2006

Outstanding Poster Presentation Award, Toledo Ohio 2006

Professional Memberships

American Chemical Society

American Institute of Chemical Engineers

Association for Women in Science

441323 Integrated design and operation for energy security and environmental protection

Abstract:

The convergence of recent advances in computer-aided engineering and theoretical developments for design and operation of industrial processes has provided unprecedented opportunities to explore new solutions for energy security and environmental protection. Integration of process design and control[1] offers enormous advantages in terms of profitability and flexibility in the presence of uncertainties, and is the leading research in Process Systems Engineering. However, this approach also results in highly combinatorial mathematical formulations that pose tough challenge for current optimization technology. During my PhD studies[1-4], I developed a new method for simultaneous optimization of process design and its control system, which is computationally manageable. It also ensures that the designed process is economically optimal and features controllability as well as self-optimizing properties. Furthermore, I have been conscious about my research impacts and during my postdoctoral studies, I applied my knowledge to novel applications for energy security and environmental protection[5-15]. I have conducted a comprehensive study on biofuel production via fast pyrolysis[6-9], including reaction kinetics, process synthesis, integration and retrofit, in addition to supply chain optimization. The other application of my interest is carbon capture and storage. In this area, I investigated[10-12] multi-scale modelling and optimization of absorption/desorption systems and their flexible operation, when integrated with power plants under a wide range of uncertainties in the electricity load. While these demonstrating studies provide the *proof of concept*, my global research objective is to enhance sustainability of energy resources and protect the environment by changing the processing paradigms and adaptation of renewable energies. My strategies to achieve this goal will include application of mathematical programming and big-data analytics for quantification and optimization of the energetic and environmental footprints of new technologies, developing enabling software tools that make new methods accessible to a wide spectrum of beneficiaries, and delivering the research impact via industrial collaborations.

Publications

[1] **Sharifzadeh M***, Integration of process design and control: a review, *Chemical Engineering Research and Design*, 91 (12), 2515–2549, (Link).

[2] **Sharifzadeh M***, (2013). Implementation of an inversely controlled process model for integrated design and control of an ETBE reactive distillation, *Chemical Engineering Science*. 92, 21–39. (Link)

[3] **Sharifzadeh M***, Thornhill NF, (2013). Integrated design and control using a dynamic inversely controlled process model. *Computers & Chemical Engineering,* 48, 121–134. (Link)

[4] **Sharifzadeh M***, Thornhill NF, (2012).Optimal selection of control structures using a steady-state inversely controlled process model. *Computers & Chemical Engineering*, 38 (5), 126–138. (Link)

[5] **Sharifzadeh M***, Rashtchian D, Pishvaie M R, Thornhill NF, (2011). Energy induced separation network synthesis of an olefin compression section: a case study. *Industrial & Engineering Chemistry Research,* 50 (3), 1610–1623, (Link).

[6] **Sharifzadeh M***, Richards CJ, Liu K, Hellgardt K, Chadwick D., Shah N. (2015). An integrated process for biomass pyrolysis oil upgrading: A synergistic approach. *Biomass & Bioenergy*. 76, 108–117, (Link).

[7] **Sharifzadeh M*,** Wang L., Shah N., (2015). Decarbonisation of olefin processes using biomass pyrolysis oil. *Applied Energy*, 149, 404–414, (Link).

[8] **Sharifzadeh M*,** Wang L, Shah N, (2015). Integrated bio-refineries: CO2 utilization for maximum biomass conversion. *Renewable and Sustainable Energy Reviews*, 47, 151–161, (Link).

[9] **Sharifzadeh M***, Cortada Garcia M., Nilay Shah, (2015). Supply chain network design and operation: Systematic decision-making for centralized, distributed, and mobile biofuel production using mixed integer linear programming (MILP) under uncertainty. Biomass and Bioenergy, 2015, 81, 401–414, (Link).

[10] **Sharifzadeh M***, Shah, N, (2015). Comparative studies of CO2 capture solvents for gas-fired power plants: Integrated modelling and pilot plant assessments. (accepted for publication at International Journal of Greenhouse Gas Control).

[11] **Sharifzadeh M***, Shah, N, (2015). Carbon capture from natural gas combined cycle (NGCC) power plants: solvent performance comparison at industrial scale (invited paper, minor revisions required by *AIChE Journal*)

[12] **Sharifzadeh M***, Shah, N, (2015). Carbon capture from pulverised coal power plants (PCPP): solvent performance comparison at industrial scale (under review at Applied Energy)

[13] Wang L, **Sharifzadeh M**, Templer R, Murphy RJ*, (2012). Technology performance and economic feasibility of bioethanol production from various waste papers. *Energy & Environmental Science*, 5, 5717-5730, (Link).

[14] Wang L, **Sharifzadeh M**, Templer R, Murphy RJ*, (2013). Bioethanol production from various waste papers: Economic feasibility and sensitivity analysis. *Applied Energy,* 111, 1172–1182, (Link).

[15] Chen L¥, **Sharifzadeh M¥**, Dowell NM, Welton T, Shah N, Hallett JP*, (2014). Inexpensive ionic liquids: [HSO4]−based solvent production at bulk scale. *Green Chemistry*, 16, 3098-3106, (Link).

441331 Conducting (flowable) suspension electrodes for water and energy technologies

Sunday, November 8, 2015
Exhibit Hall 1 (Salt Palace Convention Center)

Kelsey Hatzell, Lawrence Berkeley National Laboratory, New York, NY and Yury Gogotsi, Department of Materials Science and Engineering, Drexel University, Philadelphia, PA

Suspension-based (or semi-solid) electrodes have recently gain increased attention for large-scale applications such as grid energy storage, water deionization, and wastewater treatment. A suspension electrode is a multiphase material system comprised of an active (charge-storing) material suspended in an electrolyte. This principle enables the scalability of small-scale energy storage devices (batteries, supercapacitors,and hybrid-redox systems). Effective material utilization in a suspension electrode is highly dependent on achieving electrical percolation within a flowable electrode. Herein, we describe a novel method for visualizing electrical percolation pathways in a suspension electrode and describe the ongoing material's and rheological challenges facing flow-electrode systems.

441431 Flowable Slurry Electrodes for Electrochemical Processes

Abstract:

Flowable Slurry Electrodes for Electrochemical Processes

Enoch A. Nagelli
Advisors: Prof. Jesse S. Wainright and Prof. Robert F. Savinell
Department of Chemical and Biomolecular Engineering
Case Western Reserve University

Research Interests:

For majority of the last century, slurry and fluidized bed electrodes have been studied for electrochemical applications such as waste water treatment and energy storage systems. Recently, there has been renewed interest for the use of slurry electrodes in three specific applications: redox flow batteries (RFBs) with flowable slurry electrodes, electrochemical flow capacitors (EFCs), and flow electrode capacitive deionization (FCDI) systems. In order to successfully implement slurry electrodes in electrochemical applications requires fundamental understanding of the factors controlling electronic conduction through the Faradaic and non-Faradaic charge transfer processes occurring during flow. Moreover, recent advances in nanotechnology and materials science offers a selection of metal or carbon particles of any size, geometry and surface area to incorporate as slurry electrodes. Slurry consisting of particles with high surface area and surface to volume ratio can reduce kinetic losses and enhance mass transfer rates, respectively. For example, leveraging the benefits of utilizing slurry electrodes decouples energy storage capacity from power requirements in hybrid flow batteries for achieving low cost energy storage and broad versatility for electrochemical applications. There is also potential for broad applications beyond energy storage such as when processing dilute streams for separation/extraction of ions for chemical treatment and/or when in need of electrode replacement or reprocessing. Therefore, the application of slurry electrodes as a platform electrochemical technology can address environmental concerns, reduce energy costs, and enhance sustainability of many industrial processes.

Postdoctoral Project: ARPA-E Awarded Project - "Large Scale High Energy Capacity All-Iron Flow Batteries"
Under supervision of Professor Robert F. Savinell and Professor Jesse S. Wainright, Department of Chemical and Biomolecular Engineering, Case Western Reserve University

PhD Dissertation: "Controlled Functionalization and Assembly of Graphene Nanostructures for Sensing and Energy Storage Applications"
Under supervision of Professor Liming Dai, Department of Chemical and Biomolecular Engineering, Case Western Reserve University

Awards:
AFRL/DAGSI Ohio Research Fellowship
Project Awarded: "DNA-Directed Self-Assembling of Nanomaterials for Electronics"
Description: Novel chemical modification approach to asymmetrically modify nanomaterials with two different functional groups for supermolecular structures.
Sponsor: Dr. Rajesh Naik, Air Force Research Laboratory, Wright-Patterson Air Force Base, OH

Research Experience:
My graduate and postdoctoral training and research experience have furthered my commitment to pursue a career in academia. I am interested in an academic career that will allow me to expand my research knowledge of organic and inorganic nanomaterials into studying the electrochemical application of these materials in both large and small scale energy storage and conversion devices. My graduate research included the development of novel functionalization and assembly methods for spatial micro- and nanopatterning of carbon nanomaterials in both covalent and noncovalent functionalization schemes crucial for furthering methods of nanolayer assembly of programmable architectures for sensors, flexible electronics, and energy storage. Moreover, my research interests expand across materials science and chemistry of carbon nanomaterials into the electrochemical processes involved in the device applications of these materials. Consequently, my current postdoctoral research focuses on the design, testing, implementation, and electrochemical analysis of redox flow battery systems and the fundamental understanding of electrochemical carbon slurry electrodes for the all-iron redox flow battery. An in-depth understanding of the properties which influence how flowable carbon slurry electrodes create a current distribution that favors plating on the slurry particles is essential for the performance of the all-iron redox flow battery. Currently at CWRU, my research focus is to further understand the influence of physicochemical properties on the electron transfer mechanism and overall conductivity of carbon slurry particles during electrochemical cycling in redox flow batteries.

Teaching Experience:
In addition to my research career, I also have extensive teaching experience as a TA and guest-lecturer in core undergraduate chemical engineering courses (Transport Phenomena and Chemical Reaction Engineering) and measurements laboratory. My duties included guest-lecturing and preparing class material when needed, holding office hours, and grading course materials. As a graduate student and postdoc, I have mentored eleven undergraduate students and five high school students to develop and execute independent research projects by leveraging my lab's expertise with the student's own interests.
My time in academia has taught me the importance of fundamental chemical engineering concepts to perceive the balance between theory and practice, analytical rigor and intuition. My teaching interests include undergraduate core chemical engineering courses such as transport phenomena, thermodynamics, chemical reaction engineering, and process control. I would also like to teach graduate level courses in electrochemical engineering, materials for electrochemical energy storage and conversion systems, and carbon nanotechnology.

Future Direction:
My goal is to establish a research program that will continue to contribute to innovative research at the junction of electrochemistry and materials science and engineering. Specifically, I would like to combine the knowledge that I obtained from the research areas that I worked on during my Ph.D. and postdoc to investigate flowable slurry electrodes for electrochemical processes in energy storage and conversion, chemical purification and recovery, and electrode regeneration and protection. My research would be focused on the theory of electron transfer in slurries of

particles,
investigations of size, morphology, inherent particle properties, and concentration of particles on effective electronic conductivity, mass transport properties, and stability of the slurry for broad electrochemical applications. Understanding the influence of the intrinsic physicochemical properties of particles on overall conductivity can help predict and design flowable slurry electrodes for specific electrochemical processes. Previous efforts have focused on micron-sized or larger particles in slurry but recent advances in nanotechnology can enable the study and implementation of sub-micron sized particles of different shape, size, and surface area in slurry for electrochemical processes. Thus, the application of slurry electrodes in electroextraction processes can lead to larger range of recovery/extraction of fine organic or inorganic particles from solutions and lower electrode replacement costs for industrial processes.Moreover, incorporating the variety of particles available as flowable slurry electrodes with very high surface area and conductivity can decouple energy and power ratings to build novel energy storage and conversion systems.

Selected Publications:

E. Nagelli, R. Naik, Y. Xue, Y. Gao, M. Zhang, and L. Dai "Sensor arrays from multicomponent micropatterned nanoparticles and graphene" Nanotechnology 2013, 24, 444010.
D. Yu, **E. Nagelli**, R. Naik, and L. Dai "Asymmetrically Functionalized Graphene for Photodependent Diode Rectifying Behavior" Angew. Chem. Int. Ed. 2011, 50, 6575.
D. Yu, **E. Nagelli**, F. Du, and L. Dai "Metal-Free Carbon Nanomaterials Become More Active than Metal Catalysts and Last Longer" J. Phys. Chem. Lett. 2010, 1, 2165-2173
Selected Conference Abstracts:
E. Agar, **E. A. Nagelli**, N. S. Sinclair, N. C. Hoyt, E. A. Striker, R. F. Savinell, J. S. Wainright. "Cost-Effective Redox Flow Batteries using a Flowable Slurry Electrode". 2015 MRS Fall Meeting & Exhibit. November 29 – December 4, 2015, Boston, MA.
E. Agar, **E. A. Nagelli**, N. S. Sinclair, N. C. Hoyt, E. A. Striker, R. F. Savinell, J. S. Wainright. "Cost-Effective All-Copper Flow Battery Using Flowable Slurry Electrode for Large-Scale Energy Storage". 228th Electrochemical Society (ECS) Meeting. October 11-16, 2015, Phoenix, AZ.
N. C. Hoyt, E. Agar, **E. A. Nagelli**, R. F. Savinell, J. S. Wainright, "Characterization of Carbon Black Particles for Use in Aqueous Slurry Electrodes". 228th ECS Meeting. October 11-16, 2015, Phoenix, AZ.

441433 Development of Advanced Polymeric Membranes for Water-Energy Nexus Challenges

Abstract:

Water and energy systems are inextricably linked. Water is used in all phases of energy production. Energy is required to extract, convey, and deliver water of appropriate quality for diverse human uses, and then again to treat wastewaters prior to their return to the environment. Both resources are limiting the other—and both may be running short.

Water treatment technologies can enhance energy efficiency of water systems and enable the productive and safe use of non-traditional water resources for energy and non-energy applications. Such improvements in water treatment and management have particular use for treating oil- and gas-produced waters, as well as saline aquifers, brackish groundwater, brines, seawater, and municipal wastewater.

Membrane separations represent sustainable and often more cost-efficient alternatives to classical separation methods such as distillations, crystallizations, extractions, preparative chromatography, etc. However, existing membrane technology is far from providing optimal sustainability, particularly due to performance decline caused by compaction, fouling, repeated cleaning to alleviate fouling, and resulting gradual deterioration of the membrane material.

My research has been focused on development of novel membrane materials; fabrication techniques; and advanced membrane-based processes. I have been particularly active in the following following research areas: (1) Development of polyelectrolyte multilayer membranes for water and organic solvent purification. (2) Development of mixed matrix nano- and meso-composite membranes with enhanced performances. (3) Development of charged mosaic membranes for separation of water-soluble organic substances from complex mixtures. (4) Development of novel approaches to control membrane fouling. (5) Development of interfacially polymerized catalytic membranes for simultaneous biomass hydrolysis and sugar separation.

The future research that I intend to implement promises to advance multiple areas that currently dominate the field of membrane separations: (1) Development of novel approaches to enhance performance of functional membranes for specific (reactive) separation applications by incorporating functional materials. (2) Development of highly selective and permeable membranes yielding superior permeate flux performance. (3) Providing important insights into fundamental mechanisms of membrane fouling under a wide range of operating conditions. (4) Development of solvent resistant nanofiltration (SRNF) membranes for substituting conventional separation processes of organic solutions and crude oil.

441844 Tissue Engineering: From Microfluidic Devices to Biopreservation

Sunday, November 8, 2015
Exhibit Hall 1 (Salt Palace Convention Center)

Berk Usta, Center for Engineering in Medicine, Harvard Medical School - Massachusetts General Hospital - Shriners Hospitals for Children, Boston, Boston, MA

With the ever increasing cost of medical therapies and the nearly impossible odds of developing one successful drug that can pass the regulatory screenings, the time is right to search for cheaper, more efficient and higher throughput solutions for our screening and therapeutic needs. The micro-nano fabrication revolution in engineering and the ever improving computational capabilities are enabling technologies that now make this search more practical and feasible. Tissue engineering, a field that can take advantage of these advances, is still in its infancy despite being viewed as the future of medicine. As an engineer who has been trained extensively in computational modeling and has recently become experienced in bio-preservation and microfabrication; my ultimate goal is to contribute to tissue engineering finally bearing complete practical solutions and fundamental understanding by integrating all three areas of my current expertise; tissue engineering, biopreservation and computational modeling.

My long term vision is one where we can provide live tissue engineered products (TEPs), for clinical, research and industrial use, to anyone and anywhere in the world in a reliable and reproducible fashion. Such a vision requires a multidisciplinary research program which encompasses 1) tissue engineering 2) biopreservation 3) computational and theoretical modeling to accelerate experimental work. Accordingly my long term goal, as an independent faculty member, is to create a 3 thrust research program where I a) build micro/macro fabricated TEPs as organ models and assist/therapy devices (tissue engineering) b) devise optimized preservation methods to make the "off-the-shelf" dream a reality (bio-preservation) c) conduct coarse grained and detailed modeling to understand and accelerate experimental work (modeling).

During this poster session I will also focus on recent work where we are developing novel micro-tissue devices to actively induce and control in-vitro zonation of liver tissues as well as patterning of other tissues. The motivation in this work stems from the realization that despite a plethora of advances in the "organs-on-a-chip" field homogeneity of tissues is still a widely overlooked phenomenon. This work was recently funded by both the NIH via the R21 mechanism **(NIH 1R21EB020192, PI: Usta)** and MGH ECOR Interim Funding **(MGH #226569, PI: Usta).**

441849 Dynamic Properties of Interfaces in Soft Matter: Synthesis and New Characterization Techniques

Abstract:

The dynamic properties of soft materials are central to a wide range of engineering and biomedical applications. While the synthesis and structural characterization of soft materials have come a long way, there are still many big gaps in our understanding of soft materials, especially in the realm of dynamics, quantification of intermolecular forces, interconnection among different length scales, and non-equilibrium, time- and rate- dependent physical and engineering properties. The knowledge filling these gaps will allow rational design of functional soft matter interfaces, and has profound implications across a wide range of disciplines, including colloidal and interface science, polymer chemistry and physics, nanotechnology, biomaterials, and biology.

My research expertise is in the fields of soft materials, surface forces, and intermolecular interactions. My PhD research under the supervision of Prof. Jacob Israelachvili at the University of California, Santa Barbara, concerns the nanomechanics of biomaterials and biomimetics. I have performed the experimental and theoretical study on the mussel and gecko adhesive systems and development of bio-inspired polymer surfaces and nano-structured materials. I have conducted a systematic study to demonstrate the importance of interfacial redox of 3,4-dihydroxy-*L*-phenylalanine (Dopa) to the adhesion and mechanical properties of mussel adhesive proteins. I have also investigated the friction of rough and structured polymer surfaces, and boundary lubrication properties of surface tethered polyelectrolytes.

My postdoc research with Prof. Matthew Tirrell at the University of Chicago/Argonne National Laboratory focuses on the characterization of the structure and functionality of polyelectrolyte brushes. I have combined surface forces apparatus and X ray/neutron scattering techniques to elucidate the structure of polyelectrolyte brushes in the presence of multivalent counterions, the configuration of surface tethered DNA oligomers, and the interactions between lipid membranes and polymers in solution. The knowledge gained is critical for many polyelectrolyte brushes related applications, such as lubrication, antifouling coating for water infiltration process, colloidal stability, biosensor, and drug delivery.

In my independent research, I would like to develop an interdisciplinary research program centered on molecular design of soft matter surfaces for energy and biomedicine applications. The overall goal of my research is to understand the connection between a soft matter system's molecular interactions and architecture, and its surface properties. I would like to search for the physical laws that govern the structure, dynamics, and cooperative behavior of soft materials at the nano and molecular scale and to develop functional soft matter surfaces through rational molecular design and chemical synthesis.

438670 Microwave Assisted Heating of Human Blood at 2450 MHz Frequency Using Various Composite Supports

Abstract:

Abstract — The growing interest in the use of microwaves to process materials is primarily due to 'volumetric heating'. Microwave energy has been beneficial for a spectrum of applications in engineering and medical fields especially for heating, thawing, drying, warming and material processing. Warming/heating of blood sample is a very common medical practice and microwave heating could be potentially very useful due to the advantages such as rapid, controlled, selective and uniform heating effects. In Blood Bank, blood samples (whole blood) are usually preserved at low temperatures (1 to 6 °C) with permitted storage periods of 2-4 weeks and for longer storage periods (more than months) blood samples need to be preserved at very low temperatures (-30 to -65 °C). Warming/heating of blood sample collected from blood bank is a very essential medical practice before blood transfusion. Conventional method of blood sample heating/warming may not be useful for emergency situations as it takes hours to heat the blood sample. However, microwave heating could be extremely useful for blood sample warming/heating especially during emergency situations as it takes very short time (few minutes only). The present work has been carried out to study efficient heating due to microwaves for one-dimensional (1-D) human blood samples placed on Teflon, various ceramic (Al_2O_3, SiC), and composite supports. Maxwell's equation for electromagnetic wave propagation (electric field equation) and the non-linear heat conduction equation (energy balance equation) with the appropriate boundary conditions are simultaneously solved using Galerkin's finite element method to obtain power absorption and transient temperature profile for the blood sample-support assembly. A preliminary study has been carried out via average power vs sample thickness diagram to estimate microwave power absorption within blood samples for various cases. In addition, support thickness sensitivity analysis has also been carried out and suitable support thickness has been recommended. The maxima in average power, also termed as 'resonances', are observed for specific blood sample thicknesses and the two consecutive resonances of significant magnitudes are termed as R_1 and R_2 modes. For all cases, it is observed that microwave power absorption is enhanced in presence of metallic and composite supports during both R_1 and R_2 modes. The efficient heating strategies characterized by 'large heating rates' with 'minimal thermal runaway' i.e. uniform temperature distributions within the sample have been assessed for both small and large blood sample thicknesses. Based on the detailed spatial distributions of power and temperature for various cases, suitable combinations of supports have been recommended as optimal heating strategies for blood samples corresponding to both R_1 mode and R_2 mode. Present study recommends the efficient way to use microwaves in a single mode waveguide and the heating scheme can be suitably extended for heating of any other animal blood samples with known/ measurable dielectric properties.

438671 Multiphase Flow Research for Sustainable Production and Use of Energy, Chemicals and Water

Abstract:

<align="center" style="text-align: left;">**4th Year Postdoctoral Fellow**

Research Interests:

My research interests span the broad areas of experimental and computational fluid dynamics and its applications, with a focus on multiphase flow reactor systems. My current research mainly focuses on developing advanced models for polydisperse multiphase flow and multiphase turbulence. I firmly believe that a judicious balance of theory, experiments and computation can yield critical insight into multiphase flow physics that enables the development of physics-based models that can be predictive and robust, which in turn can lead to more efficient and sustainable production and use of energy, chemical and water, thus generate broad impact to industries and society. Looking forward, I am expecting to apply my research experience and talent to advance both fundamental research and applications of multiphase flow science.

Postdoctoral Project:

1. *"A novel Taylor-Couette photobioreactor for energy efficient micro-algae cultivation." –Conoco-Phillips*

Under supervision of R. Dennis Vigil and Jacqueline V. Shanks, Department of Chemical and Biological Engineering, Iowa State University

2. *"Kinetic Theory Modeling of Turbulent Multiphase Flow"—DOE-NETL*

3. *"Development of a computational fluid dynamic model for large-scale three-phase stirred-tank reactor."* — *SABIC*

Under supervision of Rodney O. Fox, Department of Chemical and Biological Engineering, Iowa State University

PhD Dissertation:

"*Experimental and computational study of turbulent mixing in a confined rectangular reactor*."

Under supervision of Michael G. Olsen and James C. Hill, Department of Mechanical Engineering, Iowa State University

Research Experience:

My academic career path has been a blend of many fields of science and engineering. I am an experimentalist by training. During my Ph.D. years at Iowa State, I performed PIV, LDV, PLIF and reactive PLIF measurements to study turbulent mixing and reacting flows, and accumulated extensive experience on various laser-based fluid diagnostic techniques. After I earned my Ph.D., I worked with Dr. Dennis Vigil and Dr. Jacqueline Shanks as a post-doctoral research associate to develop a novel Taylor-Couette photobioreactor for energy efficient micro-algae cultivation. I also successfully developed a multi-dimensional numerical method for simulating the spectral light distribution in algal photobioreactors based on the characteristics of algae cells, which attracted wide interest from algae bio-refinery community. Furthermore, I designed and conducted various experimental and computational studies on the gas-liquid hydrodynamics and mass transfer in algal bioreactors. Since then, I worked under Dr. Rodney Fox to develop an advanced polydispersed multiphase model based quadrature-based moment method (QBMM) for energy applications, such as bio-mass gasification, which is a project funded by DOE. The model was successfully developed and implemented in both OpenFOAM and MFIX, and validated with Euler-Lagrange simulations. Currently, I am working on advanced gas-solid multiphase turbulence models, which is not well understood or predicted in spite of their critical importance to engineering design and energy policy. What is more, in the last two years, I worked on three industrial projects sponsored by major petrochemical companies, BP, SABIC, and Chevron-Phillips, which helped me accumulate some good experiences in working with industries using open-sourced CFD software. In summary, my background in fundamental research in multiphase flow science, code development, high performance simulation, industrial applications, and experimental studies of fluid systems brought me extraordinary broad and deep knowledge and skills to allow me work closely with people with different background to solve some important and complicated issues of industries and society today.

Teaching Experience:

Aside from my research career, I also have extensive teaching experience. I TAed and guest-lectured undergraduate lecture and laboratory courses in both Mechanical and Chemical Engineering at the Iowa State University; I was repeatedly chosen by my department to be the "Fluent Expert" (duties included lecture undergrads, holding office hours and developing course material); and, lastly, I mentored many graduate students in my group.

Future Direction:

As faculty I would like to continue applying my extensive background experimental and computational fluid dynamics to achieve higher efficiency and sustainability in producing and use of energy, chemicals and water. In particular, there are three areas I am very interested to work on. First, with my experiences on bioreactor design and optimization, I would like to conduct more research on using microalgae to treat wastewater and produce high value chemicals. For example, the wastewater from crude oil desalting process is high in inorganic salts, which can't be removed by using traditional bio-degenerate method, but it can be used to culture high pH and high salt tolerated microalgae or cyanobacteria, such as *Dunaliella* and *Euhalothece* spp. Second, I want to develop a better model for gas liquid mass transfer based on a deeper understanding of this phenomenon. The hybrid model between penetration theory and eddy-cell model we developed is quite ad-hoc, which performs well in weak and highly turbulent flow regimes, but not in the moderately turbulent flow conditions. But in real industrial reactors, the flow conditions are always complex, ranging from laminar to highly turbulent, which we didn't find any published models so far can handle. Thus, I would like to propose a well-thought experiment study to collect information on this issue, gain better fundamental understanding on it, and develop a more accurate and more general gas-liquid mass transfer model. Third, I would like to explore the application of QBMM method to the area of atomization and sprays. Either volume of fluid (VOF) or level-set method can be used to simulate the fuel droplet formation near the nozzle, or then the results can be fed into a variable droplet size model with size-conditioned velocity to model the flow development and combustion further downstream. Overall, I am looking forward to get an opportunity to conduct more interesting researches which can contribute more of my effort to the society.

Selected Publications:

M. Ramezania, B. Kong, X. Gao, M. G. Olsen and R. D. Vigil, ``Experimental measurement of oxygen mass transfer and bubble size distribution in an air-water multiphase Taylor-Couette vortex bioreactor", Chemical Engineering Journal 270 (2015): 508-518.

X. Gao, B. Kong and R. D. Vigil, ``CFD investigation of bubble effects on Taylor-Couette flow patterns in the weakly turbulent vortex regime", Chemical Engineering Journal, vol. 270 pp. 508-518, 2015.

C. Yuan, B. Kong, A. Passalacqua, and R. O. Fox, ``An extended quadrature-based mass-velocity moment model for polydisperse bubbly flows", The Canadian Journal of Chemical Engineering, vol. 92, no. 12, pp. 2053-2066, 2014.

K. M. Nilsen, B. Kong, R. O. Fox, J. C. Hill, and M. G. Olsen, ``Effect of inlet conditions on the accuracy of large eddy simulations of a turbulent rectangular wake", Chemical Engineering Journal, vol. 250, pp. 175-189, Aug. 2014.

B. Kong and R. D. Vigil, ``Simulation of photosynthetically active radiation distribution in algal photobioreactors using a multidimensional spectral radiation model", Bioresource Technology, vol. 158, pp. 141-148, Apr. 2014.

B. Kong and R. D. Vigil, ``Light-limited continuous culture of Chlorella vulgaris in a Taylor vortex reactor", Environ. Prog. Sustainable Energy, vol. 32, no. 4, pp. 884-890, Dec. 2013.

B. Kong, J. V. Shanks, and R. D. Vigil, ``Enhanced algal growth rate in a Taylor vortex reactor", Biotechnology and Bioengineering. vol. 110, no. 8, pp. 2140–2149, Aug. 2013.

B. Kong, M. G. Olsen, R. O. Fox, and J. C. Hill, ``Predictive capability of Large Eddy Simulation for point-wise and spatial turbulence statistics in a confined rectangular jet", Chemical Engineering Science, vol. 69, no. 1, pp. 240-256, Feb. 2012.

B. Kong, M. G. Olsen, R. O. Fox, and J. C. Hill, ``Population, characteristics and kinematics of vortices in a confined rectangular jet with a co-flow", Exp Fluids, vol. 50, no. 6, pp. 1473-1493, Jun. 2011.

438673 Advanced Membrane Materials for Energy Efficient Separations

Sunday, November 8, 2015
Exhibit Hall 1 (Salt Palace Convention Center)
Rajinder P. Singh, Los Alamos National Laboratory, Los Alamos, NM

Ability to precisely control molecular transport in materials is the next paradigm in the separation technology development. Design and implementation of the controlled functionality materials would transformationally improve the energy efficiency of the membrane-based separation processes much needed to meet the future clean energy and water demands. In energy applications, the proposed carbon capture and sequestration approach would tremendously gain from materials having high H_2/CO_2 and CO_2/N_2 perm-selectivities. Such membrane materials would enable an energy efficient CO_2 capture from the industrial gas streams without incurring the huge energy penalty currently estimated for using the industry standard cryogenics and sorption gas separation methods. Exploiting abundant shale gas resources requires next generation high efficiency membranes for the CO_2/CH_4separations, and treatment of the produced high salinity water prior to disposal or re-injection. Similarly, sea water desalination and wastewater treatment will also benefit from the development of advanced membrane materials having high water transport capacity together with anti-fouling, chlorine resistant, and high salt rejection characteristics.

My research is focused on the advanced material design and cutting edge fabrication techniques development aimed at obtaining high performance membranes for the small molecule separations. During my PhD at Colorado School of Mines, I applied molecular self-assembly (MSA) for designing the organic-inorganic hybrid membrane materials. I demonstrated that by manipulating the surface functionality, membrane separation selectivity was tuned from molecular sieving to reverse selective. For my postdoctoral work at Los Alamos National Laboratory (LANL), I systematically altered the molecular structure of the polybenzimidazole (PBI)-based H_2selective membranes and developed an extensive structure-permselectivity correlation for this unique class of the polymer membranes.

Advancements in the selective membrane materials development often require material specific fabrication techniques to enable their successful industrial deployment. I have also developed novel methods and techniques for the membrane material transformation into industrially attractive hollow fiber and thin film platforms. PBI hollow fiber membranes development project, I am currently working on as a Staff Scientist at LANL, is focused on understanding the underlying liquid-liquid demixing phenomena for the PBI phase inversion used in the fiber spinning process. Recently, ground-breaking spinning methods were developed to achieve an unprecedented H_2/CO_2 separation performance at synthesis gas process conditions. These PBI fibers are very attractive for efficient H_2production from coal, natural gas or biomass for the fuels, power and chemicals production. As a LANL Co-PI for DOE ARPAe funded project, I have applied ultrasonic atomization assisted polymer deposition technique for thin (ca. ~100 nm) selective layer deposition on the porous polymer substrates. Reducing selective layer thickness directly improves the membrane permeance, resulting in reduced membrane surface area requirements. Therefore, this novel selective layer deposition technique enables the use of expensive high functionality materials for the large scale applications.

My future research work will be focused on the design and development of advanced materials and deployment platforms for next generation membrane-based separation processes. I plan to harness the extraordinary separation characteristics of the recently identified 2D and 3D materials including GOs, MOFs, ZIFs, ionic liquids, planar zeolites and polymers with intrinsic micro-porosity (PIMs) for gas, vapor and liquid separations, diffusion and reaction. It is difficult to process these materials into the desired robust thin film platforms due to limited control on the grain and crystal size, defect density and morphology during their synthesis. I would explore novel nano-composite, hybrid and MSA material fabrication design routes together with advanced membrane fabrication techniques to transform these materials into industrially attractive membranes. I will develop a comprehensive understanding for controlling and effectively manipulating the interfacial, chemical and molecular interactions among membrane material constituents for molecular scale mixing to achieve the desired high separation performance.

438674 Controlling the Metabolic Activity of Bacterial Cells By Physico-Chemical Factors

Abstract:

Bacterial colonization on medical and environmental surfaces is a common phenomenon encountered in various settings. During an infection, for instance, bacterial cells could withstand the challenge of shear stress along the blood vessels, and electrostatic discharges of the medical implants in order to develop mature aggregates, known as biofilms. Similar behaviors are exhibited by oil-degrading bacteria, which can overcome strong interfacial energies to establish films of bacteria at oil-water interfaces. Thus, the understanding of physiological response of bacteria to physical forces is essential to elaborate effective strategies that can prevent the proliferation of bacteria on medical and industrial surfaces as well as alleviate the persistent ecological impacts of oil spills. The premise underlying my research plan is to elucidate the growth mechanism of bateria as function of physicochemical properties of interfaces and design sustainable means for controlling the cells. My previous focus on the electrophysiology of highly drug-tolerant bacterial cells, known as persister cells, has allowed me to develop a mode of sensitization of *P. aeruginosa* PAO1 persisters to antibiotics using electrochemical currents. The discovery that electrochemical treatments disrupted the membrane property, and caused profound physiological damage in persister cells contributed to the design of a more effective way of eradicating the persistent infections *in vitro* and *in vivo* using safe electrochemical factors. My current focus in biotechnology and soft nanomaterials allows me to develop a new research program utilizing the physical challenges as a means for controlling desired physiological response and directing metabolic activity of cell towards creation of new materials and bioproducts.

438683 Development of Calcium-Based Liquid Metal Batteries for Grid Scale Energy Storage

Sunday, November 8, 2015
Exhibit Hall 1 (Salt Palace Convention Center)
Takanari Ouchi, Massachusetts Institute of Technology, Cambridge, MA

A liquid metal battery (LMB) is well-positioned to satisfy the demands of large-scale electrochemical energy storage devices due to its ability to obviate classic solid state failure mechanisms and do so with earth abundant materials and easily scalable means of construction. As the negative electrode of LMBs, calcium is an attractive material due to its low electronegativity (consequent high cell voltage) and earth abundance, and its resulting low electrode material costs. However, most researchers have taken Ca off the list for rechargeable batteries due to its high melting temperature and high reactivity combined with its high solubility in molten salt electrolyte, which causes an unacceptably high self-discharge current and poor electrochemical performance. To address these challenges, I initiated a single-electrode study in a three-electrode electrochemical cell to evaluate the electrode performance of Ca-based electrodes including the solubility of Ca into the molten salt. Based on the electrode performance of the single electrode, Ca-based liquid metal batteries were successfully designed.

(6bh) Rational Materials Design for Energy Conversion and Storage Applications

Shen, J., Northwestern University

I am interested in faculty positions in the field of material design for energy conversion and storage applications. My research direction is in the realm of inorganic/materials chemistry, and related to energy conversion and storage systems. I have worked with homogeneous complex catalysts for ethylene oligomerization/polymerization (my graduate work in University of Delaware), rational designed heterogeneous catalysts using synthesized polymer (my postdoc work in Northwestern U), materials for Li-batteries (postdoc work in Northwestern U and GM), and metal-organic frameworks (MOFs) for natural gas storage (postdoc work in GM). I have always been trying to find connections between small molecules chemistry and complex materials chemistry. I am currently working on applying MOFs in several energy related systems, such as battery and natural gas storage. When I worked in Northwestern U, by applying the knowledge I have earned from homogeneous catalysts to complex heterogeneous catalysts design, I've studied the physical properties, and catalysis performance of single cite Co ions confined in a Nano space by various techniques, including Infrared spectroscopy (IR), Gas chromatography (GC), X-ray photoelectron spectroscopy (XPS) and so on. I've been able to state that the nano confinement effect can stabilize metal ions with rare oxidation state, but does not show much in the catalysis performance. The most important finding in this work is that traditional small molecule characterization techniques can in fact be employed. This sets the stage for my own successful research agenda in the future which I anticipate will be at the catalysis design and complex materials and will provide valuable training for undergraduates and graduate students. My teaching experience to date includes general chemistry labs, and thus centers on helping undergraduate students to learn chemistry in a more practical, hands-on way. I am interested in gaining more teaching experience.

438689 Engineering of the Corneal Epithelium

Sunday, November 8, 2015
Exhibit Hall 1 (Salt Palace Convention Center)
Bernardo Yáñez Soto, Instituto de Fisica, Universidad Autonoma de San Luis Potosi, San Luis Potosi, SLP, Mexico

Every year, two million people worldwide suffer from corneal blindness. While the traditional therapy (the use of corneal transplants) is highly successful (72% retention after five years), the number of donors is insufficient, especially on developing countries. Besides this, corneal transplants are not suitable for some pathologies such as dry eye syndrome or alkali burns. These facts highlight the need to develop suitable artificial corneas. However, the artificial corneas currently on the market suffer from integration problems such as extrusion, infection, leakage or epithelial downgrowth. In general, these devices address the integration to the host only through the stroma of the cornea (the middle portion of the cornea), therefore we propose that their success rate would improve if we promote the formation and maintenance of a healthy corneal epithelium, which is a highly specialized stratified tissue with a very high regenerating potential that covers the cornea and acts as a barrier to protect from mechanical injury and infection.

The healthy functioning of the corneal epithelium depends mainly on three attributes:

1) Barrier properties, dependent on the formation of tight junctions between cells. The corneal epithelium must have impermeable characteristics to prevent the hydration of the stroma (corneal edema) and the penetration of pathogens.

2) Adherence of the basal cells to the underlying basement membrane. The basement membrane of the corneal epithelium is a laminar structure that provides desmosomal attachments to the corneal epithelium. The physicochemical characteristics of the basement membrane of the cornea have been demonstrated to affect the adhesion, shape, migration, differentiation and proliferation of the corneal epithelial cells.

3) Hydration. The cornea has the unique characteristic of being an avascular tissue. The nourishment of the corneal epithelium is achieved mainly through the adequate formation of a tear film covering the ocular surface.

The epithelialization of corneal prosthetics must replicate the process of epithelial wound healing, which occurs in four steps: 1) Lag phase, where cells reorganize their attachments to adjacent cells and basement membrane; 2) Migration phase, where the leading edge of the wound migrates toward the center of the defect; 3) Differentiation phase, where epithelial cells proliferate and stratify, and 4) Maintenance phase, where basement membrane components are synthesized, assembled and remodeled. Since the basement membrane of the corneal epithelium is the microenvironment providing cues to the epithelial cells, it is fundamental to have a surface that supports cells adhesion, is permeable to nutrients and metabolites, and presents physical and biochemical cues to the cells. Therefore, we propose that by engineering an artificial basement membrane of the corneal epithelium we can promote the epithelialization of corneal prosthetics.

The design of an artificial basement membrane must incorporate three different types of signals: 1) Biochemical: ligand-receptor interactions; 2) Mechanical: cells respond to the stiffness the substrate by modifying the cytoskeletal elements, and 3) Topographical: The 3-D topography of the extracellular matrix affect the behavior of the cells. We developed a hydrogel substrate with controlled compliance that can be functionalized with specific biomolecules. This substrate can be molded with topographic features ranging from the micro to the nanoscale that can mimic the rich topography found in the native basement membrane.

By using nonfouling hydrogel substrates, such as poly (ethylene glycol), functionalized with specific molecules such as the cell-adhesive peptide RGD, we were able to successfully isolate the confounding influence of nonspecific protein adsorption from the medium onto the surfaces from other soluble factors and provided evidence that nanoscale topography is a fundamental biomimetic cue impacting cell phenotype. Furthermore, our biomimetic topographic substrates improved the rates of healing due to increased migration of the wound border towards the center of an *in vitro* wound. These substrates may be fabricated as laminates and be incorporated to the surface of artificial corneas to increase their success rate.

However, as mentioned earlier, the maintenance of a healthy corneal epithelium depends on the nourishment of the corneal epithelial cells through the aqueous tears; and some pathologies of the ocular surface such as the dry eye syndrome impedes the formation of a film that hydrates the epithelial cells.

Dry eye syndromes are a heterogeneous group of pathologies of the ocular surface, affecting up to 20% of the adult population. In spite of its name, dry eye are not exclusively due to drying or lack of aqueous tears, but also to insufficient wettability of the epithelial surface.

The tear film is a stratified structure, forming three principal phases: mucous phase, aqueous phase and lipid phase. Traditionally, the treatment of dry eye is focused on replacing the aqueous tears with the use of eye drops or reducing the evaporation by treating the lipid layer of the tears. However, the interaction between the different components and phases of the tear film is largely understudied. We hypothesize that those surface properties are relevant for the stability of the tear film and propose the engineering of the epithelial surface to modify its wetting properties and improve the retention of the fluid on the ocular surface.

The wettability of the ocular surface has traditionally been attributed to the existence of a thick glycocalyx covering the cell surface, composed mainly of membrane-associated mucins and other glycosylated proteins. However, the dependence of the surface properties of the corneal epithelium on the expression of membrane-associated mucins has never been characterized. We have investigated the influence of the glycocalyx on the surface properties of the ocular epithelium by inducing the expression of mucins on the apical surface of corneal epithelial cells. We measured the contact angle of the cell surfaces using a two fluid method, with phosphate buffer saline as the medium and perfluorocarbons (perfluorodecalin, perfluorooctane or tetradecafluorohexane) as the testing droplets, and found a significant increase of the contact angle hysteresis when ocular mucins were expressed on the surface. This contact angle hysteresis is most likely due to the differential expression of mucins on the cell surface (chemical heterogeneity), where some cells show low expression of surface glycans and other cells show high surface glycosylation. This high contact angle hysteresis is closely related to adhesion of the droplets onto the surface, impeding the growth of dry patches and may be key for the stability of the tear film.

Although our current aim is to improve the outcomes of artificial corneas through the application of chemical engineering principles; the general approaches and efforts used in our research may also be used to enhance the integration of other medical devices, improve the outcomes of wound healing, or develop novel therapies to treat dry mucosal syndromes.

438691 From Vapors to Films: Creating Smart Surfaces Via Vapor-Phase Depositions

Abstract:

From Vapors to Films: Creating Smart Surfaces *via* Vapor-Phase Depositions

Do Han Kim

Postdoctoral Associate

Chemical Engineering

Massachusetts Institute of Technology

Research Interests:

In techniques of vapor-phase depositions, multiple vapor-phase precursors are introduced into a vacuum chamber and react on a surface, generating thin solid films. In particular, initiated Chemical Vapor Deposition (iCVD) and Atomic Layer Deposition (ALD) are powerful techniques to deposit ultra-thin polymeric and inorganic films, respectively. In contrast to conventional solution-based deposition such as spin-casting, the key features of iCVD and ALD facilitate exceptionally conformal, defect-free coatings even on nano-porous structures and the precise thickness control in a molecular scale. In addition, the film properties are easily tuned by adjusting the feed gases and processing conditions. Therefore, those depositions open up the real nanoscale level which was even inaccessible in the past. As such, vapor-phase depositions satisfy strong demands for emerging multi-disciplinary fields in a nanoscale level; for example, energy, biotechnology, and electronics. In my research for Ph.D. and Postdoc., iCVD polymeric films are used to fabricate sub 10 nm patterns in conjunction with directed self-assembly (DSA) as a means to replace conventional photolithography. ALD films were also extensively exploited to engineer the interfaces inside dye-sensitized solar cells (DSSCs) and photoelectrochemical cells (PECs) for the efficient charge transfer kinetics.

Postdoctoral Project:

"Scalable Nano-manufacturing for Sub-10 nm Patterns Using initiated Chemical Vapor Deposition (iCVD) and Directed Self-Assembly (DSA)" under supervision of Karen K. Gleason, Chemical Engineering, Massachusetts Institute of Technology

Ph.D. Dissertation:

"Atomic Layer Deposition for Dye-Sensitized Solar Cells." Under Supervision of Gregory N. Parsons, Chemical and Biomolecular Engineering, North Carolina State University

Research Plans:

I would like to continue my research on vapor deposition techniques for polymer and inorganic films. Vapor-phase depositions, such as iCVD and ALD capable of coating ultra-thin, uniform, and pinhole-free films at low temperature, go beyond solution-based deposition for fundamental studies on surface chemistry as well as for high-performance catalysis, chemical sensing, energy, and bio- applications. For my future research as faculty, I conceive a novel and innovative deposition technique to create organic-inorganic hybrid films enabling to control the thickness and tailor the composition in a molecular level. The approach employing key features of iCVD and ALD facilitates to control the properties between two original phases or to bring up new properties with the precise thickness control on complex structures such as membranes, nanotubes, and particles. The applications of these materials in the fields of energy, optics, mechanics, biology, electronics and others are expected.

Teaching Experience:

I have had experiences as a TA for several years in experimental and lecturing classes related to core-classes of chemical engineering; process control, kinetics, transport phenomena lab. These experiences gave me insight on how teachers should encourage and involve students into understanding core and challenging classes of chemical engineering.

As an effort to be an effective teacher, I took intensive teaching classes and workshops during my Ph.D. program and post-doctoral training. Recently, I completed an intensive class aimed at improving teaching skill, provided by the MIT Teaching and Learning laboratory. I have mentored one undergraduate and two graduates so far, who successfully participated to publish two papers during doctoral and postdoctoral programs.

As outreach activities, I had the opportunity to participate in an annual state-wide outreach, 'Nanodays' coordinated by North Carolina State University, in which I demonstrated representative my results and samples of my research. To encourage K-12 students, I also volunteered in the National Mathematics and Science Competition organized by KSEA (Korean-American Scientists Engineers Association).

Patents

1. P. Hoertz, G. Parsons, Q. Peng, J. Liu, J. Glass, B. Kalanyan, D. H. Kim, M.D. Losego, L. Alibabaei, T. J. Meyer "Advanced semiconductor-conductor composite particle structures for solar energy conversion" U.S. patent, US20140261646 A1, 2014.

2. K. G. Henson, T. J. Meyer, G. N. Parsons, M. D. Losego, B. Kanlanyan, and D. H. Kim "Assemblies and methods of stabilization", U.S. patent, WO2014081921 A1, 2014.

Selected Publications

1. N. Chen*, D. H. Kim*, P. Kovaick, M. Wang, H. Sojoudi, and K. K. Gleason, "CVD polymeric thin films for surface modification and device fabrication: recent progress" (2015) (submitted) *equally contributed

2. D. H. Kim, H. S. Suh, P. Moni., S. Xiong, L. C. Ocola, P. Nealey, and K. K. Gleason, " iCVD Top-coat for sub-10 nm patterning with directed self-assembly of block copolymer film" (2015) (in preparation)

3. B. Reeja-Jayan, P. Kovacik, R. Yang, H. Sojoudi, A. U. Katmis, D. H. Kim, C. D. Petruczok, X. Wang, A. Liu, and K. K. Gleason, "A Route towards sustainability through engineered polymeric interfaces", *Advanced Materials Interfaces*, 1, 1400117 (2014). (featured on frontispiece)

4. D. H. Kim, Q. Peng, M. D. Losego, and G. N. Parsons, "Atomic layer deposition for sensitized solar cells" (2015). (submitted)

5. D. H. Kim, S. E. Atanasov, P. Lemaire, K. Lee, and G. N. Parsons, "Platimum-free cathode for dye-sensitized solar cells using poly(3,4-ethylenedioxythiophene) (PEDOT) formed via oxidative molecular layer deposition" (2015), ACS Applied Materials & Interfaces, 17, 3866 (2015)

6. D. H. Kim, M. D. Losego, K. Hanson, L. Alibabaei, K. Lee, T. J. Meyer, and G. N. Parsons, "Stabilizing chromophores for dye-sensitized solar cells using multicomponent, sub-nanometer atomic layer deposition", Physical Chemistry Chemical Physics, 16 (18), 8615 (2014).

7. D. H. Kim, M. Woodroof, K. Lee, and G. N. Parsons, "Atomic layer deposition of high performance ultra-thin (< 10 nm) TiO_2 blocking layers for dye-sensitized solar cells", ChemSusChem, 6 (6), 1014, (2013). (featured on front cover)

438692 Enabling Technologies for High-Throughput Synthetic Biology and Metabolic Engineering: From Engineering Genomes and Pathways to Genetic Circuits

Abstract:

Cells can be programmed to dynamically respond to their surroundings and selectively manufacture useful biomolecules using the tools of synthetic biology and metabolic engineering. However, engineering these intricate biological systems remains challenging due to time-consuming DNA construction, toxicity of biomolecules, and a lack of algorithms to guide genetic design. I have developed technologies to address these limitations that leverage next-generation sequencing, directed genome-wide mapping, and statistical design algorithms. In combination, these approaches allow for model-guided optimization of massively multipart genetic systems and complex phenotypes. Additionally, these novel methods enabled approximately 1,000-fold more genetic variants to be built over previous approaches and reduced the engineering cycle from a month or more to a week. I have applied these techniques to engineer bacteria to perform a variety of tasks including: manufacture biofuel to high titer with improved productivity, acquire the ability to fix atmospheric nitrogen towards more sustainable agriculture, and perform complex digital logic circuitry in response to multiple chemical inputs. These technologies pave the way for a new scale in synthetic biology and provide a platform to rapidly engineer functional biological systems for new pharmaceutical and industrial applications.

441866 Layered Double Hydroxides as Anion Intercalation Electrodes for Battery-Inspired Water Desalination

Abstract:

Layered double hydroxides (LDHs) are a class of inorganic anion exchange materials with atomic structures similar to naturally occurring clays. Here we study LDH nanoparticles as a model for electrochemically cycleable anion intercalation materials. This work aims to build a fundamental basis for understanding anion intercalation electrochemistry in LDHs through a combined experimental and computation approach to provide direction to the development of cheaper, more efficient, and longer-lifetime materials for anion extraction in battery-inspired water desalination.

LDH nanoparticles are synthesized by coprecipitation followed by hydrothermal treatment, and studied electrochemically in electrolytes containing halide anions. Equilibrium potentials for anion insertion/extraction in the LDH are evaluated theoretically using ab initio quantum mechanical calculations. The combination of these experimental and theoretical techniques provides a mechanistic picture for anion intercalation based on the electronic properties of the anion host material. Future work will focus on experimentally observing insertion and extraction of anions in the LDH nanoparticles during electrochemical cycling using in situ high energy X-ray diffraction (HE-XRD) and X-ray absorption fine-structure spectroscopy (XAFS).

441966 Colloid and Nanoparticle Growth and Assembly Dynamics for Materials Design and Molecular Analogues

Abstract:

Research Interests: Colloids have the unparalleled ability to assemble into hierarchical structures and act as molecular analogues due to the multitude of possible particle-particle interactions, ranging from electrostatic to hydrodynamic forces. Approaches for assembling colloidal building blocks are often empirical in nature, especially for nanoscale colloids in which the assembly dynamics are not easily observed. Direct visualization of assembly dynamics for micron scale colloids leads to phenomenological understanding of the dominant physical mechanisms. *In situ* electron microscopy allows for similar real-time visualization of nanoscale systems in liquid, and promises to yield a depth of understanding to nanoparticle assembly as has been established in micron scale systems.

Successful Proposals: NRC Research Associateship Program Postdoctoral Fellowship

Postdoctoral Projects: "*Imaging Biomimetic Magnetite Nanocrystals in Live Magnetotactic Bacteria and Protein Templates with Liquid Cell Electron Microscopy*"

Under supervision of Tanya Prozorov, Division of Materials Science and Engineering, Ames Laboratory

"New Electron and Ion Transmission Techniques for Nanomaterial Characterization in and out of Liquid Environments"

Under supervision of Robert Keller, Advanced Chemicals and Materials Division, National Institute of Standards and Technology

Ph.D. Dissertation: *"Direct Observations of Colloidal and Nanoparticle Behavior in the Presence of External Fields"*

***Winner of 2014 Zuhair A. Munir award for best dissertation in the College of Engineering**

Under the supervision of William D. Ristenpart and Nigel D. Browning at University of California, Davis, Department of Chemical Engineering and Materials Science

Research Experience: My research experience has been multidisciplinary but has always followed the common theme of applying chemical engineering principles to materials science problems. My formal Ph.D. training was in transport, fluid mechanics, and colloidal science, but through my joint adviser in Materials Science I was introduced to the field of *in situ* electron microscopy. Optical video microscopy experiments for observing and modelling electrokinetically induced micron scale colloidal assembly motivated me to apply similar approaches to nanoscale systems using *in situ* electron microscopy. Using this technique, I developed electron video microscopy techniques to observe colloidal nanoparticle growth, aggregation, and self-assembly, and modelled these observations using kinetic and scaling analyses inspired by my transport and fluid mechanics background. Through intensive training at several national laboratories, I have become proficient in several high resolution electron and ion microscopy techniques, which I continue to apply to reveal and explain hidden nanoscale transport processes.

Teaching Experience: During my undergraduate degree I was a peer-group tutor for engineering physics, a program that emphasized a Socratic method of teaching where tutors helped students answer problems by asking critical questions. In graduate school I TAed several laboratory courses, which involved teaching students experimental techniques, lecturing on course material, and critiquing lab reports and oral presentations. Additionally, I mentored several undergraduate researchers in graduate school as well as one graduate student during my postdoctoral studies. I plan to continue these teaching themes as a faculty member, for example using experimental demonstrations to visualize transport phenomena.

Future Direction: As faculty I plan to continue applying electron and optical video microscopy to different areas of nanoparticle and colloidal assembly. In particular I propose to apply *in situ* electron microscopy to an unsolved mystery I addressed during my Ph.D.: electric field induced aggregation of nanoparticles near electrodes. While many researchers suggest this occurs due to induced charge electrokinetic fluid flows, there are no direct observations of the aggregation dynamics. *In situ* electron microscopy is an ideal technique for visualizing this phenomenon, which once understood could be exploited to form self-assembled superstructures with tunable order and material properties across many length scales. I plan to continue my work on electrokinetic manipulation of micron scale colloidal particles near electrodes, specifically developing new colloidal systems with analogous potential energy landscapes to atomic and molecular systems, such as superconductivity and surface sorption.

Besides assembly and growth of colloids and nanoparticles, my research interests include biomimetics and developing new electron and ion microscopy techniques with new contrast mechanisms. In the future I foresee myself developing *in situ* electron microscopy techniques capable of imaging solvated ions in liquids, with applications for observing electric double layers and ion binding in biomimetic matrices.

Selected Publications:

T.J. Woehl and T. Prozorov, The Mechanisms for Nanoparticle Surface Diffusion and Chain Self-Assembly Determined from Real-Time Nanoscale Kinetics in Liquid. *J. Phys. Chem. C*, **2015**, 119, 21261-21269.

T.J. Woehl, B.J. Chen, K.L. Heatley, N.H. Talken, S.C. Bukosky, C.S. Dutcher, W.D. Ristenpart, Bifurcation in the equilibrium height of colloidal particles near an electrode in oscillatory electric fields: Evidence for a tertiary potential minimum. *Phys. Rev. X*, **2015**, 5, 011023.

T.J. Woehl, S. Kashyap, E. Firlar, T. Perez-Gonzalez, D. Faivre, D. Trubitsyn, D.A. Bazylinski, T. Prozorov, Correlative Electron and Fluorescence Microscopy of Magnetotactic Bacteria in Liquid: Toward In Vivo Imaging. *Sci. Rep.*, **2014**, 4.

P. Abellan, **T.J. Woehl**, L.R. Parent, N.D. Browning, I. Arslan, Factors controlling quantitative liquid (scanning) transmission electron microscopy. *Chem. Comm.* **2014**, 50, 4873-4880.

***Featured on the back cover**

T.J. Woehl, C. Park, J.E. Evans, I. Arslan, W.D. Ristenpart, N.D. Browning, Direct Observation of Aggregative Nanoparticle Growth: Kinetic Modeling of the Size Distribution and Growth Rate. *Nano Lett.* **2013**, 14, 373-378.

C.S. Dutcher†, **T.J. Woehl**†, N.H. Talken, W.D. Ristenpart, Hexatic-to-Disorder Transition in Colloidal Crystals Near Electrodes: Rapid Annealing of Polycrystalline Domains. *Phys. Rev. Lett.* **2013**, 6, 128302.

†Equal author contribution

T.J. Woehl, J.E. Evans, I. Arslan, W.D. Ristenpart, N.D. Browning, Direct *in situ* determination of the mechanisms controlling nanoparticle nucleation and growth. *ACS Nano* **2012**, 6, 8599-8610.

***Featured on weekly ACS Nano podcast**

Book Chapters:

T.J. Woehl and T. Prozorov, Future prospects for biomolecular, biomimetic, and biomaterials research enabled by new liquid cell electron microscopy techniques. In *Liquid Cell Electron Microscopy*, edited by Frances Ross, **2015**.

442145 Fundamental Modeling of Gas-Solid and Granular Flows

Sunday, November 8, 2015
Exhibit Hall 1 (Salt Palace Convention Center)

Aaron Morris, Chemical and Biological Engineering, The University of Colorado Boulder, Boulder, CO

Research Interests

My research will use and develop fundamental models to better understand and predict gas-solid and granular multi-phase flows, which are pervasive in nature and industry. Simulation tools are gaining traction in industry as companies aim to decrease R&D costs, but require further model validation and development for confidence in the accuracy of the predictions. Fundamental modeling is necessary to develop tools that can be applied to a large range of disciplines, as particle flows often exhibit complex phenomena that cause empirical correlations to be unreliable as a general modeling tool. My research can be applied to many fields, examples of which include: optimizing heat transfer to flowing particles for emerging energy technologies, predicting particle segregation and elutriation in fluidized bed reactors to improve safety and performance, and modeling the role that particle shape has on fluidization behavior for biomass particles.

Successful Proposals

"MFIX DEM Enhancement for Industry-Relevant Flows"

Role: co-author
Collaborators: Prof. Hrenya, Univ. Colorado (PI), Dr. Cocco, PSRI (co-PI), Dr. Grout, NREL (co-PI), and Dr. Hauser, Univ. Colorado (co-PI)
Funding Agency: U.S. Department of Energy, National Energy Technology Laboratory
Amount: $827,421
Dates: Aug 2015 – Feb 2017

"Simulating Multiphase Heat Transfer in a Novel Receiver for Concentrating Solar Power (CSP) Plants"

Role: co-PI
Collaborators: Prof. Hrenya, Univ. Colorado (PI), Dr. Pannala, ORNL (co-PI)
Funding Agency: U.S. Department of Energy, Office of Science
Amount: 5M Titan-core hours at OLCF ($166,667 equivalent)
Dates: Jul 2015 – Jun 2016

"Heat Transfer in Granular Flows: Understanding Similarities and Differences with Molecular Fluids"

Role: co-PI
Collaborators: Prof. Christine Hrenya, Univ. Colorado (PI)
Funding Agency: National Science Foundation – Thermal Transport Processes
Amount: $320,000
Dates: May 2015 – Apr 2018

"Simulating Multiphase Heat Transfer in a Novel Receiver for Concentrating Solar Power (CSP) Plants"

Role: co-PI
Collaborators: Prof. Hrenya, Univ. Colorado (PI), Dr. Pannala, ORNL (co-PI)
Funding Agency: Department of Energy Office of Science
Amount: 15M Titan-core hours at OLCF ($500,000 equivalent)
Dates: Jul 2014 – Jun 2015

Research Experience

I have used many simulation methods to study fluid dynamics in a wide range of disciplines. In my current postdoctoral work at the National Energy Technology Laboratory (July 2015 – present), I use simulations to aid development of fluidized bed reactors that will process up to 900,000 gallons of sodium bearing nuclear waste. The large volume of nuclear waste is reduced by dual fluidized-bed steam reformers to a smaller amount of waste material that can be safely stored. The simulations are used to "look" inside each vessel to identify regions of attrition, channeling, or poor mixing that can adversely affect system performance.

From March 2013 – July 2015, I worked with Prof. Hrenya at The University of Colorado as a postdoctoral researcher. My project studied how solid particles perform as a heat transfer fluid for use in concentrating solar power. I performed discrete element simulations to test the validity of previous continuum heat transfer models and found that the previous models failed to yield accurate predictions. After identifying the shortcomings of previous models, I then developed a new fundamental continuum heat transfer model that was valid over a wide range of solar receiver configurations.

My PhD and MS training was done in the aerospace engineering department at The University of Texas at Austin, under the guidance of Professors David Goldstein and Philip Varghese. I studied how the rocket exhaust plume from a lunar lander interacts with the surface, eroding and entraining lunar regolith into a dangerous high velocity spray, where particle speeds can exceed 1 km/s. The direct simulation Monte Carlo method was used to simulate the rocket plume and the entrainment of dust for various landing scenarios. The maximum particle speeds and trajectories were predicted for single- and multi-engine landers, which will help ensure the safety of future landers and potential lunar outposts.

Teaching Experience

During graduate school, I worked as a teaching assistant for two semesters, guiding undergraduates through a low-speed fluid dynamics laboratory. I also guest-lectured for an undergraduate fluid dynamics course on multiple occasions.

During my postdoctoral appointment at The University of Colorado, I mentored two undergraduate workers through the undergraduate research program, one on a project to study hydrodynamic instabilities for falling particles, and another doing simulations. I guided one student as he prepared a poster for the Rocky Mountain Advanced Computing Consortium, where he earned third place in the student poster competition.

Sample Publications

Morris, A., Pannala, S., Ma, Z., and Hrenya, C., (2015) "A conductive heat transfer model for particle flows over immersed surfaces," *International Journal of Heat and Mass Transfer*, Vol. 89, pp 1277-1289.

Morris, A., Goldstein, D., Varghese, P., and Trafton, L., (2015) "Approach for modeling rocket plume impingement and dust dispersal on the Moon," *Journal of Spacecraft and Rockets*, Vol. 52, pp 362-374.

Morris, A., Varghese, P., and Goldstein, D., (2011) "Monte Carlo solution of the Boltzmann equation via a discrete velocity model," *Journal of Computational Physics*, Vol. 230, pp 1265-1280.

442146 Innovative applications in tissue engineering and regeneration

Sunday, November 8, 2015
Exhibit Hall 1 (Salt Palace Convention Center)

Sumati Sundaram, Department of Biomedical Engineering, Yale University, New Haven, CT

SUMATI SUNDARAM, PhD
Yale University, Department of Biomedical Engineering, New Haven, CT 06510
(c) 908-217-9631 (e) sumati.sundaram@yale.edu

RESEARCH OVERVIEW:

Over the past 5--6 years, I have been striving to work at the intersection of two very exciting fields - stem cells and tissue engineering; as I believe together they have the potential to transfor m medicine by providing tailored therapies and custom tissue replacements. As a postdoc toral scientist at Yale University, I have worked on various aspects of stem cell differenti ation, focused on understanding developmental aspects of lineage specification, formulating strategies for the derivation of cells in vitro, and designing appr oaches to utilize them effectively in engineering applications. A key focus of my work ha s been in the area of vascular tissue engineering, where I have been successful in creating blood vessel mimics using human induced pluripotent stem (iPS) cells. Currently as a Research Scientist in the Niklason laboratory, I have had the opportu nity to engage in a wide range of on-going projects including lung tissue engineering, creating artificial tracheal replacements and esophageal engineering working with surgeons, biologists, chemists and engineers. Overall, these experiences have provided me with a strong foundation and exposure to a wide breadth of unique applications in which tissues can be regenerated in innovative ways. My immediate goal is to utilize my expertise in stem cells along with my engineering training to creating physiological models of disease ex vivo that will enable the design of personalized therapies. I propose to do this by gaining a critical understanding of the biochemical cues, and mechanical interactions with the matrix environment that enable these cells to organize into distinct sophisticated functional structures. Alongside this initiative, I am also interested in utilizing my Chemical Engineering background to design reactors and optimize processes to accelerate the translation of laboratory scale cell and tissue regeneration experiments.

EDUCATION:

Rutgers University, USA, PhD Chemical and Biochemical Engineering Dec 2007
Institute of Chemical Technology, India, B Chem Eng, Chemical Engineering June 2000

RESEARCH/WORK EXPERIENCE:

• Yale University, Department of Biomedical Engineering Advisor: Dr. Laura Niklason

Associate Research Scientist 2013-Present

Postdoctoral Research Associate 2008-2013

• Rutgers University, Department of Chemical Engineering, Advisor: Dr. Charles Roth
Graduate Research Assistant 2001-2007

SELECTED PUBLICATIONS:

1.
Zhao L*, Sundaram S*, Huang AH, Zhang J, Caty, MG, Yi T, Le A, Leiby K, Rocco K, Urbani L, Maghsoudlou P, Burns A, De Coppi P, Niklason L, Engineered tissue-tent biocomposites as tracheal replacements. * Zhao L and Sundaram S contributed equally to this work (Submitted to Biomaterials)

2.
Sundaram S, One J, Siewert J, Teodosescu S, Zhao L, Dimitrievska S, Qian H, Huang AH, Niklason L. Tissue--engineered vascular grafts created from human induced pluripotent stem cells. Stem Cells Transl Med. 2014 Dec;3(12):1535-43

3. Sundaram S, Echter A, Qiu C, Niklason L. Small--diameter vascular graft engineered using human embryonic stem cell--derived mesenchymal cells. Tissue Eng Part A. 2014 Feb;20(3-4):740-50

4.
Sundaram S, Niklason L. Smooth muscle and other cell sources for human blood vessel engineering. Cells Tissues Organs. 2012; 195(1-2): 15-25

5.
Sundaram S, Roth CM. Interplay of polyethyleneimine molecular weight and oligonucleotide chemistry in the dynamics of antisense activity. Nucleic Acids Research 2007; 35(13): 4396-4408

AWARDS/HONORS:

- Two time recipient of Young Investigator Funding Award by Connecticut Innovations and CT Stem Cell Foundation, $200,000, 2013-2015, 2011-2013
- Travel award to attend Gordon Conference on Lung Tissue Engineering

442365 High energy batteries: materials design and optical diagnostic tool development

Abstract:

Future Research Directions:

My research interests lie at the interfaces of nanomaterials, electrochemistry, and optics, in the big context of energy, environment, and health. My approach will be the combination of (i) understanding of fundamental materials problems featuring the use of advanced optical techniques, and (ii) engineering materials and solve the problems featuring the use of nanotechnologies. Topics of interests include high energy batteries, electrochemical water splitting, electrochemical CO_2 reduction & sequestration, water desalination and air filtering. Bio-related sciences and technologies will be pursued in the long-term.

Awards:

2015 Division of Inorganic Chemistry Young Investigator Award (American Chemical Society)

2014 Daniel Cubicciotti Award (The Electrochemical Society)

2014 Chinese Government Award for Outstanding Students Abroad (China Scholarship Council)

Postdoctoral Project: “Live imaging of lithium batteries using advanced optical techniques”.

Advisors: Steven Chu (Department of Physics) and Yi Cui (Department of Materials Science & Engineering), Stanford University

PhD Dissertation: "Structure design of silicon anodes for high energy lithium batteries".

Advisor: Yi Cui, Department of Materials Science & Engineering, Stanford University

Research Experience:

My academic career path has been a blend of many fields of science and engineering. My PhD training was in nanomaterials and electrochemistry in the context of high energy lithium-ion batteries, during which I identified the fundamental materials problems, designed & fabricated new materials, and improved the performance. This powerful strategy can be extended to broader contexts and materials systems.

My postdoc training with Steven Chu added two new dimensions, optics and biology, to my research experience. Optical microscopy has been heavily used in biology but very little in materials science, mainly due to its lower resolution than electron microscopy. However, because of its ambient operation, low sample damage, and different contrast mechanism, optical microscopy is ideal for studying complex and delicate materials system under real time, real environment, without radiation damage. This tool becomes more powerful integrating recent advances such as superresolution microscopy, nonlinear optics, and single molecule techniques. Using optical techniques, I have made new discoveries on batteries and other electrochemical systems (two manuscripts in preparation). During postdoc, my eyes were also opened to biology, to see tremendous opportunities of using nanomaterials to solve problems in biotechnology and biomedicine.

Teaching Experience:

Aside from my research career, I also have extensive experience teaching Stanford undergrads and graduates. From 2013 to 2015, I have been invited three consecutive years to give guest lectures in an introductory Stanford freshman class on nanoscience and nanotechnology. Even though my PhD is in Chemistry, I've guest-lectured and TAed (duties included making & grading problem sets, and holding office hours) two graduate-level classes at Stanford School of Engineering. One class was on nanotechnology, the other on batteries. In addition to lecture classes, I have TAed three undergrad lab classes in Stanford Department of Chemistry, which I received 4.6 points out of 5.0 in students' evaluation. I have taken some Chemical Engineering core classes in undergrad and am willing to study and teach any existing classes, as well as design new classes. Lastly, I mentored graduate students in my group.

Selected Publications (50 journal articles in total. Citations: 3773. H-index 25):

(* Denotes equal contribution)

1. **Liu, N.***; Lu, Z.*; Zhao, J.; McDowell, M.; Lee, H.; Zhao, W.; Cui, Y. A Pomegranate-Inspired Nanoscale Design for Large-Volume-Change Lithium Battery Anodes. ***Nature Nanotechnology*** (**Cover article**), 2014, 9, 187-192.

Featured in ***Nano Today***, 2014, 9, 161.

2. **Liu, N.***; Wu, H.*; McDowell, M. T.; Yao, Y.; Wang, C.; Cui, Y. A Yolk-shell Design for Stabilized and Scalable Li-ion Battery Alloy Anodes. ***Nano Letters*** 2012, 12, 3315–3321.

3. **Liu, N.**; Hu, L.; McDowell, M. T.; Jackson, A.; Cui, Y. Prelithiated Silicon Nanowires as an Anode for Lithium Ion Batteries. ***ACS Nano*** 2011, 5, 6487–6493.

4. **Liu, N.**; Huo, K.; McDowell, M.; Zhao, J.; Cui, Y. Rice Husks as a Sustainable Source of Nanostructured Silicon for High Performance Li-ion Battery Anodes. ***Scientific Reports*** (Nature Publishing Group). 2013, 3, 1919.

5. **Liu, N.**; Yao, Y.; Cha, J.; McDowell, M.; Han, Y.; Cui, Y. Functionalization of Silicon Nanowire Surfaces with Metal-organic Frameworks. ***Nano Research*** 2012, 5, 109–116.

6. Lu, Z.*; **Liu, N.***; Lee, H.-W.; Zhao, J.; Li, W.; Li, Y.; Cui, Y. Non-filling carbon coating of porous silicon micronsized particles for high performance lithium battery anodes. ***ACS Nano***, 2015, 9, 2540–2547.

7. Misra, S.*; **Liu, N.***; Nelson, J.; Hong, S. S.; Cui, Y.; Toney, M. F. In Situ X-ray Diffraction Studies of (De)lithiation Mechanism in Silicon Nanowire Anodes. ***ACS Nano*** 2012, 6, 5465–5473.

8. Zhu, B.*; **Liu, N.***; McDowell, M. T.; Jin, Y.; Cui, Y.; Zhu, J. Interfacial Stabilizing Effect of ZnO on Si Anodes for Lithium Ion Battery. ***Nano Energy***, 2015, 13, 620–625.

9. Hu, L.*; **Liu, N.***; Eskilsson, M.*; Zheng, G.; McDonough, J.; Wågberg, L.; Cui, Y. Silicon-conductive Nanopaper for Li-ion Batteries. ***Nano Energy*** 2013, 2, 138–145

Invited Review Article:

Liu, N.; Li, W.; Pasta, M.; Cui, Y. Nanomaterials for Electrochemical Energy Storage (Featured as Cover Article). Frontiers of Physics. 2014, 9, 323–350.

Invited Book Chapter:

Liu, N.; Zheng, G.; Cui, Y. Nanocarbon hybrids with silicon, sulfur, or paper/textile for high-energy lithium ion batteries. Nanocarbons for Advanced Energy Storage. Wiley-VCH, 2015.

Patent:

Cui, Y.; **Liu, N.**; Lu, Z.; Zhao, J. Large-volume-change lithium battery electrodes. US 14/506,345 (2014)

443387 Computationally Assisted Biofuel Production: Hydrodynamics, Optimization, and Heuristics

Abstract:

Microorganisms, such as microalgae, have recently been a popular research field because of their promise as a biofuel feedstock with advantages over first generation biomass sources, such as wood, corn, and sugarcane. Such advantages include algae's ability to be grown on low value land and to be grown using low quality, brackish, or even salt water. Algae also produce much more oil per acre than traditional energy crops. Currently, however, the production cost of microalgae biofuels can be uneconomical, which impedes its wide scale implementation as a feedstock. One way to improve the economics of algae-based fuels is to maximize the amount of algae that can be grown within a given bioreactor. This would have the effect of not only reducing the number of reactors needed to produce a given amount of algae, but would also reduce the land area needed for the algae growth process.

One method of maximizing algal biomass production is to develop better bioreactors, with reactor geometry and flow conditions optimized for maximal algae production. Computational fluid dynamics (CFD) models excel at capturing the important hydrodynamics of this problem, including turbulence and nutrient and carbon dioxide distribution within the growth substrate. These models can be employed along with surrogate modeling and optimization solvers to determine the reactor conditions for maximum algae growth [1,2]. Unfortunately, these methods can be very computationally expensive. To this end, I have developed several alternative methods to respond to these challenges by reducing the computational expense associated with individual simulations, reducing the number of simulations necessary, and reducing the costs in the "downstream" algae supply chain, such as transportation. I have developed a heuristic for simulation of bioreactors which reduces the computational expense of applying CFD in testing of many different growth conditions. Additionally, I have conducted a linear stability analysis of an algae bioreactor and have determined hydrodynamic conditions which would lead to maximum algae growth. Finally, I have also developed a pair of optimization directed sampling algorithms which allow for better selection of sample points when developing a surrogate model involving CFD studies. These algorithms have been used to investigate algae growth, and all show promise in assisting with maximizing the amount of algae produced. In this poster session, I will present my work to this point as well as continuing work in this subject area.

Reference:

1. Smith, Justin, Cremaschi, Selen and Crunkleton, Daniel W. (2012), "CFD Based Optimization of a Flooded Bed Bioreactor for Algae Production," Computer Aided Chemical Engineering **31**, pp. 910-914

2. Smith, Justin D., Neto, Amadeu, Cremaschi, Selen and Crunkleton, Daniel W. (2013), "CFD-Based Optimization of a Flooded Bed Algae Bioreactor," Industrial & Engineering Chemistry Research **52**(22), pp. 7181-7188, doi:10.1021/ie302478d

438703 Development of Novel Poly(Ionic Liquid) Membranes for Electrodialytic Separations and Desalination

Abstract:

Novel energy and water recovery systems are crucial to meeting global water and energy demands. Research on ionic liquids have demonstrated unique chemical and physical properties, and significant work on their use has resulted in advances for water and energy applications, making these chemicals ideal for the development of novel membrane systems. Specifically, electrodialysis and electrodeionization are mature membrane technologies that can be improved substantially through the use of ionic liquids in ion exchange membranes. *My research will focus on the development of water production and recovery through electrodialytic desalination systems as well as improved ion separations through ionic liquid supported membrane synthesis.* I will create novel ion-exchange membranes using polymerized ionic liquids (i.e. poly(ionic liquid)) composites for improved ion permeability and selectivity with good membrane mechanical properties and tunability. I *will investigate the transport properties and molecular interactions expressed by ionic liquid/solvent systems and their applicability in electrochemical applications.* Synthesized membranes will be tested for ionic properties as well as chemical and mechanical stability. Previous work has shown that ionic liquid impregnated membranes can operate at high ion conductivities (> 10^{-3} S/cm) and high temperatures (> 80 °C). Synthesized membranes will operate at these performance levels with tuned characteristics for high selectivity and ion exchange capacity, leading to substantial cost and energy savings compared to current state-of-the-art. For energy reduction in the separations industry, I will develop novel membranes for improved industrial separations with a focus on specialty chemicals and pharmaceuticals containing ionic groups. Through my work, I plan to foster scientific debate and progress, higher education, and academic/industrial collaborations.

438704 Applied Statistics and Data Analytics for Advanced Process Systems Engineering

Abstract:

Since the formal inception of Process Systems Engineering (PSE) in 1961, PSE has rapidly evolved at the interface between chemical engineering, applied mathematics, statistics and computer science. Recent advancements in computing capabilities and big-data analytics, along with emerging trends in information technology have enabled PSE ideas to be extrapolated and applied to systems beyond chemical processes, such as molecular, health-care, neuroscience, financial, social, internet, and behavioral systems.

Today, process industries face unique challenges in terms of growing process complexity, shift towards process integration and miniaturization, dynamic operating conditions, hard economic constraints and stricter environmental regulations. In the face of rising uncertainty in process operations, data becomes an indispensable asset for smart economic decision-making and safe plant operations. The use of process operation and process control computers along with process information systems have enabled process industries to collect and archive massive databases from equipment, process, operation and customer data [1]. Furthermore, advancements in process sensing technologies ranging from conventional process sensors to images, videos, and virtual-sensing technologies have made data acquisition (DAQ) not only accurate and cheap, but also contributed to the diversity and complexity of process data [1]. A recent AIChE Journal Perspective article [2] characterizes the massiveness of process data as "drowning in data." Despite the richness of process data available at our disposal, PSE research has not been able to keep pace with developments in advanced statistical inferencing, machine-learning, data mining and analytics techniques.

My current research focus is on the use of statistics and data-analytics to develop novel data-based solutions for complex process systems engineering problems. The thrust of my research is on developing data-based process methods for inferential sensor development, process monitoring and control, product optimization, and safety and compliance management. With my strong background in chemical engineering, applied statistics, and computational sciences, we have been successful in developing data-based solutions for various applications in process, health-care, oil and gas, pharmaceutical and aerospace industries. Some of the key problems addressed in the current research plan includes:

• Mining and fusion of process information for improved process identification **[8]**

• Data quality and optimal experiment design for efficient non-linear identification **[8, 9]**

• Real-time adaptive Bayesian identification of non-linear processes with irregular data **[7, 12]**

• Performance assessment, diagnosis, and selection of non-linear filtering methods **[4, 10, 11, 18]**

• Inferential sensor for online monitoring of Type-II diabetic patients using clinical data **[3]**

• Model-based delay-timer alarm design for non-linear industrial processes **[16]**

• Reachability study for multivariate chemical systems subject to flow-rate disturbances **[13, 14, 15]**

• Global parameter estimation in chemical reaction networks using kinetic data **[17]**

• Multi-model approach to non-linear identification of distillation columns using EM algorithm **[5]**

• Theoretical statistical inferencing and filtering limits in ballistic target tracking problems **[6]**

As a faculty member, my research plan will be to continue to build on the fundamental and applied research in statistics and data analytics, and to provide improved data-based solutions to challenging problems in process, energy and health-care systems. Using my interdisciplinary research experience along with my expertise in applied statistics, some of the key research directions we would like to explore in future includes:

• Data Mining, Machine Learning and Big Data Analytics for Advanced Process Systems Engineering;

• Inferential Sensors for Process, Energy and Health-Care Systems; and

• Process Monitoring, Safe Operations, and Compliance Management in Process Industries.

References

1. Qin, S.J. "Process data analytics in the era of big data." AIChE Journal. Vol. 60, No. 9, pp. 3092-3100, 2014.

2. Venkatasubramanian, V. "DROWNING IN DATA: Informatics and modeling challenges in a data rich networked world." AIChE Journal. Vol. 55, No. 1, pp. 2-8, 2014

3. Barazandegan, M., Ekram, F., Kwok. E., Gopaluni, R.B., **Tulsyan, A.**"Assessment of type II diabetes mellitus using irregularly sampled measurements with missing data". Bioprocess and Biosystems Engineering. Vol. 38, No. 4, pp. 615–629, 2015

4. **Tulsyan, A.,** Huang, B., Gopaluni, R.B., Forbes, J.F. "Performance assessment, diagnosis, and optimal selection of non-linear state filters". Journal of Process Control, Vol. 24, No. 2, pp. 460-478, 2014

5. Chen, L., **Tulsyan, A.,** Huang, B., Liu, F. "Multiple model approach to non-linear system identification with uncertain scheduling variables using EM algorithm". Journal of Process Control, Vol. 23, No. 10, pp. 1480-1496, 2013

6. **Tulsyan, A.,** Huang, B., Gopaluni, R.B., Forbes, J.F. "A particle filter approach to approximate posterior Cram´er-Rao lower bound: The case of hidden states". IEEE Transactions on Aerospace and Electronic Systems, Vol. 49, No. 4, pp. 2478-2495, 2013

7. **Tulsyan, A.**, Huang, B., Gopaluni, R.B., Forbes, J.F. "On simultaneous on-line state and parameter estimation in non-linear state-space models". Journal of Process Control, Vol. 23, No. 4, pp. 516-526, 2013.

8. **Tulsyan, A.**, Huang, B., Forbes, J.F. "Designing prior for robust Bayesian optimal experimental design". Journal of Process Control, Vol. 22, No. 2, pp. 450-462, 2012

9. **Tulsyan, A.**, Huang, B., Gopaluni, R.B., Forbes, J.F. "Bayesian identification of non-linear state-space models: Part I- Input design". IFAC 10th International Symposium on Dynamics and Control of Process Systems, December 18-20, 2013, Mumbai, India

10. **Tulsyan, A.**, Khare, S.R., Huang, B., Gopaluni, R.B., Forbes, J.F. "Bayesian identification of nonlinear state-space models: Part II- Error Analysis". IFAC 10th International Symposium on Dynamics and Control of Process Systems, December 18-20, 2013, Mumbai, India

11. **Tulsyan, A.**, Huang, B., Gopaluni, R.B., Forbes, J.F. "Performance assessment of non-linear state filters", IFAC International Symposium on Advanced Control of Chemical Processes, July 10-13, 2012, Singapore

12. **Tulsyan, A.**, Gopaluni, R.B. "Sequential Monte-Carlo methods for inferencing in non-linear chemical systems- A tutorial". Processes, 2015.

13. **Tulsyan, A.**, Barton, P.I. "Interval enclosures for reachable set of chemical kinetic flow systems. Part 1: Developing sparse transformation". Chemical Engineering Science, 2015.

14. **Tulsyan, A.**, Barton, P.I. "Interval enclosures for reachable set of chemical kinetic flow systems. Part 2: Direct bounding method". Chemical Engineering Science, 2015.

15. **Tulsyan, A.**, Barton, P.I. "Interval enclosures for reachable set of chemical kinetic flow systems. Part 3: Indirect bounding method". Chemical Engineering Science, 2015.

16. **Tulsyan, A.**, Gopaluni, R.B. "Industrial alarm design for nonlinear dynamical systems". Journal of Process Control. Working paper.

17. **Tulsyan, A.**, Schaber, S., Barton, P.I. "PERKS: Software for parameter estimation in reaction kinetic systems". AIChE Journal. Working paper.

18. **Tulsyan, A.**, Khare, S.R., Huang, B., Gopaluni, R.B., Forbes, J.F. "An optimal filter switching strategy for on-line state and parameter estimation". IEEE Transactions on Signal Processing. Working paper.

443406 Design of Hierarchical 3D Architectures for Energy, Electronic Applications

Sunday, November 8, 2015
Exhibit Hall 1 (Salt Palace Convention Center)
Po-Yen Chen, Chemical Engineering, Massachusetts Institute of Technology, Cambridge, MA

We live in age of great challenges such as the energy crisis. The society has been trying to develop new strategies 1) to efficiently convert renewable energy to usable electricity, 2) to store energy into high performance rechargeable batteries, and 3) to fabricate advanced electronic devices in environmentally benign manners. To develop high-performance energy, electronic technology, the design of three-dimensional (3D) nanostructures and nanocomposites have become central issues.

My doctoral research at MIT (with Prof. Angela M. Belcher and Prof. Paula T. Hammond) focuses on the design of 3D nanostructured composites by using biological materials (*e.g.*, M13 viruses) as templates. M13 viruses can be genetically engineered to display the specific functionalities for the organization of targeted materials. The capability of M13 virus to manipulate nanomaterials provides an attractive route to creating nanostructured hybrid composites for the electrodes of solar cells and batteries. First, via self-assembling M13 virus into 3D viral hydrogels, the virus-based hydrogel serves as a multifunctional platform capable of synthesizing core-shell Au-TiO_2 nanocomposites to produce plasmon-enhanced sensitized solar cells with substantial improvement on power conversion efficiency. Second, M13 viruses are genetically programmed to cluster SWNTs along the phage without aggregation. The SWNTs-virus complexes are crosslinked into hydrogels and coated with conducting polyaniline. The virus-templated SWNTs-polyaniline composites can largely improve the electrical conductivity and stored capacity of battery electrodes. The effective incorporation of SWNT results the electrochemical electrode with the lowest effective cost (material cost/capacity). During my doctoral research, I also invented a practical solution to reusing an old car battery to manufacture high-efficiency perovskite solar cells to power ~30 households.

As an independent postdoctoral research fellow at Brown University, I have established my own small lab, mentored masters and undergraduates, and built substantive collaboration with faculty members. My postdoctoral research involves the creation of high-ordered and multiscale buckled surface topologies by extreme compression of 2D nanomaterials (e.g., graphene), and the hierarchical surfaces can achieve superhydrophobic surfaces ($\theta_w > 160°$) and exhibit efficient electrochemical performance.

Selected Publications:

1. **P.-Y. Chen**, et al., Multiscale buckled surface topographies created by extreme compression of graphene surface films. *(Under review)*
2. **P.-Y. Chen**, et al., Carbon nanotube-polyaniline core-shell nanostructured hydrogel for electrochemical energy storage. *RSC Adv.* **2015**, *5*, 37970.
3. **P.-Y. Chen**, et al., M13 virus-enabled synthesis of titanium dioxide nanowires for tunable mesoporous semiconducting networks. *Chem. Mater.* **2015**, *27*, 1531.
4. **P.-Y. Chen**, et al., Environmentally responsible fabrication of efficient perovskite solar cells from recycled car batteries. *Energ. & Environ. Sci.* **2014**, *7*, 3659. *(Highlighted in Washington Post)*
5. **P.-Y. Chen**, et al., Assembly of viral hydrogels for three-dimensional conducting nanocomposites. *Adv. Mater.* **2014**, *26*, 5101. *(Front cover)*
6. N.-M. D. Courchesne, M. T. Klug, **P.-Y. Chen**, et al., Assembly of a bacteriophage-based template for the organization of materials into nanoporous networks. *Adv. Mater.* **2014**, *26*, 3398. *(Back cover)*
7. **P.-Y. Chen**, et al., Versatile three-dimensional virus-based template for dye-sensitized solar cells with improved electron transport and light harvesting. *ACS Nano* **2013**, *7*, 6563.
8. **P.-Y. Chen**, et al., Layer-by-layer assembled porous photoanodes for efficient electron collection in dye-sensitized solar cells. *J. Mater. Chem. A* **2013**, *1*, 2217.
9. **P.-Y. Chen**, et al., A quasi solid-state dye-sensitized solar cell containing binary ionic liquid and polyaniline-loaded carbon black. *J. Power Sources* **2010**, *195*, 3933.
10. C.-P. Lee, **P.-Y. Chen**, et al, Iodine-free high efficient quasi solid-state dye-sensitized solar cell containing ionic liquid and polyaniline-loaded carbon black. *J. Mater. Chem.* **2010**, *20*, 2356.

Thermochemical conversion of biomass and organic wastes into renewable fuels and high value co-products

Umakanta Jena, Desert Research Institute

Current Position: Assistant Research Professor –Bioenergy

Research Interests:

My primary research interest is in the generation of alternative fuels and co-products from biomass and wastes via thermochemical conversion (TCC) routes and catalysis. Pyrolysis and hydrothermal liquefaction are two important TCC processes that convert biomass components into a liquid fuel (called, bio-oil or biocrude) in a single step. Bio-oil can be catalytically modified and co-processed into hydrocarbon fuels in the presence of hydrogen and co-solvents. Integration of a 'well defined' TCC process with a suitable low-cost catalytic method is important for production of commodity biofuels (gasoline/diesel). Co-product development and utilization in biomass processing has significance to the process sustainability and cost-economics. My research interests are in developing new processes and optimizing process parameters to maximize the primary fuel production (example, bio-oil) and co-product utilization. For example, in hydrothermal liquefaction and carbonization processes, the aqueous co-product streams have significant amounts of organics and nutrients that can be separated: the nutrients can be reused and recycled for plant growth, and the organics can be catalytically upgraded into high-value chemicals (levoglucosans, sugars, hydroxymethyl furfurals etc.), or converted into energy-dense gases via hydrothermal gasification process.

My secondary research interest in the integration of thermochemical and biochemical/biological conversion pathways. Using most of the biomass will be a key aspect in a sustainable biofuel production process. For example, in algal biodiesel or renewable diesel production, large amount of solid is left-behind (60-80% of the staring algae feedstock). Similarly, in the fermentation of agricultural grains and cellulosic crops, significant unused or residual mass is obtained. The utilization of the above residue by a second conversion process (catalytic or non-catalytic TCC) into fuels/chemicals will be important for the overall biofuel process viability. My overall goal will be to design integrated process and evaluate the process mass and energy balance, and its environmental and economic impacts.

Successful Grant Proposals:

Qatar National Research Foundation (2015; $900,000, International grant– collaborative research).

United States Department of Agriculture –NIFA (2014; $90,000)

Department of Energy – EERE (2012; $698,000)

DRI's Institute Project Assignments (2013 and 2015; $14,500 and $14, 879)

Indian Council of Agricultural Research (2006; INR 1,250,000)

Postdoctoral Research Projects:
1. "Catalytic hydrodeoxygenation of fast pyrolysis bio-oil."

Mentors: Prof. James R. Kastner and Prof. K.C. Das, College of Engineering, University of Georgia

2. "Development of bio-oil commodity fuel as a refinery feedstock from high impact algae feedstock"

Mentor: Prof. James R. Kastner, College of Engineering, University of Georgia

3. "Recovering valuable products through hydrothermal carbonization of whole and lipid-extracted algae"

Mentor: Prof. S. Kent Hoekman

PhD Dissertation: "Thermochemical conversion of microalgal biomass for production of biofuels and co-products"

Under supervision of Prof. K.C. Das, College of Engineering, University of Georgia

Research Experience:

I have over 12 years' experience working in thermochemical and catalytic conversion of biomass and intermediate products. My doctoral research demonstrated hydrothermal liquefaction (HTL) and pyrolysis of algae, comprehensively evaluated process mass and energy balance and developed the concept of algae biorefinery. Understanding the process chemistry and reaction mechanism are important for new process development and its optimization. My laboratory studies have evaluated reaction kinetics of biomass decomposition (algae, woody biomass, and coal) using thermogravimetric techniques and mass spectrometry. My research has involved development of processes for HTL, pyrolysis, and gasification of dry and wet biomass (lignocellulosic, algae, and animal wastes) and conduct of laboratory and field evaluation. For example, my doctoral and postdoctoral studies evaluated batch HTL and pyrolysis of algae and woody biomass. One of the previous studies involved the field evaluation of downdraft gasification system in agricultural processing operation.

My current research is focused on integrating thermochemical conversion (HTL, pyrolysis, gasification) with the biological processes (fermentation, anaerobic digestion etc.) for production fuels/chemicals and waste remediation. For example, one of my recent studies reported on the renewable diesel production via HTL of yeast biomass generated from fermentation process using delactosed dairy effluent. Another study conducted under a DOE study has explored a novel two-step HTL process for converting algae into low-nitrogen fuel. I am presently exploring the integration of ethanol production from sweet sorghum while converting the residues into an energy-dense hydrochar (a coal-like solid fuel). Both of the studies involve environmental life cycle and/or techno-economic assessments (LCA and TEA) for evaluation of process sustainability.

Teaching Experience:

Aside from my research career, I also have extensive teaching experience. I spent a semester co-teaching a Chemical Engineering Course (Process Separations) at University of Nevada Reno (over 50% of the course); I TAed and guest-lectured undergraduate lecture and laboratory courses in Engineering Unit Operations and Environmental Engineering-I at the University of Georgia; and, lastly, I mentored graduate and undergraduate students in my group. As a post-doc I have worked with undergraduate summer students and new graduate students from chemistry, biochemistry, biological systems engineering and chemical engineering from outside universities (University of Nevada Reno, University of Nevada Las Vagas, Fort Valley State University-GA etc.). Their unique backgrounds have helped me generalize my teaching methods. I have helped them learn how to work in the laboratory, analyze data, and present results. The students helped identify process evaluation and optimization, develop experimental set-ups, develop product separation protocol in HTL, and identify methods for collecting liquid and gas samples. I enjoyed watching students succeed with my input which has always motivated me furthering my career in the academic profession.

Future Direction:

As faculty I would like to continue investigating fundamental and applied aspects of thermochemical conversion technologies: process development, evaluation, and optimization. In particular I would like to apply my skills that I obtained from various research works that I worked on during my Ph.D. and postdoc: hydrothermal reactions, integrated thermochemical and biological processes, and sustainability evaluation (TEA and LCA). For strengthening our understanding of the process, my future research projects will designed for conducting basic and applied studies investigating the reaction chemistry, mechanism, and the process as a whole (through process modeling). I would also conduct their life cycle and techno-economic assessments for sustainable biofuel production, waste utilization and environmental remediation. Listed below are the areas of my future research interests.

Research Areas of Interests

- Catalytic upgrading of bio-oil from HTL and pyrolysis processes
- Hydrothermal carbonization of biomass into hydrochar and pelletization
- Biomass pretreatments and pyrolysis
- New process development and evaluation
- Process techno-economics and life cycle assessments (LCA)

Selected Publications:

Jena, U., A. McCurdy, R. Ledbetter, H. Summers, Rhesa Ledbetter A, Warren, L. Seefeldt, S.K. Hoekman, J. Quinn. 2015. Oleaginous Yeast Platform for Producing Biofuels via Co-Solvent Hydrothermal Liquefaction. *Biotechnology for Biofuels, 8:167.*

Costanzo, W., **U. Jena**, R.N. Hilten, K.C. Das, J.R. Kastner. 2015. Evaluation of low temperature hydrothermal liquefaction of algae for low nitrogen biocrude generation and nutrient recycling, *Algal Research, 12(2015): 377-387.*

Broch, A. **U. Jena,** S.K. Hoekman. 2014. Analysis of solid and aqueous phase products from hydrothermal carbonization of whole and lipid-extracted algae. *Energies*, 7(1): 62-79.

Jena, U., K.C. Das and J.R. Kastner. 2012. Comparison of the effects of Na_2CO_3, $Ca_3(PO_4)_2$, and NiO catalysts on the thermochemical liquefaction of microalga *Spirulina Platensis. Applied Energy, 98(2012): 368-375.*

Jena, U. and K.C. Das. 2011. Comparative evaluation of BioOil production from microalgae by thermochemical liquefaction and pyrolysis. *Energy &Fuels,* 25, 5472–5482.

Jena, U., K.C. Das and J.R. Kastner. 2011. Effect of operating conditions of thermochemical liquefaction on biocrude production from *Spirulina platensis. Bioresource Technology* 102(10): 6221-6229.

Jena, U., N. Vaidyanathan, S. Chinnasamy and K.C. Das. 2010. Evaluation of microalgae cultivation using recovered aqueous solution from thermochemical liquefaction of algal biomass. *Bioresource Technology,* 102(3): 3380-3387.

Singh, R.N., **U. Jena,** J.B. Patel and A.M. Sharma. 2006. Feasibility study of cashew nut shells as an open core gasifier feedstock. *Renewable Energy* 31: 481–487.

Book Chapters:

Biomass energy from revegetation of landfill sites, B. Seshadri, N. S. Bolan, **U. Jena**, R. Thangarajan, K.C. Das, H. Wang and R. Naidu. Chapter 5, In Bioremediation and Bioeconomy, Edited by M.N. Prasad, Elsevier, pp: 99-109.

AUTHOR INDEX

AUTHOR INDEX

AUTHOR INDEX

AUTHOR INDEX